Collins

Revision

NEW GCSE SCIENCE

Science and Additional Science

Higher

For OCR Gateway B

Series editor: Chris Sherry

Authors: Colin Bell
Brian Cowie
Ann Daniels
Maureen Elliot
Ian Honeysett

Revision Guide +
Exam Practice Workbook

Contents

Fitness and health

Blood pressure

- **Blood pressure** is measured in millimetres of mercury. This is written as mmHg.
- Blood pressure has two measurements: **systolic pressure** is the maximum pressure the heart produces and **diastolic pressure** is the blood pressure between heart beats.
- Different factors can cause a person's blood pressure to increase or decrease:
  – It can be increased by stress, high **alcohol** intake, smoking and being overweight.
  – It can be decreased by regular exercise and eating a balanced **diet**.
- High blood pressure can cause blood vessels to burst. This can cause damage to the brain, which is often called a **stroke**. It can also cause damage to the kidneys.
- Low blood pressure can lead to dizziness and fainting as the blood supply to the brain is reduced, and poor circulation to other areas such as the fingers and toes.

Fitness and health

- There is a difference between fitness and health:
  – Fitness is the ability to do physical activity.
  – Health is being free from diseases such as those caused by **bacteria** and viruses.
- Your general level of fitness can be measured by your cardiovascular **efficiency**.
- Your fitness can also be measured for different activities:
  – strength, by the amount of weights lifted
  – flexibility, by the amount of joint movement
  – stamina, by the time of sustained exercise
  – agility, by changing direction many times
  – speed, by a sprint race.
- Therefore, you can be very fit for a sprint race but not perform well in a marathon.
- Ways of measuring fitness should be evaluated to check effectiveness in particular situations.

Smoking

- Smoking can increase blood pressure in a number of ways:
  – **Carbon monoxide** in cigarette smoke causes the blood to carry less oxygen. This means that the heart rate increases so that the tissues receive enough oxygen.
  – Nicotine in cigarette smoke directly increases heart rate.
- Carbon monoxide decreases the oxygen-carrying capacity of blood. It combines with **haemoglobin**, preventing it from combining with oxygen, so less oxygen is carried.

Diet and heart disease

- Heart disease is caused by a restricted blood flow to the heart muscle. The risk of getting heart disease is increased by:
  – a high level of saturated fat in the diet, which leads to a build-up of **cholesterol** (a **plaque**) in **arteries**
  – high levels of salt, which can increase blood pressure.
- It is essential to be able to interpret data showing links between the amount of saturated fat eaten, the build-up of cholesterol and the incidence of heart disease.
- The narrowing of the arteries caused by plaques in the coronary arteries can reduce blood flow to heart muscle. The plaques also make blood clots or **thrombosis** more likely to happen, which will also block the artery.

Remember!
It is the blood vessel into the heart muscle that is blocked, not the blood flowing into the heart.

 Improve your grade

Heart disease

Excess saturated fat or excess salt in the diet can increase the risk of heart disease.
Explain why. *AO1* [4 marks]

Human health and diet

A balanced diet

- It is important for good health to eat a balanced **diet** containing the correct amounts of the chemicals found in food. Three of these are:
  - carbohydrates, which are made up of simple sugars such as glucose
  - proteins, which are made up of **amino acids**
  - fats, which are made up of fatty acids and glycerol.
- A balanced diet varies according to factors including age, gender, level of activity, religion, being **vegetarian** or **vegan**, or because of medical issues such as food allergies.

- If you eat too much fat and carbohydrate, they are stored in the body.
  - Carbohydrates are stored in the liver as glycogen or converted into fats.
  - Fats are stored under the skin and around organs as adipose tissue. Although proteins are essential for growth and repair, they cannot be stored in the body.

glucose units → starch is a complex carbohydrate

amino acids → protein

glycerol   fatty acids → fat

Chemistry of foods

D–C

B–A*

Protein intake

- Proteins are needed for growth and so it is important to eat the correct amount. This is called the estimated average daily requirement (**EAR**) and can be calculated using the formula:

  EAR in g = 0.6 × body mass in kg

- Sue has a mass of 72.5 kg. Her EAR is 0.6 × 72.5 = 43.5 g/day.

- Too little protein in the diet causes the condition called **kwashiorkor**. This is more common in developing countries due to overpopulation and lack of money to improve agriculture.

D–C

- The EAR is only an estimated figure based on an average person. The EAR for protein might be affected by factors such as body mass, age, pregnancy or breast-feeding (lactation).

- Although proteins cannot be stored in the body, some amino acids can be converted by the body into other amino acids.

- Proteins from meat and fish are called **first-class proteins**. They contain all the essential amino acids that cannot be made by the human body.

- Plant proteins are called **second-class proteins** as they do not contain all the essential amino acids.

B–A*

Overweight or underweight?

- To work out if a person is overweight or underweight, calculate their **body mass index** (BMI):

  $$BMI = \frac{\text{mass in kg}}{(\text{height in m})^2}$$

  Tom is 170 cm tall and has a mass of 80 kg. Calculate his BMI.
  170 cm = 1.7 m
  $$BMI = \frac{80}{1.7^2} = 27.7$$

> ### EXAM TIP
> The height must be in metres and the mass in kilograms.

D–C

- A BMI of more than 30 means the person is **obese**, 25–30 is overweight, 20–25 is normal, less than 20 is underweight. With a BMI of 27.7, Tom is overweight.

- Some people may become ill as they choose to eat less than they need. This may be caused by low self-esteem, poor self-image or a desire for what they think is perfection.

Improve your grade

EAR for protein
Explain the importance of knowing your EAR for protein. *AO1/2* [4 marks]

Staying healthy

Malaria

- Malaria is caused by a protozoan called *Plasmodium*, which feeds on human **red blood cells**.
- *Plasmodium* is carried by mosquitoes, which are **vectors** (i.e. not affected by the disease), and transmitted to humans by mosquito bites.
- *Plasmodium* is a **parasite** and humans are its host. A parasite is an organism that feeds on another living organism, causing it harm.

- Knowledge of the mosquito's life cycle has helped to stop the spread of malaria (by draining stagnant water, putting oil on the water surface and spraying **insecticide**). This knowledge has also helped to develop new treatments for malaria.

Cancers

- Changes in lifestyle and **diet** can reduce the risk of some **cancers**:
  – Not smoking reduces the risk of lung cancer.
  – Using sunscreen reduces the risk of skin cancer.

Remember!
Different antibodies are required to deal with different pathogens.

- Benign **tumour** cells, such as in warts, divide slowly and are harmless.
  Cancers are malignant tumours: the cells display uncontrolled growth and may spread.
- Ways of interpreting data on cancer and survival/mortality rates should be considered.

The fight against illness

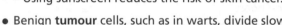

- **Pathogens** (disease-causing organisms) produce the symptoms of an infectious disease by damaging the body's cells or producing poisonous waste products called **toxins**.
- The body protects itself by producing **antibodies**, which lock onto **antigens** on the surface of pathogens such as a **bacterium**. This kills the pathogen.
- Human **white blood cells** produce antibodies, resulting in **active immunity**. This can be a slow process but has a long-lasting effect. Vaccinations using antibodies from another human or animal result in **passive immunity**, which has a quick but short-term effect.

- Each pathogen has its own antigens, so a specific antibody is needed for each pathogen.
- The process of immunisation is also called vaccination:
  – It starts with injecting a harmless pathogen carrying antigens.
  – The antigens trigger a response by white blood cells, producing the correct antibodies.
  – Memory cells (a type of T-lymphocyte cell) remain in the body, providing long-lasting immunity to that disease.
- Immunisation carries a small risk to the individual, but it avoids the potentially lethal effect of the pathogen, as well as decreasing the risk of spreading the disease.

Treatments and trials

- **Antibiotics** (against bacteria and fungi) and **antiviral drugs** (against viruses) are specific in their action.
- An antibiotic destroys a pathogen; an antiviral drug slows down the pathogen's development.
- New treatments, such as vaccinations, are tested using animals, human tissue and computer models before human **trials**. Some people object to causing suffering in animals in such tests.

- A **placebo** is a harmless pill. Placebos are used as a comparison in drug testing so the effect of a new drug can be assessed.
- In a **blind trial**, the patient does not know whether they are receiving a new drug or a placebo. In a double-blind trial, neither the patient nor the doctor know which treatment is being used. These types of trials avoid a 'feel-good factor' and a biased opinion.
- Excessive use of antibiotics has resulted in resistant forms of bacteria being more common than non-resistant forms. For example, resistant MRSA has thrived, causing serious illness.

Improve your grade

Immunisation
Immunisation carries a small risk. Despite this risk, why is immunisation important? *AO1* [4 marks]

The nervous system

How do eyes work

- The main parts of the eye have special functions.
- Light rays are **refracted** (bent) by the cornea and lens.
- The retina contains light receptors. Some are sensitive to different colours.
- **Binocular vision** helps to judge distance by comparing the images from each eye; the more different they are, the nearer the object.
- The eye can focus light from distant or near objects by altering the shape of the lens. This is called **accommodation**.
- To focus on distant objects, the ciliary muscles relax, and the suspensory ligaments tighten, so the lens has a less rounded shape.
- To focus on near objects, the ciliary muscles contract and the suspensory ligaments slacken, so the lens regains a more rounded shape due to its elasticity.

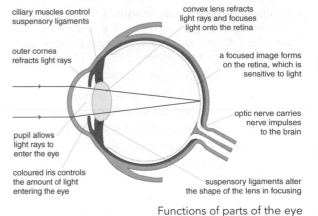

ciliary muscles control suspensory ligaments

outer cornea refracts light rays

pupil allows light rays to enter the eye

coloured iris controls the amount of light entering the eye

convex lens refracts light rays and focuses light onto the retina

a focused image forms on the retina, which is sensitive to light

optic nerve carries nerve impulses to the brain

suspensory ligaments alter the shape of the lens in focusing

Functions of parts of the eye

 D–C

 B–A*

Faults in vision

- Red-green colour blindness is caused by a lack of specialised cells in the retina.
- Long and short sight are caused by the eyeball or lens being the wrong shape. In long sight, the eyeball is too short or the lens is too thin, so the image is focused behind the retina. In short sight, the eyeball is too long or the lens is too rounded so the lens refracts light too much, so the image would be focused in front of the retina.

D–C

- Corneal surgery or a lens in glasses or contact lenses corrects long and short sight. A convex lens is used to correct long sight, a concave lens to correct short sight.

B–A*

Nerve cells

- Nerve cells are called neurones. Nerve impulses pass along the **axon**.
- What happens in a **reflex** action is shown by a reflex arc. The links in a reflex arc are:

  stimulus→receptor→**sensory neurone**→**central nervous system**→**motor neurone**→effector→response

- The pathway for a spinal reflex is:

  receptor→sensory neurone→relay neurone→motor neurone→effector

D–C

- Neurones are adapted by being long, having branched endings (dendrites) to pick up impulses and having an insulator sheath.
- The gap between neurones is called a **synapse**. The arrival of an impulse triggers the release of a transmitter substance, which **diffuses** across the synapse. The transmitter substance binds with receptor molecules in the membrane of the next neurone causing the impulse to continue.

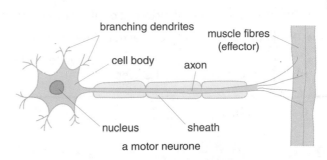

branching dendrites

cell body

axon

muscle fibres (effector)

nucleus

sheath

a motor neurone

Parts of a motor neurone

 B–A*

Improve your grade

Vision

In old age, muscles lose their ability to contract and relax quickly. Ligaments become less flexible. The lens becomes less elastic. Explain the effects of these changes on vision. *AO2* [4 marks]

Remember!
In the eye the light rays are refracted, *not* reflected.

Drugs and you

Types of drugs

- Drugs have a legal classification. Class A drugs are the most dangerous and have the heaviest penalties. Class C drugs are the least dangerous, with the lightest penalties.

- There are different types of drugs:
  - **depressants** (**alcohol**, solvents, temazepam)
  - **painkillers** (aspirin, paracetamol)
  - **stimulants** (nicotine, MDMA ('ecstasy'), caffeine)
  - **performance enhancers** (anabolic steroids)
  - **hallucinogens** (LSD).

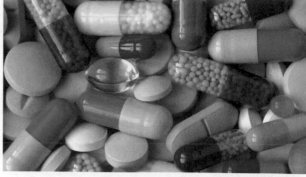

Different prescription drugs

- Depressants block the transmission of nerve impulses across **synapses** by binding with receptor molecules in the membrane of the receiving neurone.

- Stimulants cause more neurotransmitter substances to cross synapses.

Effects of smoking

- Cigarette smoke contains many chemicals that stop cilia moving.

- Cilia (tiny hairs) are found in the epithelial lining of the trachea, bronchi and bronchioles.

- A 'smoker's cough' is a result of
  - dust and particulates in cigarette smoke collecting and irritating the epithelial lining
  - mucus not being moved by the cilia.

> ### EXAM TIP
> Avoid the common error of writing 'the rate of lung cancer is affected' without explaining that it increases or decreases.

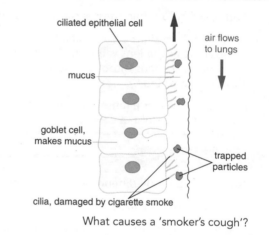

ciliated epithelial cell

air flows to lungs

mucus

goblet cell, makes mucus

trapped particles

cilia, damaged by cigarette smoke

What causes a 'smoker's cough'?

Effects of alcohol

- The alcohol content of alcoholic drinks is measured in **units of alcohol**.

- Drinking alcohol increases **reaction times** and increases the risk of accidents.

- The liver is damaged when it breaks down **toxic** chemicals such as alcohol. This is called cirrhosis of the liver.

How science works

You should be able to:

- identify complex relationships between variables such as in the interpretation of data related to alcohol content, accident statistics and reaction times.

Improve your grade

Affects of alcohol

Drinking alcohol increases the risk of accidents. Explain why. *AO2* [4 marks]

Staying in balance

Homeostasis

- Keeping a constant internal environment is called homeostasis.
- Homeostasis involves balancing bodily inputs and outputs.
- Automatic control systems keep the levels of **temperature**, water and **carbon dioxide** steady. This makes sure all cells can work at their optimum level.

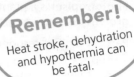

Remember!
Heat stroke, dehydration and hypothermia can be fatal.

D–C

- Negative feedback controls are used in homeostasis. Negative feedback systems act to cancel out a change such as a decreasing temperature level.

B–A*

Temperature control

- The body temperature of 37 °C is linked to the **optimum temperature** for many **enzymes**.
- A high temperature can cause:
  - **heat stroke** (skin becomes cold and clammy and pulse is rapid and weak)
  - **dehydration** (loss of too much water).
- Both heat stroke and dehydration can be fatal if not treated.
- To avoid overheating, sweating increases heat transfer from the body to the environment.
- The **evaporation** of sweat requires body heat to change the liquid sweat into water vapour.
- A very low temperature can cause **hypothermia** (slow pulse rate, violent shivering), which can be fatal if not treated.

Vasoconstriction. When the body is too cold small blood vessels in the skin constrict and so less blood flows through them, reducing heat loss.

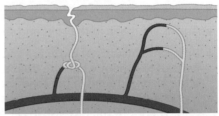

Vasodilation. When the body is too hot small blood vessels in the skin dilate and so blood flow increases, bringing more blood to the surface, where it loses heat.

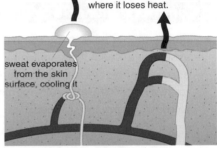

sweat evaporates from the skin surface, cooling it

D–C

- Blood temperature is monitored by the **hypothalamus** gland in the brain. Reaction to temperature extremes are controlled by the nervous and hormonal systems, which trigger vasoconstriction or vasodilation.
- Vasoconstriction is the constriction (narrowing) of small blood vessels in the skin. This causes less blood flow and less heat transfer.
- Vasodilation is the dilation (widening) of small blood vessels in the skin. This causes more blood flow near the skin surface resulting in more heat transfer.

Vasoconstriction and vasodilation

B–A*

EXAM TIP

In vasodilation and vasoconstriction, the *size* of the small blood vessels is altered. The blood vessels do *not* move.

Control of blood sugar levels

- A **hormone** called **insulin** controls **blood sugar levels**.
- Hormone action is slower than nervous reactions as the hormones travel in the blood.
- Type 1 diabetes is caused by the pancreas not producing any insulin, so must be treated by doses of insulin. Type 2 diabetes, which is caused either by the body producing too little insulin or the body not reacting to it, can be controlled by **diet**.

D–C

- Insulin converts excess glucose in the blood into glycogen, which is stored in the liver. This regulates the blood sugar level.
- The insulin dosage in Type 1 diabetes needs to vary according to the person's diet and activity. Strenuous exercise needs more glucose to be present in the blood, so a lower insulin dose is required.

B–A*

Improve your grade

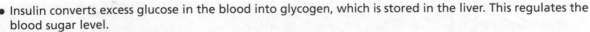
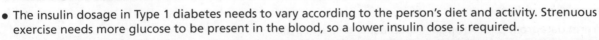

Types of diabetes

Joe has Type 1 diabetes and Charlie has Type 2 diabetes. Explain why they require different treatments. *AO1* [4 marks]

Controlling plant growth

Plant responses

D–C

- **Phototropism** is a plant's growth response to light. **Geotropism** is a plant's growth response to **gravity**.

- Parts of a plant respond in different ways:
  - Shoots are positively phototropic (they grow towards light) and negatively geotropic (they grow away from the pull of gravity).
  - Roots are negatively phototropic (they grow away from light) and positively geotropic (they grow with the pull of gravity).

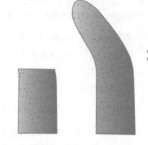

Remember!
A positive reaction means that the root or shoot grows towards the stimulus.

Plant hormones (auxins)

D–C

- **Auxins** are a group of **plant hormones**. They move through the plant in solution.

- Auxins are involved in phototropism and geotropism.

- Auxins are made in the root and shoot tip.

B–A*

- Different amounts of auxin are found in different parts of the shoot when the tip is exposed to light. More auxin is found in the shady part of shoots. A higher amount of auxin will increase the length of cells. Therefore the increase in cell length on the shady side of the shoot causes curvature of the shoot towards the light.

direction of light

When the shoot tip is removed the shoot does not grow

Commercial uses of plant hormones

D–C

- Plant hormones have many commercial uses. They are used:
  - as selective weedkillers, which kill specific weeds and increase crop yield
  - as rooting powder to increase root growth of cuttings
  - to delay or accelerate fruit ripening to meet market demands
  - to control dormancy in seeds.

Spraying crops with selective weedkiller

How science works

You should be able to:

- interpret data from phototropism experiments on auxin action.

Improve your grade

Phototropism

Plant shoots grow straight upwards in the dark but will grow towards a light source.
Explain how and why they do this. *AO1* [4 marks]

Variation and inheritance

Inherited characteristics

- Some human characteristics, such as facial features, can be inherited. They can be **dominant** or **recessive**.

- **Alleles** are different versions of the same **gene**.

- There is debate over how much 'nature or nurture' (genetic or environmental factors) affect intelligence, sporting ability and health.

- Dominant and recessive characteristics depend on dominant and **recessive alleles**.

- Dominant alleles are expressed when present but recessive alleles are expressed only in the absence of the dominant allele.

Remember!
A gene can have two different alleles, one dominant and the other recessive.

 D–C

 B–A*

Chromosomes

- Most body cells have the same number of **chromosomes**. The number depends on the **species** of organism. Human cells have 23 pairs.

- **Sex chromosomes** determine sex in **mammals**. Females have identical sex chromosomes called XX, males have different sex chromosomes called XY.

- A sperm will carry either an X or a Y chromosome. All eggs will carry an X chromosome.

- There is a **random** chance of which sperm **fertilises** an egg. There is therefore an equal chance of the offspring being male or female.

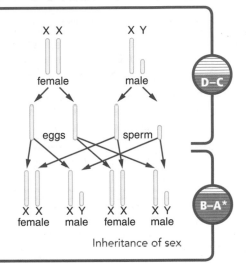

Inheritance of sex

 D–C

B–A*

Genetic variation

- Genetic **variation** is caused by:
  - **mutations**, which are random changes in genes or chromosomes
  - rearrangement of genes during the formation of **gametes**
  - fertilisation, which results in a zygote with alleles from the father and mother.

D–C

A monohybrid cross

- A **monohybrid cross** involves only one pair of characteristics controlled by a single gene, one allele being dominant and one recessive.

- **Homozygous** means having identical alleles, **heterozygous** means having different alleles.

- A person's **genotype** is their genetic make-up. Their **phenotype** is which alleles are expressed.

B–A*

Inherited disorders

- Inherited disorders are caused by faulty genes.

- Many personal and ethical issues are raised:
  - in deciding to have a genetic test (a positive result could alter lifestyle, career, insurance)
  - by knowing the risks of passing on an inherited disorder (whether to marry/have a family).

- Inherited disorders are caused by faulty alleles, most of which are recessive.

- It is possible to predict the probability of inheriting such disorders by interpreting genetic diagrams.

D–C

B–A*

Improve your grade

Inherited disorders

Rabeena finds out that her mother has cystic fibrosis, an inherited disorder caused by a faulty allele (c). Her father does not have this faulty allele. Explain why Rabeena hopes her future husband will not be heterozygous for cystic fibrosis. *AO2/3* [4 marks]

B1 Summary

Blood pressure has two readings, diastolic and systolic pressure, in mmHg.

A balanced diet will vary according to age, gender, activity, religion and personal choice.

Diet and exercise

Smoking, a high alcohol intake and a diet rich in saturated fats and salt increase blood pressure

The body mass index (BMI) can be used to indicate being over or underweight.

Being fit is the ability to do exercise, being healthy is being free from disease.

High blood pressure can damage the brain (stroke) and kidneys.

The EAR can be used to calculate protein requirements. It depends on age, pregnancy and lactation.

Harmful drugs are classified as Class A, B and C, Class A being the most harmful.

Changes in lifestyle can reduce the risk of some cancers.

Drugs and disease

Depressant and stimulant drugs affect the nervous system by affecting the transmission across synapses.

The mosquito is a vector that carries malaria.

Immunisation protects against certain diseases by using harmless pathogens.

Plasmodium is the pathogen that causes malaria. It is a parasite and humans are its host.

High levels of alcohol can cause cirrhosis of the liver.

Homeostasis is maintaining a constant internal environment.

Auxins are a group of plant hormones. They cause shoot curvature by cell elongation.

Homeostasis and plant hormones

Automatic systems in the body keep water, temperature and carbon dioxide levels constant.

Auxins are involved in phototropism (response to light) and geotropism (response to gravity).

The hormone insulin controls blood sugar levels. It converts excess blood glucose into glycogen.

Vasodilation and vasoconstriction control heat transfer from the body.

Plant hormones have many commercial uses (selective weedkiller, rooting powder, control of fruit ripening).

Light rays are refracted as they pass through the cornea and lens.

A nerve impulse travels along the axon of a neurone.

The nervous system

The eye accommodates by altering the shape of the lens.

A neurotransmitter substance diffuses across a synapse, so the nerve impulse can pass to the next neurone.

Monocular vision has a wider field of view but poorer distance judgement than binocular vision.

Long and short sight is caused by the eyeball or lens being the wrong shape.

A spinal reflex involves a receptor, sensory, relay and motor neurones and an effector.

Alleles are different versions of the same gene.

Dominant alleles are expressed if present, recessive alleles are expressed in the absence of a dominant allele.

Variation and inheritance

Sex is determined by sex chromosomes, XX in female, XY in male.

Being homozygous is having two identical alleles, being heterozygous is having two different alleles,

Human body cells have 23 pairs of chromosomes.

Inherited disorders are caused by faulty alleles.

Most faulty alleles are recessive.

Classification

Grouping organisms

- All organisms are classified into a number of different groups, starting with their kingdom and finishing with their **species**.

- The groups are: kingdom, phylum, class, order, family, genus and species.

- As you move down towards 'species', there are fewer organisms within each group and they share more similarities.

Remember! Try remembering the order of the groups by using the first letter of each one.

- Organisms can be classified in two ways:
  - An *artificial system* is based on one or two characteristics that make identification easier, for example birds that always live by or on the sea can be called seabirds.
  - A *natural system* is based on **evolutionary** relationships and is much more detailed. Animals that are more closely related are more likely to be in the same group.

- Sequencing the bases in **DNA** has enabled scientists to know much more about how closely related organisms are, and has often meant that organisms can be reclassified.

Species

- A species is a group of organisms that can interbreed to produce fertile offspring.

- All organisms are named by the **binomial system**. The system works like this:
  - There are two parts to the name, the first is the genus and the second the species.
  - The genus part starts with a capital letter; the species part starts with a lower-case letter.

Problems with classifying

- Living things are at different stages of evolution, and new ones are being discovered all the time. This makes it difficult to place organisms into distinct groups. An example of this is *Archaeopteryx*. This creature had characteristics that would put it into two different groups:
  - It had feathers, like a bird.
  - It also had teeth and a long, bony tail, like a **reptile**.

- Some organisms present specific problems:
  - **Bacteria** do not interbreed, they reproduce **asexually**, so they cannot be classified into different species using the 'fertile offspring' idea.
  - Mules are **hybrids**, produced when members of two species (a donkey and a horse) interbreed. Hybrids are infertile, so mules cannot be classed as a species.

Classification and evolution

- Organisms that are grouped together are usually closely related and share a recent common ancestor. However, they may have different features if they live in different **habitats**. `

- When classifying organisms, it is important to bear in mind that similarities and differences between organisms may have different explanations:
  - Dolphins have similarities to fish because they live in the same habitat (ecologically related). However, they are classified differently – dolphins are **mammals**.
  - Dolphins and bats have evolved to live in different habitats, but both are mammals – they are related through evolution.

How science works

You should be able to:
- explain how a scientific idea has changed as new evidence has been found.

Improve your grade

Classifying newly discovered organisms

Two similar types of animals have been discovered living close together in a jungle.
Describe how scientists could find out how closely related the two animals are.
AO2 [3 marks]

Energy flow

Pyramids of biomass

- Pyramids of numbers and pyramids of **biomass** can both be used to represent feeding relationships between organisms in a food chain or web.

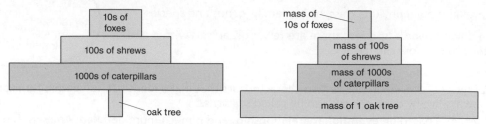

Pyramids of numbers and biomass for a woodland

- Pyramids of biomass show the *dry mass* of living material at each stage of a food chain.

- They may look different to pyramids of numbers if:
  - **producers** are very large
  - a small **parasite** lives on a large animal.

- Although pyramids of biomass are a better way of representing **trophic levels** they are difficult to construct. This is because:
  - some organisms feed on organisms from different trophic levels
  - measuring dry mass is difficult as it involves removing all the water from an organism, which will kill it.

Energy flow

- As energy flows along a food chain some is used in growth. However, at each trophic level much of the energy is transferred into other, less useful forms, such as:
  - heat from **respiration**
  - **egestion**
  - **excretion**.

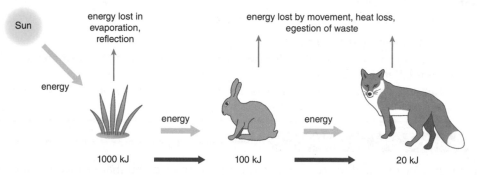

Energy flow in a food chain

- The material that is lost at each stage of the food chain is not wasted. Most of the waste is used by **decomposers** that can then start another food chain.

- Because each trophic level 'loses' up to 90 per cent of the available energy, an animal at the end of a long food chain does not have much food available to it.

- The **efficiency** of energy transfer can be calculated between trophic levels:

> What is the efficiency of energy transfer between the rabbit and the fox?
> $$\frac{\text{energy used for growth}}{\text{energy input}} = \frac{20}{100} = 0.2 \text{ or } 20\%$$

Improve your grade

Pyramids of biomass

Explain one advantage and one disadvantage of using pyramids of biomass to show feeding relationships. *AO2* [3 marks]

  B2 Understanding our environment

Recycling

The carbon cycle

- **Carbon** is one of a number of elements that are found in living organisms.
- Carbon needs to be **recycled** so it can become available again to other living organisms.
- **Carbon dioxide** is removed from the air by **photosynthesis** in plants.
- Feeding passes carbon compounds along a food chain or web.
- Carbon dioxide is released into the air by:
  – plants and animals **respiring**
  – soil bacteria and fungi acting as **decomposers**
  – the burning of **fossil fuels (combustion)**.
- Carbon dioxide is also absorbed from the air by oceans. Marine organisms make shells made of carbonate, which become **limestone** rocks.
- The carbon in limestone can return to the air as carbon dioxide during volcanic eruptions or weathering.

Remember!
Carbon dioxide can be locked up in limestone for a long time. The oceans are often called a carbon sink.

 D–C

B–A*

The nitrogen cycle

- Plants take in nitrogen as nitrates from the soil to make protein for growth.
- Feeding passes nitrogen compounds along a food chain or web.
- The nitrogen compounds in dead plants and animals are broken down by decomposers and returned to the soil.

- A number of **microorganisms** are responsible for the recycling of nitrogen:
  – Decomposers are soil **bacteria** and fungi and they convert proteins and urea into ammonia.
  – **Nitrifying bacteria** convert the ammonia to nitrates.
  – **Denitrifying bacteria** convert nitrates to nitrogen gas.
  – **Nitrogen-fixing bacteria** living in root nodules (or in the soil) fix nitrogen gas – this also occurs by the action of lightning.

EXAM TIP

Diagrams of the carbon or nitrogen cycle may not look exactly like this one. Just look for where the arrows go to and from.

D–C

B–A*

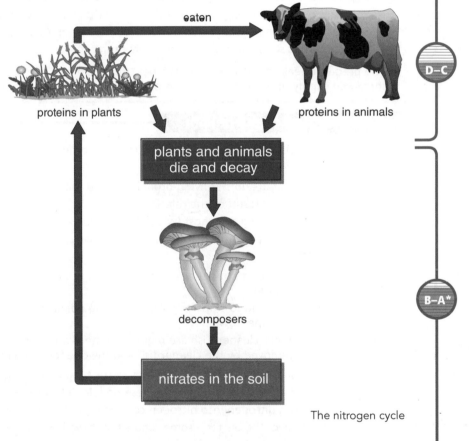

eaten

proteins in plants   proteins in animals

plants and animals die and decay

decomposers

nitrates in the soil

The nitrogen cycle

Keeping decomposers working

- For decomposers to break down dead material in soil, they need oxygen and a suitable **pH**.
  – **Decay** will therefore be slower in waterlogged soils as there will be less oxygen.
  – Acidic conditions will also slow down decay.

D–C

Improve your grade

The nitrogen cycle

Explain how nitrogen in a protein molecule in a dead leaf can become available again to a plant. *AO1* [4 marks]

Interdependence

Competition

- Similar animals living in the same **habitat** compete with each other for resources (e.g. food).
- If they are members of the same **species** they will also compete with each other for mates so they can breed.

- An **ecological niche** describes the habitat that an organism lives in and also its role in the habitat. For example, ladybirds live on trees such as sycamore and eat greenfly.
- Organisms that share similar niches are more likely to compete, as they require similar resources. The harlequin ladybird arrived in Britain in 2004 and competes strongly with native ladybirds.
- Competition can be *interspecific* and *intraspecific*:
  - Interspecific is between organisms of different species.
  - Intraspecific is between organisms of the same species and is likely to be more significant as the organisms share more similarities and so need the same resources.

Predator–prey relationships

- Both **predator** and **prey** show cyclical changes (ups and downs) in their numbers. This is because:
  - When there are lots of prey, more predators survive and so their numbers increase.
  - This means that the increased number of predators eat more prey, so prey numbers drop.
  - More predators starve and so their numbers drop.
- The predator peaks occur soon after the peaks of the prey. This is because it takes a little while for the increased supply of food to allow more predators to survive and reproduce.

Parasitism and mutualism

- As well as competing with each other or eating each other, organisms of different species can also be dependent on each other in other ways.
- **Parasites** feed on or in another living organism called the host.
  - The host suffers as a result of the relationship.
  - Fleas are parasites living on a host (which may be human).
  - Tapeworms are also parasites feeding in the digestive systems of various animals.

Remember!
A successfully adapted parasite does not kill its host quickly, as it would then need to find another one.

- Sometimes both organisms benefit as a result of their relationship. This is called **mutualism**.
  - Insects visit flowers and so transfer pollen, allowing **pollination** to happen. They are 'rewarded' by sugary nectar from the flower.
  - On some coral reefs 'cleaner' fish are regularly visited by larger fish. The large fish benefit by having their parasites removed by the cleaner fish and the cleaner fish gain food.
- Pea plants and certain types of bacteria also benefit from mutualism. Pea plants are **legumes** with structures on the roots called nodules. In these nodules live **nitrogen-fixing bacteria**.
  - The bacteria turn nitrogen into nitrogen-containing chemicals and give some to the pea.
  - The pea plant gives the bacteria some sugars that have been produced by **photosynthesis**.

How science works

You should be able to:
- Identify some arguments for and against a scientific or technological development in an unfamiliar situation, in terms of its impact on different groups of people or the environment.

Improve your grade

Niches

The scientist Gauss put forward a theory that said organisms of two different species cannot share an identical ecological niche.

(a) Explain what is meant by the term 'ecological niche'. *AO1* [2 marks]

(b) Suggest why Gauss said that two species cannot share the same niche. *AO2* [2 marks]

Adaptations

Adapting to the cold

- Some animals are adapted to living in very cold conditions. They keep warm by reducing heat loss. Some have anatomical **adaptations** to help reduce heat loss.

- So:
  - They have excellent **insulation** to cut down heat loss. The arctic fox has thick fur that traps plenty of air for insulation. Seals have thin fur but a thick layer of fat under the skin.
  - These animals are usually quite large, with small ears. This helps to decrease heat loss by decreasing the surface area to volume ratio.

- Animals may try to avoid the cold by changing their behaviour. Some migrate long distances to warmer areas. Others slow down all their body processes and hibernate.

- Penguins have a counter-current heat exchange mechanism to help reduce heat loss. The warm blood entering the flippers warms up the cold blood leaving, to stop it cooling the body.

- Other organisms that live in cold climates may have biochemical adaptations, such as antifreeze proteins in their cells.

Adapting to hot, dry conditions

- Organisms such as camels and cacti live in deserts, in very hot, dry conditions.

- To increase heat loss, animals adapt in a variety of ways.
  - Some are anatomical adaptations, for example camels increase the loss of heat by having very little hair on the underside of their bodies. Animals that live in hot areas are usually smaller and have larger ears than similar animals that live in cold areas. These factors give them a larger surface area to volume ratio, so that they can lose more heat.
  - Other adaptations to lose more heat are behavioural, such as panting or licking their fur.

- To reduce heat gain, animals may change their behaviour, for example they seek shade during the hotter hours around the middle of the day.

- To cope with dry conditions, organisms have behavioural, anatomical and physiological adaptations. For example:
  - Camels can survive with little water because they can produce very concentrated urine.
  - Cacti reduce water loss because their leaves have been reduced to spines. They also have deep roots and can store water in the stem.

- Organisms that can survive in hot conditions are called extremophiles. Some bacteria can live in hot springs as they have **enzymes** that do not denature at **temperatures** as high as 100°C.

Spines on a cactus

Specialists or generalists

- Some organisms, like polar bears, are called *specialists*, as they are very well adapted to living in specific **habitats**. They would struggle to live elsewhere.

- Others, for example rats, can live in several habitats.
  - These organisms are called *generalists*.
  - They will lose to the specialists in certain habitats.

EXAM TIP

You may have to write about different organisms. Just apply the same principles.

Improve your grade

Living in hot, dry conditions

An elephant has a large body, large ears, skin with few hairs and the ability to produce concentrated urine. Explain which of these features are advantages or disadvantages when living in hot, dry areas. *AO2* [4 marks]

Natural selection

Charles Darwin and natural selection

- Over 150 years ago Charles Darwin wrote his theory of **natural selection** to explain how **evolution** might happen. It says that if animals and plants are better adapted to their environment, they and the following generations are more likely to survive.

- He did not know exactly how **adaptations** were passed on. We now know that when organisms reproduce, their **genes** are passed on to the next generation.

- The modern version of natural selection can be summarised like this:
  - Within any **species** there is **variation**.
  - Organisms produce far more young than will survive, so there is competition for limited resources such as food.
  - Only those best adapted will survive, which is called *survival of the fittest*.
  - Those that survive pass on successful adaptations to the next generation in their genes.

- Over time, the changes produced by natural selection may result in a new species.
  - This only happens if different groups of organisms cannot mate for a long time.
  - The organisms might be prevented from mating because they live in different areas. This is geographical isolation. They might be prevented from mating because of behavioural isolation.
  - If each group evolves differently they might over time become different enough to be classified as separate species.

Modern examples of natural selection

- Natural selection is difficult to study because it usually takes thousands of years to see the effect. Some examples have been studied over shorter time spans:
  - More and more **bacteria** are developing resistance to **antibiotics**.
  - Peppered moths are dark or pale in colour. Dark moths are better camouflaged in polluted areas, so more of them survive.

Light and dark peppered moths on a tree trunk

Arguments over natural selection

- At first many people disagreed with Darwin's ideas:
  - Some people thought he did not have enough evidence to back up his theory.
  - Many people disagreed because they thought God had created all species.

- Now Darwin's theory is much more widely accepted. This is because:
  - it explains lots of observations
  - it has been discussed and tested by a wide range of scientists.

- There have been other attempts to explain evolution. Before Darwin, Jean Baptiste de Lamarck had a different theory, called the law of acquired characteristics. This said, for example, that giraffes acquired long necks to feed, and this characteristic was passed on.

- As we have discovered more about how genes are passed on, theories like Lamarck's have been proved incorrect and Darwin's theory has become more widely accepted.

Improve your grade

Explanations for evolution

Human ancestors had more hair than modern humans. This could be explained by saying that scratching the skin due to parasites has gradually over many generations made some of the hair fall out.

(a) Explain why this theory uses Lamarck's ideas. *AO2* [2 marks]

(b) How might Darwin's ideas be used to explain why modern humans have less hair? *AO2* [2 marks]

Population and pollution

Pollution

- There are many different types of **pollution**. Three that have caused much concern are:
  - **carbon dioxide**, from increased burning of **fossil fuels**, which may increase the greenhouse effect and **global warming**
  - **CFCs**, from aerosols, which destroy the **ozone layer**
  - sulfur dioxide, from burning fossil fuels, which causes **acid rain**.

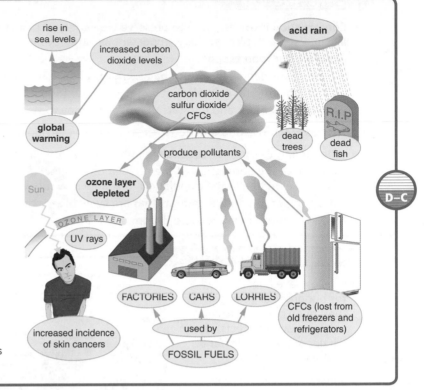

The source and effects of three pollutants

Population and pollution

- The human **population** of the world is growing at an ever-increasing rate. This is called **exponential growth**.

- This growth in population is happening because the birth rate is exceeding the death rate.

- The greatest rise in world population figures is occurring in under-developed land masses, such as Africa and India. However, the developed world uses the most fossil fuels (principally the USA and Europe).

- The amount of pollution caused per person or organisation is called the '**carbon footprint**.' This measures the total **greenhouse gas** given off by a person or organisation within a certain time.

Measuring pollution

- Pollution in water or air can be measured using direct methods or by indicator organisms.

- Direct methods include oxygen probes attached to computers that can measure the exact levels of oxygen in a pond. Special chemicals can be used to indicate levels of nitrate pollution from **fertilisers**.

- The presence or absence of an **indicator species** is used to estimate levels of pollution. For example:
  - The mayfly larva is an insect that can only live in clean water.
  - The water louse, bloodworm and mussels can live in polluted water.
  - Lichen grows on trees and rocks but only when the air is clean. It is unusual to find lichen growing in cities, because it is killed by the pollution from motor engines.

Remember!
If many different species are present in a habitat, this is usually a sign of low levels of pollution.

- There are advantages to the different methods of measuring pollution:
  - Using indicator organisms is cheaper, does not need equipment that can go wrong and monitors pollution levels over long periods of time.
  - Using direct methods can give more accurate results at any specific time.

Improve your grade

Population and pollution
Explain the reasons for the increase in carbon dioxide levels in the atmosphere and explain why people are concerned about this. *AO1* [4 marks]

Sustainability

Conservation

D–C

- **Conservation** involves trying to preserve the variety of plants and animals and the **habitats** that they live in. People think that this is important because it can:
  - protect our food supply
  - prevent any damage to food chains, which can be hard to predict
  - protect plants and animals that might be useful for medical uses
  - protect organisms and habitats that people enjoy to visit and study.
- **Species** are at risk of **extinction** if the number of individuals or habitats falls below critical levels.

B–A*

- When trying to conserve species the important factors to bear in mind include:
  - the size of the **population** (if the population is below a critical level there is unlikely to be enough genetic **variation** in the population to enable it to survive)
  - the number of suitable habitats that are available for the organism to live in
  - how much competition there is from other species.

Whale conservation

D–C

- Whales have been hunted for hundreds of years for their body parts, which are used in many products. Live whales are also important to the tourist trade.
- Some whales are kept in captivity for research or **captive breeding** programmes or just for our entertainment. However, many people object when whales lose their freedom.

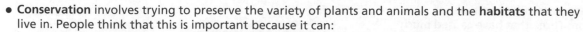

skin: used in belts, shoes, handbags and luggage

sinews: used in tennis rackets

spermacetti: used in high-grade machine oil

oil: sperm whale oil taken from bone and skin used in high-grade alcohol, shoe cream, lipstick, ointment, crayons, candles, fertiliser, soap and animal feeds

whalemeat: used in pet food and human food

teeth: used in buttons, piano keys and jewellery

liver: used in oil

bone: used in fertiliser and animal feeds

ambergris: from intestine, used in perfumes

Many uses of whale parts

B–A*

- It has been difficult to set up and police international agreements on whaling and some countries want to lift the ban on whaling.
- Scientists believe there is a need to kill some whales to help find out more about how whales can survive at extreme depths. However, they could study the whales without killing them. Migration patterns and whale communication can only be investigated if the animal is alive.

Sustainable development

D–C

- **Sustainable development** means taking enough resources from the environment for current needs, while leaving enough for the future and preventing permanent damage. For example:
  - Fishing quotas are set, so that there are enough fish left to breed.
  - Woods are replanted to keep up the supply of trees.

B–A*

- As the world population increases it is crucial to carry out sustainable development.
  - **Fossil fuels** will run out. As there is an increase in demand for **energy**, we must manage alternative fuels, such as wood.
  - We need to supply increasing amounts of food for growing populations without destroying large areas of natural habitats.
  - Large amounts waste products must be **disposed** of so as to prevent or minimise pollution.
- If all of this is achieved it should help to save **endangered** species.

Improve your grade

Saving endangered species

The Hawaiian goose is only found on the islands of Hawaii. In the mid-1900s only about 30 were left alive. The space on the islands is restricted and a number of animals have been introduced to the islands. Write about the problems facing scientists trying to save the goose from extinction. *AO2* [4 marks]

> **EXAM TIP**
> You should be able to suggest how an endangered organism could best be conserved.

B2 Summary

Classification and living together

Organisms are classified into discrete groups starting with kingdom and ending with species.

This causes some problems with:
- intermediate organisms
- hybrids
- asexual organisms.

Organisms can:
- eat each other (predators)
- gain from each other (mutualism)
- feed off each other (parasites).

Organisms are classified using natural systems.

This gives information about evolutionary relationships.

Similar organisms will compete with each other for food.

Organisms that share the same niche or are in the same species will compete more.

Energy flow and recycling

Pyramids of biomass are harder to construct but always form pyramids.

- Energy is lost from each stage of a food chain.

Food chains are limited to a small number of trophic levels.

Pyramids of biomass and numbers can show feeding relationships.

The recycling of nitrogen involves the action of four types of bacteria.

The recycling of carbon involves:
- photosynthesis
- feeding
- respiration
- decomposition.

Adaption and natural selection

Organisms in hot, dry areas have adaptations to:
- increase heat loss
- move on sand
- cope with lack of water.

Heat loss from organisms depends on their surface area to volume ratio.

Organisms in cold conditions are adapted to:
- keep warm
- move on the snow.

Darwin's theory of natural selection involves variation, competition, survival of the fittest and selective reproduction.

Darwin's theory was widely criticised at first but is now widely accepted.

Examples of natural selection occurring today are:
- warfarin resistance in rats
- antibiotic resistance in bacteria
- frequency of colour in peppered moths.

Population, pollution and sustainability

Human population has been increasing exponentially.

This has led to an increase in pollutants, such as:
- carbon dioxide causing global warming
- sulfur dioxide causing acid rain
- CFCs breaking down the ozone layer.

Removing waste, producing food and supplying energy in a sustainable way will help to conserve habitats and organisms.

Pollution can be measured using direct methods or by using indicator species.

Conservation is important to:
- protect our food supply
- prevent damage to food chains
- protect organisms for medical uses
- protect habitats for people to visit.

Making crude oil useful

Fossil fuels

- **Fossil fuels** are **finite resources** because they are no longer being made, or are being made extremely slowly.
- Fossil fuels are being used up faster than they are being formed. They are called a **non-renewable** resource.

- Specific difficulties associated with the finite nature of **crude oil** include:
  - all the readily extractable resources will be used up in the future
  - finding replacements.

Fractional distillation

- Crude oil is a mixture of many types of oil, which are all 'hydrocarbons'. A hydrocarbon is made up of molecules containing **carbon** and hydrogen only.
- Crude oil is heated at the bottom of a fractionating column.
  - Oil that doesn't boil sinks as a thick liquid to the bottom. This is **bitumen**. Bitumen has a very high **boiling point**. It 'exits' at the bottom of the column.
  - Other fractions, containing mixtures of hydrocarbons with similar boiling points, boil and their gases rise up the column. The column is cooler at the top. Fractions with lower boiling points 'exit' towards the top of the column.

- Crude oil can be separated because the hydrocarbons in different fractions have differently sized molecules.
  - The forces between the molecules are **intermolecular forces** and are broken during boiling.
  - The molecules of the liquid separate from each other as molecules of gas.
  - Large molecules, such as those of bitumen and heavy oil, have strong forces of attraction. A lot of energy is needed to break the forces between the molecules. These fractions have high boiling points.
  - Smaller molecules, such as **petrol**, have weak attractive forces between them and are easily separated. Less energy is needed to break the forces between the molecules. These fractions have low boiling points.

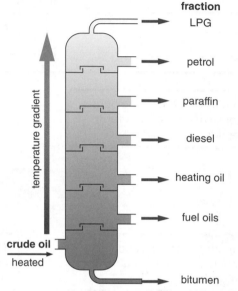

A fractional distillation column

Problems in extracting crude oil

- Transporting oil can cause problems. Oil slicks can damage birds' feathers and cause their deaths. Clean-up operations use detergents that can damage wildlife.

- There may be political problems related to the extraction of crude oil, particularly where the UK is dependent on oil and gas from politically unstable countries. Oil-producing nations can set high prices and cause problems for the future supply of non-oil producing nations.
- Because the demand for oil and its products is very high, there is a conflict between the needs for making petrochemicals and for making fuels. A fraction called naphtha is in high demand for use in medicines, plastics and dyes.

> **Remember!**
> In fractional distillation, the intermolecular forces *between* the molecules are broken during boiling. The covalent bonds *within* the molecules do not break.

Cracking

- **Cracking** is a process that turns large **alkane** molecules into smaller alkane and **alkene** molecules. An alkene molecule has a double bond, which makes it useful for making polymers.

- Cracking also helps oil manufacturers match supply with demand for products like petrol.

Improve your grade

Fractional distillation

Why can crude oil be separated using fractional distillation? *AO1* [3 marks]

C1 Carbon chemistry

Using carbon fuels

Choosing fuels

- A fuel is chosen because of its key features. For example, coal produces more **pollution** than **petrol**, as you can see in the table.

Key features	Coal	Petrol
energy value	high	high
availability	good	good
storage	bulky and dirty	volatile
cost	high	high
toxicity	produces acid fumes	produces less acid fumes
pollution caused	acid rain, carbon dioxide, soot	carbon dioxide, nitrous oxides
ease of use	easier to store for power stations	flows easily around engines

- The amount of **fossil fuels** being burnt is increasing because **populations** are increasing.
  - Governments are concerned because of the increasing **carbon dioxide** emissions that result when fossil fuels are burned.
    Countries with huge populations, such as India or China, are now using more fuel, which adds further to gas emissions.
  - Many governments have pledged to try to cut carbon dioxide emissions over the next 15–20 years. It is a global problem that cannot be solved by one country alone.

Combustion

- Burning hydrocarbon fuels in plenty of air produces carbon dioxide and water.

  methane + oxygen → carbon dioxide + water

- This can be shown using an experiment in the laboratory.

- **Complete combustion** occurs when a fuel burns in plenty of oxygen.
  - More energy is released during complete combustion than during incomplete **combustion**.
  - **Toxic gas (carbon monoxide)** and soot (carbon) is made during incomplete combustion.
  - The word equations for incomplete combustion are:

    fuel + oxygen → carbon monoxide + water

    or

    fuel + oxygen → carbon + water

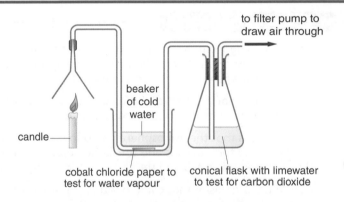

Fuels burn in oxygen to make carbon dioxide and water

Remember!
Complete combustion gives carbon dioxide and water. Incomplete combustion gives carbon monoxide and water or carbon and water.

- Given the **molecular formula** of a hydrocarbon, **balanced symbol equations** can be constructed for:
  - complete combustion

    $CH_4 + 2O_2 \rightarrow CO_2 + 2H_2O$

  - or incomplete combustion

    $2CH_4 + 3O_2 \rightarrow 2CO + 4H_2O$ or $CH_4 + O_2 \rightarrow C + 2H_2O$

EXAM TIP

When you write equations always write the molecular formulae first. Then add up the number of atoms on each side and balance by changing the number of molecules on each side.

Improve your grade

Choosing a fuel

Use the table at the top of the page to decide which fuel should be used in a car engine. Justify your answer from the evidence. *AO3* [2 marks]

Clean air

What is in clean air?

- Clean air is made up of 78% nitrogen, 21% oxygen and of the remaining 1%, only 0.035% is **carbon dioxide**.

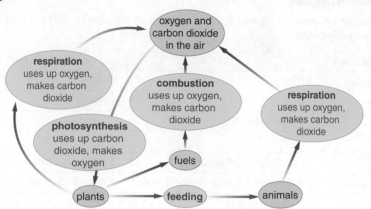

The carbon cycle

- These percentages change very little because there is a balance between the processes that use up and make both carbon dioxide and oxygen.

- Some of these processes are shown in the **carbon cycle**. The arrows in the diagram show the direction of movement of carbon compounds.

- Over the last few centuries the percentage of carbon dioxide in air has increased slightly due to a number of factors, including:
  - **deforestation** – as more rainforests are cut down, less **photosynthesis** takes place
  - increased **population** – as populations increase, the world's energy requirements increase.

The atmosphere

- Gases escaping from the interior of the Earth formed the original atmosphere. Plants that could photosynthesise removed carbon dioxide from the atmosphere and added oxygen. Eventually the amount of oxygen reached its current level.

- Gases come from the centre of the Earth through volcanoes in a process called **degassing**. Scientists analyse the composition of these gases to form theories about the original atmosphere.

- One theory is that the atmosphere was originally rich in water vapour and carbon dioxide. This vapour condensed to form oceans and the carbon dioxide dissolved in the water. The percentage of nitrogen slowly increased and, being unreactive, little nitrogen was removed.

- Over time, organisms that could photosynthesise evolved and converted carbon dioxide and water into oxygen. As the percentage of oxygen in the atmosphere increased, the percentage of carbon dioxide decreased, until today's levels were reached.

Pollution control

- It is important to control atmospheric **pollution** because of the effects it can have on people's health, the natural environment and the built environment.

- Sulfur dioxide is a **pollutant** that can cause difficulties for people with asthma. It can also dissolve in water to form **acid rain** that damages wildlife and **limestone** buildings.

- A car fitted with a **catalytic converter** changes **carbon monoxide** into carbon dioxide.

- In a catalytic converter, a reaction between nitric oxide and carbon monoxide takes place on the surface of the **catalyst**. The two gases formed are natural components of air – nitrogen and carbon dioxide.

$$2CO + 2NO \rightarrow N_2 + 2CO_2$$

Remember!
To get a B–A* grade you need to know why the high temperature inside an internal combustion engine allows nitrogen from the air to react with oxygen to make oxides of nitrogen.

Improve your grade

Evolution of the atmosphere

Describe a theory, put forward by scientists, for the formation of the atmosphere we have today.
AO1 [5 marks]

Making polymers

Hydrocarbons

- A hydrocarbon is a compound of carbon and hydrogen **atoms** only.
  - **Alkanes** are hydrocarbons that have single **covalent bonds** only.
  - **Alkenes** are hydrocarbons that have a **double covalent bond** between carbon atoms. Double bonds involve two shared pairs of electrons.
  - Propane, C_3H_8, is a hydrocarbon because it has only C and H atoms. It is an alkane because all the bonds are single covalent bonds.

  - Propanol, C_3H_7OH, is *not* a hydrocarbon because it contains an oxygen atom.

  - Propene, C_3H_6, is a hydrocarbon because it contains only C and H atoms. It is an alkene because it has a double covalent bond between carbon atoms. Propene is also a monomer. Poly(propene) is the **polymer**.

- **Bromine** is used to test for an alkene. When orange bromine water is added to an alkene it turns colourless (**decolourises**).

- The bromine and alkene form a new compound by an addition reaction. A **di-bromo compound** forms which is colourless.
- A saturated compound only has single covalent bonds between carbon atoms. Alkanes, like propane, are saturated. They have no double bond between carbon atoms.
- An unsaturated compound has at least one double covalent bond between carbon atoms. Alkenes, like propene, are unsaturated. They have a C=C double bond.

Polymerisation

- Addition polymerisation is the process in which many alkene **monomers** react to give a polymer. This reaction needs high pressure and a **catalyst**.
- You can recognise a polymer from its **displayed formula** by looking out for the following: a long chain, the pattern repeating every two carbon atoms, two brackets on the end with extended bonds through them, an '*n*' after the brackets.

> ### EXAM TIP
> When you are constructing the displayed formula for an addition polymer from a monomer, first draw the monomer without the double bond to see what the repeating unit looks like.

- This is the displayed formula of poly(ethene):

- The displayed formula of:
  - an addition polymer can be constructed when the displayed formula of its monomer is given
  - a monomer can be constructed when the displayed formula of its addition polymer is given

- This is the displayed formula of the ethene monomer:

- During an addition polymerisation reaction a long chain is made until it is stopped. This long molecule is poly(ethene). The reaction causes the double bond in the monomer to break and each of the two carbon atoms forms a new bond.
- Addition polymerisation involves the reaction of many unsaturated monomer molecules (alkenes) to form a saturated polymer.

⊙ Improve your grade

Interpreting displayed formulae

Explain why this molecule is unsaturated and whether it can be made into a polymer. *AO2* [3 marks]

Designer polymers

Breathable polymers

- Nylon is tough, lightweight, keeps water out and keeps UV (**ultraviolet**) light out but does not let water vapour through. This means that sweat condenses and makes the wearer wet and cold inside their jacket.

- GORE-TEX® has all the properties of nylon but is also breathable, so it is worn by many active outdoor people. Water vapour from sweat can pass through the membrane but rainwater cannot.

- GORE-TEX® material is waterproof and yet breathable.
  - It is made from a PTFE (polytetrafluoroethene)/polyurethane membrane.
  - The holes in PTFE are too small for water to pass through but are big enough for water vapour to pass through.

- PTFE/polyurethane membrane is too fragile on its own and so it is laminated onto nylon to produce a stronger fabric.

Disposing of polymers

- Scientists are developing new types of polymers:
  - polymers that dissolve
  - **biodegradable** polymers.

- Research into new polymers is important because there are environmental and economic issues with the use of existing polymers.
  - **Disposal** of non-biodegradable polymers means landfill sites get filled quickly.
  - Landfill means wasting land that could be valuable for other purposes.
  - Disposal by burning waste plastics makes **toxic** gases.
  - Disposal by burning or using landfill sites wastes the **crude oil** used to make the polymers.
  - It is difficult to sort out different polymers so **recycling** is difficult.

Stretchy polymers and rigid polymers

- **Atoms** in polymers are held together by strong **covalent bonds**.

- The properties of plastics can be related to simple models of their structure.
  - Plastics that have weak **intermolecular forces** between polymer molecules have low **melting points** and can be stretched easily as the polymer molecules can slide over each other.
  - Plastics that have strong forces between the polymer molecules (covalent bonds or cross-linking bridges) have high melting points, cannot be stretched and are rigid.

Remember!
Intermolecular forces are weak forces of attraction *between* molecules and are not as strong as covalent bonds *within* molecules.

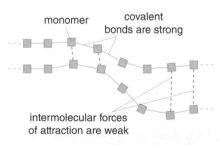

monomer
covalent bonds are strong
intermolecular forces of attraction are weak

If the intermolecular forces between two polymer molecules are weak, the plastic can easily be stretched

Improve your grade

Properties and uses of polymers
Explain why each of the polymers used for drain pipes, electrical cable covers and socks is suitable for its purpose and explain how the structure of the polymer causes it to have these properties. Use ideas about intermolecular forces. *AO2* [6 marks]

Cooking and food additives

Proteins and carbohydrates

- Protein molecules in eggs and meat permanently change shape when eggs and meat are cooked.
- This changing of shape is called 'denaturing'.
- The texture of egg or meat changes when it is cooked because the shapes of the protein molecules change permanently.
- Potato is a carbohydrate which is easier to digest if it is cooked because:
  - the starch grains swell up and spread out
  - the cell walls rupture resulting in the loss of their rigid structure and a softer texture is produced.

D–C

B–A*

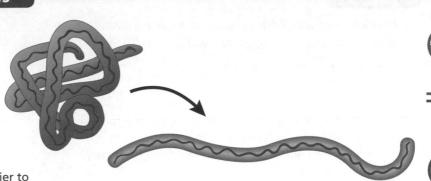

Proteins denature on heating

Baking powder

- Baking powder is sodium hydrogencarbonate.
- When it is heated it breaks down (**decomposes**) to give **carbon dioxide**.
- The word equation for the decomposition of sodium hydrogencarbonate is:

  sodium hydrogencarbonate → sodium carbonate + carbon dioxide + water
- The **balanced symbol equation** for the decomposition of sodium hydrogencarbonate is:

  $2NaHCO_3 \rightarrow Na_2CO_3 + CO_2 + H_2O$

D–C

Emulsifiers

- Emulsifiers are molecules that have a water-loving (hydrophilic) part and an oil- or fat-loving (hydrophobic) part.
- The oil- or fat-loving part (the hydrophobic end) goes into the fat droplet.
- Emulsifiers help to keep oil and water from separating:
  - The hydrophilic end bonds to the water molecules.
  - The hydrophobic end bonds with the oil or fat molecules.
  - The hydrophilic end is attracted to the water molecules which surround the oil, keeping them together.

D–C

B–A*

hydrophilic head

hydrophobic tail

An emulsifying molecule

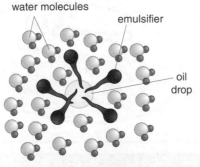

water molecules

emulsifier

oil drop

An emulsion of oil and water

Improve your grade

Using baking powder

Construct the balanced symbol equation for the decomposition of baking powder.
This is sodium hydrogencarbonate, $NaHCO_3$, which decomposes to give sodium carbonate, Na_2CO_3, carbon dioxide and water. *AO1* [2 marks]

Smells

Esters

D–C

- **Alcohols** react with **acids** to make an 'ester' and water.

  alcohol + acid → ester + water

- Esters are used to make perfumes.

- An ester can be made using a simple experiment.
  - The acid is added to the alcohol and heated for some time.
  - The condenser stops the gas from escaping and helps to cool it down again, so that it can react more.
  - The condenser allows the reaction to go on for longer.

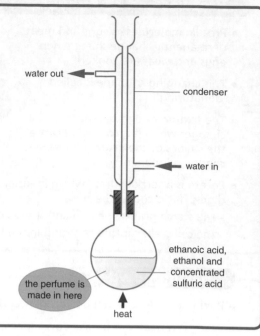

water out

condenser

water in

ethanoic acid, ethanol and concentrated sulfuric acid

the perfume is made in here

heat

Making a perfume

Perfume properties

D–C

- A perfume must have certain properties. It must:
  - evaporate easily so that the perfume particles can reach the nose
  - be non-toxic so it does not poison you
  - not react with water so the perfume does not react with perspiration
  - not irritate the skin so the perfume can be put directly onto the skin
  - be insoluble in water so it cannot be washed off easily.

Solutions

D–C

- A solution is a mixture of solvent and solute that does not separate out.
- Esters can be used as solvents.

Particles

B–A*

- The volatility, or ease of **evaporation** of perfumes, can be explained in terms of **kinetic theory**.
  - In order to evaporate, particles of a liquid need sufficient kinetic energy to overcome the forces of attraction to other molecules in the liquid.
  - Only weak attractions exist between particles of the liquid perfume so it is easy to overcome these attractions as they have sufficient kinetic energy.

- Water will not dissolve nail varnish colours.
  - The attraction between the water molecules is stronger than the attraction between the water molecules and the nail varnish molecules.
  - The attraction between the nail varnish molecules is stronger than the attraction between water molecules and the nail varnish molecules.

Remember!
Intermolecular forces are weak forces of attraction between molecules and can easily be overcome with sufficient kinetic energy.

How science works

You should be able to:

- explain why testing of cosmetics on animals has been banned in the EU and evaluate why people have different opinions about whether the testing of cosmetics on animals is ever justified.

Improve your grade

Esters and perfumes

A perfume must evaporate easily. Explain *how* it evaporates easily, using ideas on kinetic theory and why a perfume needs other specific properties. *AO1* [3 marks]

Paints and pigments

Paint drying

- Most paints dry because:
  - paints are applied as a thin layer
  - the solvent evaporates.

EXAM TIP
When you are constructing an answer that involves particle models it is often easier to draw a labelled diagram.

- Emulsion paints are water-based paints that dry when the solvent evaporates.

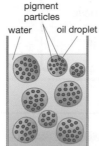

How emulsion paints dry

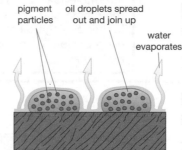 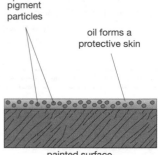

paint in the can   wet paint on a surface   painted surface

- Oil paints dry because:
  - the solvent evaporates
  - the oil is oxidised by atmospheric oxygen.

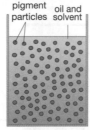 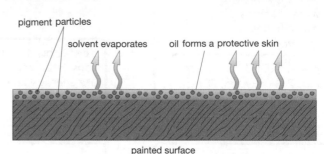

How oil paints dry   paint in the can   painted surface

Thermochromic pigments

- Thermochromic pigments change colour at different **temperatures**. Thermochromic pigments are used:
  - as thermometers because they change colour when the temperature of a body or the temperature of a fridge rises
  - in the manufacture of some cups – the colour changes to show when they are hot
  - in electric kettles to keep users safe when boiling water
  - in babies' spoons and bath toys, to warn if the spoon or toy is too hot to give to a baby.

- Thermochromic pigments can be added to acrylic paints to make even more colour changes. If a blue thermochromic pigment which turns colourless when hot is added to yellow acrylic paint, the paint will appear green when cool and yellow when hot.

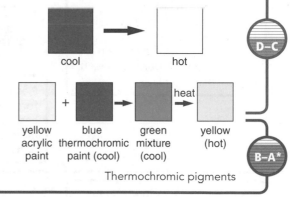

Thermochromic pigments

Phosphorescent pigments

- Phosphorescent pigments glow in the dark because:
  - they absorb and store **energy**
  - they release the energy as light over a period of time.

- Phosphorescent pigments are much safer than the older, alternative radioactive paints.

Improve your grade

Drying paint
Explain how paints dry in different ways. *AO1* [3 marks]

C1 Summary

Crude oil is a mixture of many hydrocarbons.

Fossil fuels are finite resources because they are no longer being made or are being made extremely slowly. They are non-renewable resources as they are being used up faster than they are being formed.

Making crude oil useful

Crude oil can be separated by fractional distillation. It is heated near the bottom of a fractionating column. Fractions with low boiling points 'exit' at the top. Fractions with high boiling points 'exit' at the bottom. There is a temperature gradient between the bottom and the top of the column.

Cracking helps an oil refinery match its supply of useful products such as petrol with the demand for them. Cracking converts large alkane molecules into smaller alkane and alkene molecules.

Hydrocarbons with bigger molecules have stronger intermolecular forces between their molecules, and so have higher boiling points than hydrocarbons with smaller molecules.

If hydrocarbons burn in a plentiful supply of air, carbon dioxide and water are made. In an experiment, carbon dioxide can be tested for with limewater; it turns the limewater milky.

Fuels and clean air

The original atmosphere came from the degassing of early volcanoes which were rich in water and carbon dioxide. The water condensed to form oceans. Photosynthetic organisms helped to increase the levels of oxygen through photosynthesis.

The present day atmosphere contains 21% oxygen, 78% nitrogen and 0.035% carbon dioxide.

Key factors that need to be considered when choosing a fuel are energy value, availability, storage, toxicity, pollution caused and ease of use.

Photosynthesis, respiration and combustion are processes in the carbon cycle.

A hydrocarbon is a compound of carbon atoms and hydrogen atoms only.

Addition polymers are made when alkene monomer molecules react together under high pressure and with a catalyst.

Nylon and GORE-TEX® are polymers with suitable properties for particular uses. Nylon is tough, lightweight and keeps water and UV light out. GORE-TEX® has all these properties but allows water vapour to pass out so that sweat does not condense.

Alkanes are hydrocarbons which contain single covalent bonds only. Alkenes are hydrocarbons which contain a double covalent bond between carbon atoms.

Polymers

Plastics that have weak intermolecular forces between the polymer molecules can easily be stretched as the polymer molecules can slide over one another. Rigid polymers have cross-linking bridges.

Protein molecules in eggs and meat denature when they are cooked. The change of shape of the protein molecule is permanent.

Paints are colloids because the particles are mixed and dispersed throughout a liquid but are not dissolved in it. Thermochromic pigments change colour with heat and are used, for example, in babies' toys that are immersed in water to show whether the water is too hot. Phosphorescent pigments glow in the dark and are much safer for use in products than the alternative radioactive substances.

Food, smells and paint

Emulsifiers are molecules with a water-loving (hydrophilic) part and an oil loving (hydrophobic) part. The hydrophilic end bonds to water molecules and the hydrophobic end bonds to oil molecules, keeping the oil and water from separating.

Perfumes need to be able to evaporate so that they can easily reach the nose. To evaporate, particles need sufficient kinetic energy to overcome their attraction to other molecules in the liquid.

The structure of the Earth

The lithosphere

- The outer layer of the Earth is called the **lithosphere**. This layer is (relatively) cold and rigid and comprises the **crust** and top part of the **mantle**.
- The lithosphere is made of **tectonic plates** which are less dense than the mantle below.
- The crust is too thick to drill through, so most of our information about the Earth is collected from **seismic waves** produced by earthquakes and man-made **explosions**.

More on the Earth's structure

- The mantle is the zone between the **core** and the crust. It is cold and rigid just below the crust, but at greater depths it is hot and non-rigid and able to move.

- The Earth's core transfers energy, so the temperature of the mantle increases with depth.

- Convection currents slowly move plates.

- Oceanic crust is denser than continental crust. When these plates collide, the oceanic plate (which is cooler at the margins) sinks, pulling more of the plate down, and partly melting as it reaches the hotter part of the mantle. This is called **subduction**.

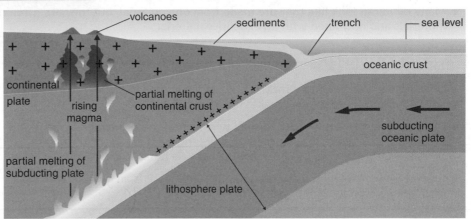

Subduction in oceanic plates

Remember!
Plate movement is due to convection currents. Collisions create mountains.

Plate tectonic theory

- Most scientists now accept the theory of plate tectonics, which suggests that Africa and South America could once have been one land mass, because:
  - it explains a wide range of evidence
  - it has been discussed and tested by many scientists.

- Wegener's continental drift theory (1914) was not accepted by scientists at the time.

- In the 1960s, new sea floor spreading evidence was found. Subsequent research led to Wegener's theory slowly becoming accepted.

Magma and rocks

- **Magma** rises up through the Earth's crust because it is less dense than the crust. This can cause volcanoes.
- Magma can have different types of composition which cause different types of eruption.
- Geologists study volcanoes to try to forecast future eruptions and reveal more about the structure of the Earth.

- Geologists are better able to predict eruptions than they used to be, but still not with 100% certainty.
- Different types of **igneous rock** are formed from lava.
- Iron-rich **basalt** rock comes from runny lava in slower volcanic eruptions. Silica-rich **rhyolite** rock comes from thick lava in explosive eruptions.

Improve your grade

Plate tectonic theory
Wegener's continental drift theory of plate tectonics was not accepted in 1914 but it is now widely accepted. Explain why. *AO1* [4 marks]

Construction materials

Raw materials

- Some raw materials used to make construction materials are found in the Earth's **crust**.

- Hardness can be compared by rubbing two materials together. **Granite** is harder than **marble**, and marble is harder than **limestone**.

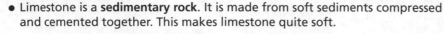

Raw material	clay	limestone and clay	sand	iron ore	aluminium ore
Building material	brick	cement	glass	iron	aluminium

- Many buildings look like they are made from one material, but they are often made with different materials and only lined with an expensive material.

Rock hardness

- Limestone is a **sedimentary rock**. It is made from soft sediments compressed and cemented together. This makes limestone quite soft.

- Marble is a **metamorphic rock** formed when limestone is changed by heat and pressure, and is typically composed of an interlocking mosaic of carbonate crystals. Marble is harder than limestone because it has been baked.

- Granite is formed when **magma** cools and solidifies. It is an **igneous rock** with interlocking crystals. This makes it very hard.

> **Remember!**
> Calcium carbonate is the chemical name for both limestone and marble.

Cement and concrete

- **Thermal decomposition** is a reaction where one substance breaks down on heating to give at least two new substances.

- Calcium carbonate (limestone) thermally decomposes when heated:

calcium carbonate → calcium oxide + carbon dioxide
$$CaCO_3 \rightarrow CaO + CO_2$$

- **Cement** is made when limestone is heated with clay.

- **Concrete** is made by mixing cement, sand and small stones with water.

- **Reinforced concrete** is a **composite material** which has steel rods or meshes running through it. Composites contain at least two materials that can still be distinguished.

> **EXAM TIP**
> For grades D–C you need to be able to construct this balanced symbol equation using the formulae you are given, but at Grades B–A* you need to be able to do it without being given any formulae.

More on reinforced concrete

- Concrete is strong under compression (squashing force) but weak under tension (pulling force).

- If heavy loads are applied to a beam, the concrete will bend. This creates tension and compression. The tension cracks the concrete.

- Reinforced concrete is harder and more flexible than concrete, so is a better construction material.

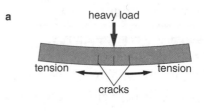

Putting steel rods or mesh in concrete stops cracks appearing in it

steel rod stops concrete from stretching and cracking

Improve your grade

Rock hardness

Explain why granite, marble and limestone have different degrees of hardness.
AO1 [4 marks]

Metals and alloys

D–C

Extracting copper

- Impure copper can be purified in the laboratory using an **electrolysis** cell.

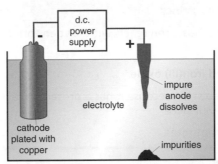

Purifying copper by electrolysis

- Advantages of **recycling** copper are that:
  - it has a fairly low **melting point** so the energy cost to melt it is low
  - it reduces the need for mining, saving reserves and the environmental problems caused by mining
  - it keeps the cost of copper down.

- Disadvantages and problems of recycling copper are that:
  - the small amounts used in electrical equipment are difficult to separate
  - valuable 'pure' copper scrap must not be mixed with less pure scrap, such as **solder**
  - less copper is mined so there are fewer mining jobs
  - the actual separating process may produce **pollution**
  - a lot of copper is thrown away as it is difficult to persuade people to recycle it.

- In the purification of copper by electrolysis, an **electrolyte** of copper(II) sulfate solution is used.

- The impure copper is the **anode**, and a sheet of pure copper is used for the **cathode**.
  - The positive anode loses mass as the copper dissolves.
  - The negative cathode gains mass as pure copper is plated onto it.

- The concentration of the copper(II) sulfate electrolyte stays the same because as the impure copper anode dissolves, pure copper is plated on to the cathode at the same rate.

- At the anode, the Cu **atoms** lose electrons to form Cu^{2+} **ions**. This is called **oxidation**. The electron half equation is:

$$Cu - 2e^- \rightarrow Cu^{2+}$$

- At the cathode, copper is plated when Cu^{2+} ions gain electrons. This is **reduction**. The electron half equation is:

$$Cu^{2+} + 2e^- \rightarrow Cu$$

Remember!
OILRIG
Oxidation Is Loss,
Reduction Is Gain.

B–A*

Alloys

- **Alloys** are mixtures containing as least one metal.

- Alloys have different properties and we match the alloy to the job we use it for. Examples include:
  - **amalgam**, which contains mercury (this is used for filling teeth)
  - **brass**, which contains copper and zinc
  - solder, which contains lead and tin.

D–C

- Some **smart alloys** return to their original shape after being heated to a certain temperature.

- **Nitinol** (nickel-titanium) is a smart alloy which is used to make spectacle frames; it returns to its original shape after being bent if it is put in hot water.

- Smart alloys are becoming more important as new ways to use them are found.

B–A*

Improve your grade

Recycling copper
Suggest advantages and disadvantages of recycling copper. *AO1* [4 marks]

Making cars

Rusting and corrosion

- Only iron and steel **rust**. Other metals **corrode**.
- **Acid rain** and **salt** water accelerate rusting.
- Rusting is an **oxidation** reaction because iron reacts with oxygen forming an oxide.
- The word equation for rusting is:

  iron + oxygen + water → **hydrated iron(III) oxide**

- Aluminium does not corrode in moist air because it has a protective layer of aluminium oxide which, unlike rust, does not flake off the surface.
- Different metals corrode at different rates.

Materials used in cars

- Different materials are used in cars because they have different properties.

Material and its use	Reasons material is used
aluminium in car bodies and wheel hubs	does not corrode, low **density**, **malleable**, quite strong
iron or steel in car bodies	malleable, strong
copper in electrical wires	ductile, good electrical conductor
lead in lead–acid batteries	chemical reaction with lead oxide produces electricity
plastic in dashboards, dials, bumpers	rigid, does not corrode, cheap
pvc in metal wire coverings	flexible, does not react with water, electrical insulator
glass and plastic/glass composite in windscreens	transparent, shatterproof (may crack)
fibre in seats	can be woven into textiles, can be dyed, hard-wearing

- **Alloys** are mixtures of elements containing at least one metal, and often have different and more useful properties than the metals they are made from. For example:
  - steel is harder and stronger than iron
  - steel is less likely to corrode than iron.
- Car bodies can be built from aluminium or steel and there are advantages and disadvantages with each. Aluminium is lighter and more resistant to corrosion than steel. However, steel costs less and is stronger.

> **Remember!**
> You need to be able to give advantages and disadvantages of both aluminium and steel for making car bodies.

- A lighter aluminium body means that fuel economy is improved. Because aluminium corrodes slowly, the car body will also last longer.

Recycling

- Advantages of **recycling** the materials from old cars include:
  - less mining saves **finite resources** needed to make metals
  - less **crude oil** is needed to make new plastics
  - less waste means less landfill
  - fewer **toxic** materials, such as lead from batteries, are dumped.
- Disadvantages of recycling the materials from old cars include:
  - fewer mines are built and fewer mining jobs created
  - difficult to separate the different materials
  - some separating techniques produce **pollution**
  - some recycling processes are very expensive.

> **EXAM TIP**
> Word equations must be given in words. If you use symbols you may not get the marks.

- There are laws which specify that a minimum percentage of all materials used to manufacture cars must be recyclable to help protect the environment.

 Improve your grade

Recycling materials from cars
Give two advantages and two disadvantages of recycling materials from cars.
AO1 [4 marks]

Manufacturing chemicals – making ammonia

The Haber process

- World food production depends on nitrogen **fertilisers**. These fertilisers are made from ammonia, which is made by the **Haber process**.
  - The word equation for the Haber process equation is:

    nitrogen + hydrogen ⇌ ammonia

  - The **balanced symbol equation** for the Haber process equation is:

    $N_2 + 3H_2 \rightleftharpoons 2NH_3$

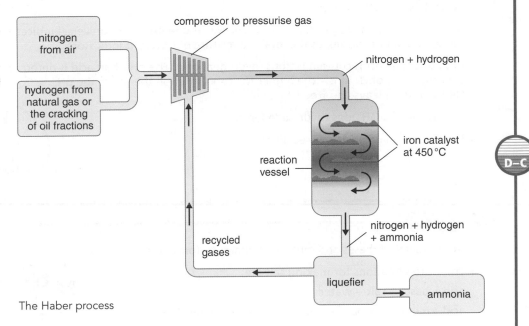

The Haber process

- The **optimum** (best) **conditions** for the Haber process are to:
  - use a **catalyst** made of iron
  - raise the temperature to about 450 °C (fairly low for an industrial process)
  - use high pressure (about 200 atmospheres/20 MPa)
  - **recycle** any unreacted nitrogen and hydrogen.

- The conditions used in the Haber process are designed to make the process as efficient as possible.
  - The iron catalyst increases the reaction rate (rate of successful collisions) but not the **percentage yield**.
  - High pressure increases the percentage yield of ammonia.
  - A high temperature increases the reaction rate.
  - A high temperature breaks down ammonia, reducing percentage yield.
  - The **optimum temperature** is around 450 °C. Although the yield is not very high, the rate is fairly quick. This is the best compromise or 'best of a bad balance'.

What affects the cost of chemical manufacture?

- Costs increase when the pressure is raised (increasing the plant costs), and the temperature is raised (increasing the energy costs).

- Costs decrease when catalysts are used, unreacted starting materials are recycled and automation is used (reducing wage bills).

Remember!
Chemical plants do not aim for the best yield – they aim for the most economical yield.

- These economic considerations determine the conditions used to manufacture a chemical.
  - Both the reaction rate and the percentage yield must be high enough to give a sufficient daily yield of product.
  - A low percentage yield can be accepted if the reaction can be repeated many times with recycled starting materials.
  - Optimum conditions give the lowest cost, rather than the fastest rate or highest yields.

Improve your grade

The Haber process
Explain how a high yield can be obtained using the Haber process. *AO1* [5 marks]

Acids and bases

Acids, bases, alkalis and neutralisation

D–C

- Metal oxides and metal hydroxides are **bases**. A few bases are **soluble** in water, and are called **alkalis**, e.g. sodium hydroxide and calcium hydroxide.

- **Neutralisation** takes place when an **acid** and a base react to make **salt** and water. The word equation for neutralisation is:

  acid + base → salt + water

- Some indicators show a sudden colour change at one **pH** value. Universal indicator shows a gradual range of colour changes, as it contains a mixture of different indicators.

- In solution, all acids contain H^+ **ions** (hydrogen ions). The pH of an acid is determined by the concentration of H^+ ions – the higher the concentration the lower the pH. Neutralisation leaves no free H^+ ions.

B–A*

- Alkalis contain OH^- ions (hydroxide ions).

- Neutralisation involves this reaction:

  $$H^+ \ + \ OH^- \rightleftharpoons H_2O$$

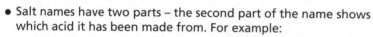

Remember!
You need to be able to work out the names of any salt made from the four common acids in the table.

Salts

D–C

- Acids react with bases and metal carbonates to form salts.

- The word equations for making salt are:

  acid + base → salt + water

  acid + metal carbonate → salt + water + carbon dioxide

- Salt names have two parts – the second part of the name shows which acid it has been made from. For example:

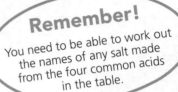

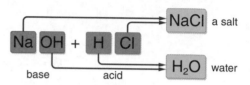

The reaction of sodium hydroxide with hydrochloric acid produces sodium chloride

Common acid	Second part of the salt name
Sulfuric	sulfate
Nitric	nitrate
Hydrochloric	chloride
Phosphoric	phosphate

B–A*

- Some common chemical equations for the neutralisation of acids by a base or metal carbonate can be constructed using:
  - sulfuric acid, nitric acid and hydrochloric acid
  - ammonia, potassium hydroxide, sodium hydroxide and copper oxide
  - sodium carbonate and calcium carbonate.

- The word equation for the reaction between hydrochloric acid and copper carbonate is:

  hydrochloric acid + copper carbonate → copper chloride + water + carbon dioxide

- The symbol equation for the reaction between hydrochloric acid and copper carbonate is:

  $$2HCl \ + \ CuCO_3 \ \rightarrow \ CuCl_2 \ + \ H_2O \ + \ CO_2$$

Improve your grade

Acids and bases

(a) Predict the name of the salt formed when nitric acid reacts with potassium hydroxide. *AO1* [1 mark]

(b) Write a word equation and a balanced symbol equation for the reaction between sulfuric acid and copper oxide. *AO1* and *AO2* [4 marks]

Fertilisers and crop yields

Growing crops

- Farmers use fertilisers to increase their crop yields. These **fertilisers** must be dissolved in water before they can be absorbed by the plant roots as only dissolved substances are small enough to be absorbed.

- Some fertilisers dissolve easily, but some are designed for slow release giving crops a small amount over a long time.

- Fertilisers are needed because the world **population** is rising and there is a greater demand for food production from the land available.

- Fertilisers increase crop yield by:
  - replacing **essential elements** used by the previous crop or providing extra essential elements
  - providing nitrogen that is incorporated into plant protein resulting in increased growth.

Eutrophication

- Fertilisers create problems if they get into ponds, lakes and rivers. They encourage algae growth, leading to **eutrophication.**

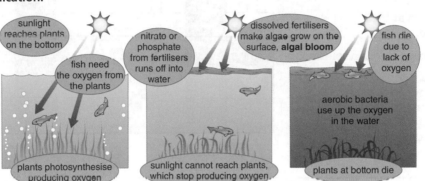

sunlight reaches plants on the bottom

fish need the oxygen from the plants

plants photosynthesise producing oxygen

1.

nitrate or phosphate from fertilisers runs off into water

sunlight cannot reach plants, which stop producing oxygen

2.

dissolved fertilisers make algae grow on the surface, **algal bloom**

fish die due to lack of oxygen

aerobic bacteria use up the oxygen in the water

plants at bottom die

3.

How eutrophication occurs

- Eutrophication happens when:
  - fertilisers are washed off fields
  - fertilisers in the water increase the nitrate and phosphate levels in ponds, lakes and rivers
  - algae grow quickly on the surface (**algal bloom**) in the presence of these chemicals
  - the algae block off the sunlight to other oxygen-producing plants which die
  - aerobic **bacteria** use up the oxygen in the water and feed on the dead and decaying plants
  - most living organisms die.

- Excessive use of fertilisers can also **pollute** water supplies.

Remember!
At Grades D–C you need to know what eutrophication is. At Grades B–A* you also need to be able to describe the sequence.

Preparing fertilisers

- Many fertilisers are **salts**, so they can be made by reacting an **acid** and an **alkali** to make salt and water. For example, the fertiliser that would be produced by combining nitric acid and ammonia solution would be ammonium nitrate.

- The process of producing a fertiliser from the reaction of an acid and an alkali (such as sulfuric acid and ammonia solution) is as follows:
  - the alkali is titrated with the acid using an indicator to find out the quantities needed before the main batch is made (this is repeated until the results are consistent)
  - although the acid and alkali have now reacted completely to produce a **neutral** solution of ammonium sulfate fertiliser, this is contaminated with indicator solution
  - the titration results are used to repeat the experiment using the correct quantities
  - the dissolved fertiliser is heated to evaporate most of the water off, then left for the remaining solution to crystallise. The crystals are then filtered off.

Improve your grade

Fertilisers
Describe how to prepare a sample of ammonia phosphate. *AO1* [3 marks]

Mining and subsidence

- **Salt** is mined in Cheshire in two different ways:
  - mining it from the ground as rock salt
  - solution mining by pumping in water and extracting saturated salt solution.
- Mining salt can lead to **subsidence**. The ground above a mine can sink causing landslips and destroying homes.
- Salt at the surface, particularly brine solution, can escape and affect **habitats**.

Electrolysis of sodium chloride solution

- Concentrated sodium chloride solution (brine) can be separated by **electrolysis**.
- Hydrogen is made at the negative **cathode**.
- Chlorine is made at the positive **anode**.
- Sodium hydroxide forms in solution.
- Hydrogen and chlorine are reactive, so it is important to use inert electrodes so that the products don't react before they are collected and the electrodes do not dissolve.

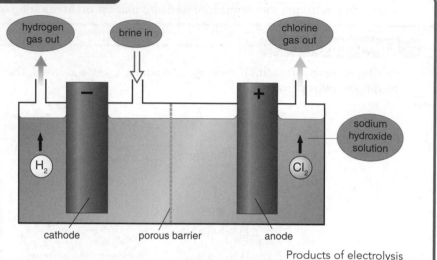

Products of electrolysis

More on electrolysis

- During the electrolysis of NaCl solution:
  - Na^+ and H^+ **ions** migrate to the negative cathode.
  - Cl^- and OH^- ions migrate to the positive anode.
  - At the cathode, hydrogen is made – electrons are gained so this is **reduction** as shown in this half-equation.
    $$2H^+ + 2e^- \rightarrow H_2$$
- At the anode, chlorine is made – electrons are lost so this is **oxidation**.
  $$2Cl^- - 2e^- \rightarrow Cl_2$$
- The ions not discharged make sodium hydroxide solution.
  $$Na^+ + OH^- \rightarrow NaOH$$

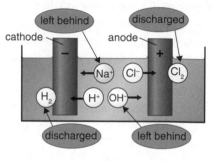

How ions move in the electrolysis of sodium chloride

The chlor-alkali industry

- Sodium hydroxide and chlorine are used to make household bleach.
- Chlorine and sodium hydroxide are important raw materials. They are involved in making about half the chemicals we use on a daily basis including solvents, plastics, paints, soaps, medicines and food additives.

How science works

You should be able to:

- identify how a scientific or technological development could affect different groups of people or the environment.

Improve your grade

Electrolysis of sodium chloride solution

Describe and explain using equations what is seen during the electrolysis of sodium chloride solution and explain what is happening. *AO1* [4 marks]

C2 Summary

Earth structure

Plate movements cause earthquakes and volcanoes.

Plate tectonic theory is now widely accepted.

Evidence needs to be repeatable and agreed for a theory to be accepted.

Rock is recycled by geological processes.

Scientists study volcanoes and earthquakes to try to predict future eruptions and seismic activity.

Materials

Rocks are used as building materials.

Concrete is not very strong, but can be reinforced with steel.

Metals are extracted from rocks.

Alloys contain at least one metal, but have different properties.

Copper is extracted by reduction, and purified by electrolysis.

Iron rusts in contact with oxygen and water, and salt speeds up rusting.

Aluminium is a metal with a low density, which does not corrode because it has a protective oxide layer.

Fertilisers

The Haber process is a reversible reaction that makes ammonia $N_2 + 3H_2 \rightleftharpoons 2NH_3$.

Conditions for the Haber process can be adjusted to maximise yield.

Ammonia is used to make nitrogen fertilisers.

Fertilisers can be made by a neutralisation reaction: acid + base → salt + water

Fertilisers can cause **eutrophication**.

Salts

Acids react with metal oxides and hydroxides, and metal carbonates to make salts.

Sodium chloride is an important raw material, obtained by mining it from the ground or by solution mining.

Electrolysis is used to split sodium chloride solution into hydrogen, chlorine and sodium hydroxide.

In electrolysis, positive ions move to the cathode, and negative ions to the anode.

Heating houses

Energy flow

D–C

- **Energy**, in the form of heat, flows from a warmer to a colder body. When energy flows away from a warm object, the **temperature** of that object decreases.

Measuring temperature

D–C

- A **thermogram** uses colour to show temperature; hottest areas are white/yellow, coldest are black/dark blue/purple.

hot end ▮▮▮▮▮ cold end

Typical range of colours used in a thermogram

B–A*

- Temperature is a measurement of hotness on an arbitrary scale. You do not need to use a thermometer. It allows one object to be compared to another.
- When the temperature of a body increases, the average **kinetic energy** of the particles increases.
- Heat is a measurement of internal energy. It is measured on an absolute scale.

Specific heat capacity

D–C

- All substances have a property called **specific heat capacity**, which is:
  - the energy needed to raise the temperature of 1 kg by 1 °C
  - measured in **joules** per kilogram degree Celsius (J/kg °C) and differs for different materials.
- When an object is heated and its temperature rises, energy is transferred.
- The equation for energy transfer by specific heat capacity is:

energy transferred = mass × specific heat capacity × temperature change

> Calculate the energy transferred when 30 kg of water cools from 25 °C to 5 °C.
> energy transferred = 30 × 4200 × (25 − 5) = 30 × 4200 × 20
> = 2 520 000 J or 2520 kJ

Specific latent heat

D–C

- **Specific latent heat** is:
  - the energy needed to melt or boil 1 kg of the material
  - measured in joule per kilogram (J/kg) and differs for different materials and each of the changes of state.
- When an object is heated and it changes state, energy is transferred, but the temperature remains constant.
- The equation for energy transfer by specific latent heat is:

energy transferred = mass × specific latent heat

> Calculate the energy transferred when 2.5 kg of water changes from solid to liquid at 0 °C
> energy transferred = 2.5 × 340 000
> = 850 000 J or 850 kJ

In a thermogram, white, yellow and red represent the hottest areas. Black, dark blue and purple represent the coldest areas.

B–A*

- When a substance changes state, energy is needed to break the bonds that hold the molecules together. This explains why there is no change in temperature.

Improve your grade

Specific heat capacity

Ed uses a stainless steel saucepan to heat his soup from 17 °C to 94 °C. The saucepan has a mass of 1.1 kg and a specific heat capacity of 510 J/kg °C. Energy is required to heat the soup.

(a) Calculate the extra energy required to raise the temperature of the saucepan.
   AO2 [2 marks]

(b) Ed reads that a 1.1 kg copper saucepan will be more energy efficient. Explain why.
   AO2 [2 marks]

P1 Energy for the home

Keeping homes warm

Practical insulation

- Double glazing reduces **energy** loss by conduction. The gap between the two pieces of glass is filled with a gas or contains a **vacuum.**
  - Particles in a gas are far apart. It is very difficult to transfer energy. There are no particles in a vacuum so it is impossible to transfer energy by conduction.

- Loft **insulation** reduces energy loss by conduction and convection:
  - warm air in the home rises
  - energy is transferred through the ceiling by conduction
  - air in the loft is warmed by the top of the ceiling and is trapped in the loft insulation
  - both sides of the ceiling are at the same temperature so no energy is transferred
  - without loft insulation, the warm air in the loft can move by convection and heat the roof tiles
  - energy is transferred to the outside by conduction.

- Cavity wall insulation reduces energy loss by conduction and convection:
  - the air in the foam is a good insulator
  - the air cannot move by convection because it is trapped in the foam.

- Insulation blocks used to build new homes have shiny foil on both sides to reduce energy transfer by radiation:
  - energy from the Sun is reflected back to keep the home cool in summer
  - energy from the home is reflected back to keep the home warm in winter.

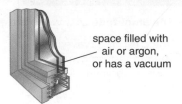

space filled with air or argon, or has a vacuum

A double glazed window

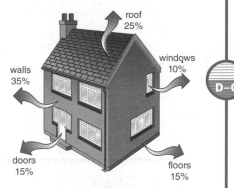

roof 25%

windows 10%

walls 35%

doors 15%

floors 15%

Energy loss from a home

Remember!
Hot air will only rise into the loft if the loft-hatch is open.

D–C

Conduction, convection and radiation

- Energy can be transferred by:
  - conduction – due to the transfer of **kinetic energy** between particles
  - convection – a gas expands when it is heated. This makes it less dense so it rises. The unit of **density** is kg/m^3 or g/cm^3.

  $$\text{density} = \frac{\text{mass}}{\text{volume}}$$

  - **radiation** does not need a material to transfer energy. Energy can be transferred through a vacuum.

B–A*

Energy efficiency

$$\text{efficiency} = \frac{\text{useful energy output} (\times 100\ \%)}{\text{total energy input}}$$

- Energy transformations can be shown by Sankey diagrams.
- Energy from the source (home) is lost to the sink (environment).
- Different types of insulation cost different amounts and save different amounts of energy.

  $$\text{payback time} = \frac{\text{cost of insulation}}{\text{annual saving}}$$

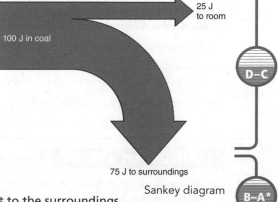

25 J to room

100 J in coal

75 J to surroundings

Sankey diagram

D–C

- Everything that transfers energy will waste some of the energy as heat to the surroundings.
- Buildings that are energy efficient are well insulated; little energy is lost to the surroundings.
- Designers and architects have to make sure that as little energy as possible is wasted.

B–A*

Improve your grade

Energy loss in a cavity wall

The Johnson's house has cavity walls. They decide to have foam injected into the cavity to reduce energy loss.

Explain how energy is transferred to the roof space from the cavity. *AO1* [3 marks]

A spectrum of waves

Wave properties

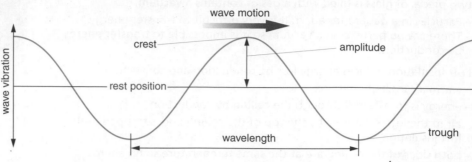

- The **amplitude** of a **wave** is the *maximum* displacement of a particle from its rest position.

- The crest of a wave is the *highest point on* a wave above its rest position.

- The trough of a wave is the *lowest point on* a wave *below* its rest position.

- The **wavelength** of a wave is the distance *between* two *successive* points on a wave having the same displacement and moving in the same direction.

A transverse wave

- The **frequency** of a wave is the number of complete waves passing a point in one second.

- The equation for the **speed** of a wave is:

  wave speed = frequency × wavelength

> When Katie throws a stone into a pond, the distance between ripples is 0.3 m and four waves reach the edge of the pond each second.
> wave speed = 0.3 × 4 = 1.2 m/s

Remember!
Always give the units in your answer:
– wavelength – metre (m)
– frequency – **hertz** (Hz)
– speed – metre per second (m/s).

> **Microwaves** travel at a speed of 300 × 10^6 m/s. A microwave oven uses microwaves with a frequency of 2.5 × 10^9 Hz. Calculate the wavelength of the microwaves.
>
> Wavelength = $\dfrac{\text{wave speed}}{\text{frequency}}$
>
> $= \dfrac{300 \times 10^6}{2.5 \times 10^9} = 0.12$ m

Remember!
At higher tier you may be expected to use scientific notation and rearrange equations.

Electromagnetic spectrum

radio   microwave   infrared   visible   ultraviolet   X-ray   gamma ray

← increasing wavelength   increasing frequency →

Getting messages across

- Some optical instruments, such as the periscope, use two or more plane mirrors.

- **Refraction** occurs because the speed of waves decreases as the wave enters a more dense medium and increases as the wave enters a less dense medium. The frequency stays the same but the wavelength changes.

- **Diffraction** is the spreading out of a wave as it passes through a gap.

- The size of a communications **receiver** depends on the wavelength of the radiation.

Diffraction effects

- The amount of diffraction depends on the size of the gap; the most diffraction occurs when the gap is a similar size to the wavelength. Larger gaps show less diffraction.

- Diffraction effects are noticeable in telescopes and microscopes.

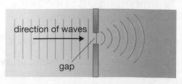

Diffraction at a narrow gap

direction of waves

gap

Improve your grade

Diffraction effects

Light is diffracted as it passes through a narrow slit. Describe how the amount of diffraction depends on the wavelength of the light and the width of the slit. *AO1* [2 marks]

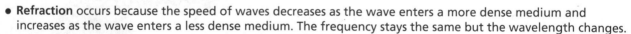

Light and lasers

Morse code

- The **Morse code** uses a series of dots and dashes to represent letters of the alphabet.
  - This code is used by signalling lamps as a series of short and long flashes of light.
  - It is an example of a **digital signal**.

D–C

Sending signals

- When a signal is sent by light, electricity, **microwaves** or radio, it is almost instantaneous.
- Each method of transmission has advantages and disadvantages:
  - can the signal be seen by others?
  - can wires be cut?
  - how far does the signal have to travel?

B–A*

Laser light

- White light is made up of different colours of different frequencies out of **phase**.
- **Laser** light has only a single frequency, is in phase and shows low divergence.
- Laser light is used to read from the surface of a compact disc (CD):
  - the surface of the CD is pitted
  - the pits represent the digital signal
  - laser light is shone onto the CD surface and the difference in the reflection provides the information for the digital signal.

B–A*

Critical angle

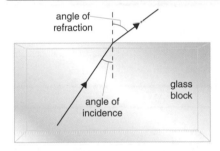
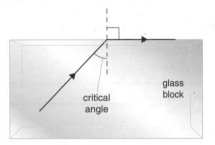
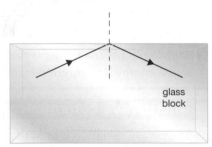
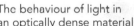

The behaviour of light in an optically dense material

D–C

- When light travels from one material to another, it is normally refracted.
- If it is passing from a more dense material into a less dense, the angle of refraction is larger than the angle of incidence.
- When the angle of refraction is 90°, the angle of incidence is called the **critical angle**.
- If the angle of incidence is bigger than the critical angle, the light is reflected:
  - this is **total internal reflection**.
- Telephone conversations and computer data are transmitted long distances along **optical fibres** at the speed of light (200 000 km/s in glass).
- Some fibres are coated to improve reflection.

Endoscopy

- An **endoscope** allows doctors to see inside a body without the need for surgery.
  - Light passes along one set of optical fibres to illuminate the inside of the body.
  - The light is reflected.
  - The reflected light passes up another set of fibres to an eyepiece or camera.

B–A*

Improve your grade

Sending signals

Adam is standing on top of a hill in line of sight and 10 km away from Becky who is on top of another hill. They can communicate either by using light, radio or electrical signals. Suggest one advantage and one disadvantage of using each type of signal. *AO1* [3 marks]

Cooking and communicating using waves

Cooking with waves

- **Infrared** radiation does not penetrate food very easily.
- **Microwaves** penetrate up to 1 cm into food.
- Microwaves can penetrate glass or plastic but are reflected by shiny metal surfaces:
  – special glass in a microwave oven door reflects microwaves
  – they can cause body tissue to burn.

Electromagnetic spectrum

- Energy is transferred by **waves**:
  – the amount of energy depends on the frequency or wavelength of the wave
  – high frequency (short wavelength) waves transfer more energy.
- Normal ovens cook food by infrared radiation:
  – energy is absorbed by the surface of the food
  – the **kinetic energy** of the surface food particles increases
  – the rest of the food is heated by conduction.
- Microwave ovens cook food by microwave radiation:
  – the water or fat molecules in the outer layers of food vibrate more.

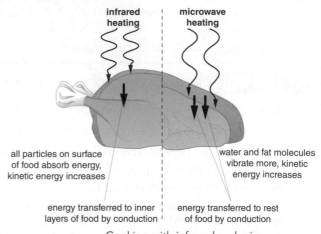

infrared heating   microwave heating

all particles on surface of food absorb energy, kinetic energy increases

water and fat molecules vibrate more, kinetic energy increases

energy transferred to inner layers of food by conduction

energy transferred to rest of food by conduction

Cooking with infrared and microwaves

Microwave properties

- Microwaves have wavelengths between 1 mm and 30 cm.
- Mobile phones use longer wavelengths than microwave ovens.
  – Less energy is transferred by mobile phones.

Microwave communication

- Microwave radiation is used to communicate over long distances.
- The **transmitter** and **receiver** must be in *line of sight*.
  – Aerials are normally situated on the top of high buildings.
- **Satellites** are used for microwave communication.
  – The signal from Earth is received, amplified and re-transmitted back to Earth.
  – Satellites are in line of sight because there are no obstructions in space.
  – Large aerials can handle thousands of phone calls and television channels at once.
- There are concerns about the use of mobile phones and where phone masts are situated.
- Scientists publish results of their studies to allow others to check their findings.
- Signal strength for mobile phones can change a lot over a short distance.
  – Microwaves do not show much **diffraction**.
  – Adverse weather and large areas of water can scatter the signals.
  – The curvature of the Earth limits the line of sight so transmitters have to be on tall buildings or close together.
- Mobile phones can **interfere** with sensitive equipment:
  – They are banned on planes and in many hospitals.

Improve your grade

Microwave transmitters
The Telecom Tower in London is one of the tallest buildings in the city. There are many microwave aerials surrounding the top of the tower.
Explain why they are sited so high up. *AO1* [2 marks]

Data transmission

Digital signals

- **Infrared** signals carry information that allows electronic and electrical devices to be controlled.

- Pressing a button on the remote control device completes the circuit. A coded signal is sent to a **light-emitting diode** or LED at the front of the remote.

- The signal includes a start command, the instruction command, a device code and a stop command.

- The LED transmits the series of pulses. This is received by the device and decoded to allow the television to change channel or volume.

- The switchover from **analogue** to **digital** started in 2009 and is planned to finish by 2015. This may be delayed until more people buy digital radios. The switchover for both radio and TV means:
  - improved signal quality for both picture and sound
  - a greater choice of programmes
  - being able to interact with the programme
  - information services such as programme guides and subtitles.

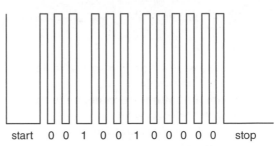

start   0   0   1   0   0   1   0   0   0   0   stop

A typical digital signal from a remote

Optical fibres

- **Optical fibres** allow data to be transmitted very quickly using pulses of light.

Remember!
An optical fibre is solid, not a hollow tube.

Advantages of digital signals

- Before an analogue signal is transmitted, it is added to a carrier **wave**.

- The **frequency** of the carrier wave is usually higher.

- The combined wave is transmitted.

- **Interference** from another wave can also be added and transmitted.

- If the wave is amplified, the interference is amplified as well.

- Interference also occurs on digital signals, but is not apparent because the digital signal only has two values.

- **Multiplexing** allows a large number of **digital signals** to be transmitted at the same time.

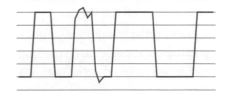

Interference on a digital wave

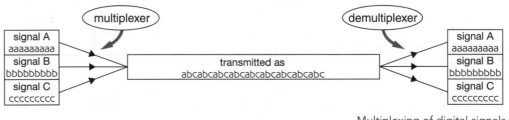

multiplexer   demultiplexer

signal A aaaaaaaaa		signal A aaaaaaaaa
signal B bbbbbbbbb	transmitted as abcabcabcabcabcabcabcabcabc	signal B bbbbbbbbb
signal C ccccccccc		signal C ccccccccc

Multiplexing of digital signals

Improve your grade

Advantages of using digital signals and optical fibres

Explain the advantages of using digital signals and optical fibres compared with analogue signals and electrical cables for data transmission. *AO1* [4 marks]

Wireless signals

Radio refraction and interference

- Wireless technology is used by:
  - radio and television
  - laptops
  - mobile phones.

- **Radio waves** are reflected and refracted in the Earth's **atmosphere**:
  - the amount of **refraction** depends on the frequency of the **wave**
  - there is less refraction at higher **frequencies**.

D–C

- Radio stations broadcast signals with a particular frequency.

- The same frequency can be used by more than one radio station:
  - the radio stations are too far away from each other to interfere
  - but in unusual weather conditions, the radio waves can travel further and the broadcasts interfere.

- **Interference** is reduced if **digital signals** are used.

- Digital Audio Broadcasting or DAB also provides a greater choice of radio stations but the audio quality is not as good as the FM signals currently used.

B–A*

- DAB eliminates interference between other radio stations.

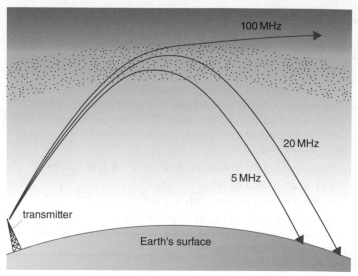

Refraction of waves in the atmosphere

Radio reflection

B–A*

- Radio waves are reflected from the **ionosphere**. They behave like light in an optical fibre and undergo **total internal reflection**.

- Water reflects radio waves but land mass does not.

- Continued reflection by the ionosphere and the oceans allows radio waves to be received from an aerial that is not in line of sight.

- **Microwaves** pass through the ionosphere.

- Microwave signals are received by orbiting **satellites**, amplified and retransmitted back to Earth.

- Communication satellites orbit above the equator and take 24 hours to orbit Earth.

Remember!
Microwave signals are not reflected from satellites.

Communication problems

B–A*

- Radio waves are diffracted when they meet an obstruction.

- Refraction in the atmosphere needs to be taken into account when sending a signal to a satellite.

- The transmitting aerial needs to send a focused beam to the satellite because its aerial is very small.

- The transmitted beam is slightly divergent.

- Some energy is lost from the edge of the transmitting aerial because of **diffraction**.

Improve your grade

Radio communication

The picture shows a transmitter and receiver on the Earth's surface, out of line of sight.

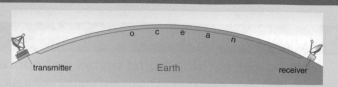

(a) Explain how long wave radio signals travel from the transmitter to the receiver. *AO1* [3 marks]

(b) Explain how microwave signals travel from the transmitter to the receiver. *AO1* [2 marks]

Stable Earth

Earthquake waves

- A seismograph shows the different types of earthquake **wave**.

- L waves travel round the surface very slowly.

- **P waves** are **longitudinal pressure waves**:
  - P waves travel through the Earth at between 5 km/s and 8 km/s
  - P waves can pass through solids and liquids.

- **S waves** are **transverse** waves:
  - S waves travel through the Earth at between 3 km/s and 5.5 km/s
  - S waves can only pass through solids.

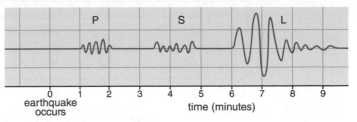

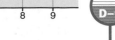

Seismograph trace

 D–C

Earth's insides

- P waves travel through the Earth:
  - they are **refracted** by the core
  - the paths taken by P waves mean that scientists can work out the size of the Earth's **core**.

- S waves are not detected on the opposite side of Earth to an earthquake:
  - they will not travel through liquid
  - this tells scientists that Earth's core is liquid.

B–A*

Tan or burn

- A tan is caused by the action of **ultraviolet** light on the skin.

- Cells in the skin produce **melanin**, a pigment that produces a tan.

- People with darker skin do not tan as easily because ultraviolet radiation is filtered out.

- Use a sunscreen with a high SPF, or sun protection factor, to reduce risks.

  maximum length of time to spend in the Sun = published normal burn time × SPF

- People are becoming more aware of the dangers of exposure to ultraviolet radiation, including the use of sun beds.

D–C

Ozone depletion

- At first scientists did not believe there was thinning of the **ozone layer** – they thought their instruments were faulty but other scientists confirmed the results and increased confidence in the findings.

 D–C

- **Ozone** is found in the **stratosphere**.

- Ozone helps to filter out ultraviolet radiation.

- **CFC** gases from aerosols and fridges destroy ozone and reduce the thickness of the ozone layer.
  - This increases the potential danger to humans.

- The ozone layer is at its thinnest above the South Pole because ozone depleting chemicals work best in cold conditions.

- Scientists monitor the thickness of the ozone layer using **satellites**.

- There is international agreement to reduce CFC emissions.

 B–A*

Improve your grade

Earthquake waves

An Earthquake occurs with its epicentre at **E**. It is detected at two monitoring stations **A** and **B**.

Describe and explain the appearance of the seismograph traces at **A** and **B**.
AO1/AO2 [4 marks]

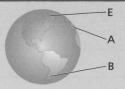

P1 Summary

Energy is transferred when a substance changes temperature.

The amount of energy transferred depends on the mass, temperature change and specific heat capacity.

energy transferred = mass × specific heat capacity × temperature change

Heat and temperature

Energy is transferred from a hotter to a colder body.

Temperature is a measure of hotness on an arbitrary scale, measured in °C.

Energy is a measure of energy transfer on an absolute scale measured in J.

Energy is transferred when a substance changes state.

The amount of energy transferred depends on the mass and the specific latent heat.

energy transferred = mass × specific latent heat

Energy transfer

Air is a good insulator and reduces energy transfer by conduction.

Conduction in a solid is by the transfer of kinetic energy.

Trapped air reduces energy transfer by convection.

Convection currents are caused by density changes.

Energy saving in the home can be achieved by:
- double glazing
- cavity wall insulation
- draught strip
- reflecting foil
- loft insulation
- curtains
- careful design.

Shiny surfaces reflect infrared radiation to reduce energy transfer.

Energy transformations can be represented by Sankey diagrams.

efficiency = $\dfrac{\text{useful energy output (×100\%)}}{\text{total energy input}}$

Waves transfer energy

Warm and hot objects emit infrared radiation.

Infrared radiation is used for cooking.

Microwaves can be used for cooking and for communication when transmitter and receiver are in line of sight.

Laser light is single colour and in phase.

Laser light, visible light and infrared are all used to send signals along optical fibres by total internal reflection.

All waves have amplitude, frequency and wavelength

wave speed = frequency × wavelength

Digital and analogue signals are used for communication.

Morse is a digital code.

Digital signals allow many signals to be transmitted at the same time.

Digital signals are clearer.

Radio waves, microwaves, infrared, visible light and ultraviolet are some of the waves in the electromagnetic spectrum.

The energy of the wave increases as the wavelength decreases.

All electromagnetic waves can be reflected, refracted and diffracted.

Radio waves are used for communication.

Longer wavelengths diffract around obstacles.

The stable Earth

Earthquake waves travel through the Earth.

Different waves help us find out about the inside of the Earth.

Exposure to ultraviolet radiation causes sun burn and skin cancer.

Sunscreen and sunblock reduces damage caused by ultraviolet radiation.

CFCs are causing the ozone layer to become thinner.

Collecting energy from the Sun

Photocells

- The advantages of **photocells** are:
  - they are robust and do not need much maintenance
  - they need no fuel and do not need long power cables
  - they cause no **pollution** and do not contribute to **global warming**
  - they use a **renewable energy** resource.
- The only disadvantage is that they do not produce electricity when it is dark or too cloudy.

How photocells work

- A photocell contains two pieces of silicon joined together to make a **p-n junction**.
- One piece has an impurity added to produce an excess of free electrons – n-type silicon.
- The other piece has a different impurity added to produce an absence of free electrons – p-type silicon.
- Sunlight contains energy packets called **photons**.
- Photons cause free electrons to move producing an electric **current**.
- The output from a photocell depends on:
  - the light intensity
  - the surface area exposed
  - the distance from the light source.

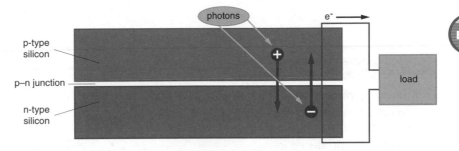

An electric field is set up across the p-n junction

Passive solar heating

- The Sun is very hot and produces infrared radiation with a very short **wavelength**:
  - glass is transparent to this short wavelength **radiation**
  - the walls and floor inside a building absorb this radiation, warm up and re-radiate infrared radiation
  - the walls and floor are not as hot as the Sun and the wavelength radiated is therefore longer
  - glass reflects this longer wavelength radiation back inside the building.

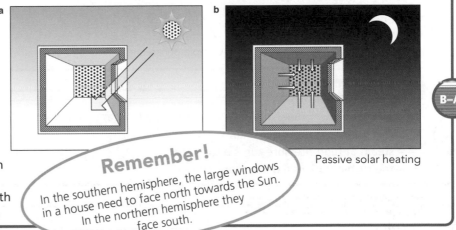

Passive solar heating

Remember!
In the southern hemisphere, the large windows in a house need to face north towards the Sun. In the northern hemisphere they face south.

Heat from the Sun

- Solar reflectors are moved by computer to make sure they are always facing the Sun.

Energy from the wind

- Wind is a renewable form of energy, but it does depend on the speed of the wind:
  - wind **turbines** do not work if there is no wind, nor if the wind speed is too great.
- Wind farms do not contribute to global warming, nor do they **pollute** the **atmosphere** but they can be noisy, take up a lot of space and people sometimes complain that they spoil the view.

Improve your grade

Photocells

A photocell contains two pieces of silicon joined together to make a p-n junction.

Explain how light falling on a photocell produces an electric current. *AO1* [3 marks]

Generating electricity

Bigger currents and voltages

- The current from a **dynamo** can be increased by:
  - using a stronger magnet
  - increasing the number of turns on the coil
  - rotating the magnet faster.

- The output from a dynamo can be displayed on an **oscilloscope**.

- An oscilloscope trace shows how the **current** produced by the dynamo varies with time.

- The time for one complete cycle is called the period of the **alternating current**.

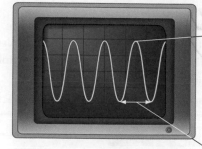

the height of the wave is the maximum (peak) **voltage**

the length of the wave represents the time for one cycle

Oscilloscope trace

D–C

> **Remember!**
> Frequency = 1 ÷ period
> **Frequency** is measured in **hertz** (Hz)
> Period is measured in seconds (s)

Practical generators

- A simple **generator** consists of a coil of wire rotating between the poles of a magnet:
  - the coil cuts through the magnetic field as it spins.
  - a current is produced in the coil.

- A current can be produced if the coil remains stationary and the magnets move.

- Generators at **power stations** work on the same principle.

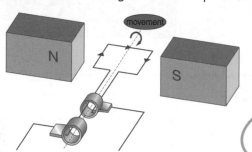

movement

N

S

A simple generating current (AC generator)

D–C

> **Remember!**
> It is the *relative* movement of magnet and coil that is important.

Power stations

- In conventional power stations, fuels are used to heat water:
  - water boils to produce steam
  - steam at high pressure turns a **turbine**
  - the turbine drives a generator.

D–C

Energy efficiency

- **Efficiency** is a measure of how well a device transfers energy.

- Energy in a power station is lost in the boilers, generator and cooling towers.

D–C

What is the efficiency of a power station if 60 MJ of fuel energy is converted into 20 MJ of electrical energy?

$$\text{efficiency} = \frac{\text{useful energy output}}{\text{total energy input}}$$
$$= \frac{20\,000\,000}{60\,000\,000}$$
$$= 0.33 \text{ or } 33\%$$

> **EXAM TIP**
> Students working at B–A* level should be able to rearrange equations.

> **EXAM TIP**
> Always check the wording of your answer is very specific and says exactly what you mean.

⊙ Improve your grade

Generators

Describe three ways to increase the output of an electrical dynamo. *AO1* [3 marks]

Global warming

Greenhouse gases

- Most **wavelengths** of electromagnetic **radiation** can pass through the Earth's **atmosphere**, but **infrared** radiation is absorbed.

- Carbon dioxide occurs naturally in the atmosphere as a result of:
  - natural forest fires
  - volcanic eruptions
  - **decay** of dead plant and animal matter
  - its escape from the oceans
  - respiration.

- Man-made **carbon dioxide** is caused by burning fossil fuels, waste incineration, **deforestation** and **cement** manufacture.

- Water vapour is the most significant **greenhouse gas**:
  - almost all of the water vapour occurs naturally
  - a mere 0.001% comes from human activity
  - half of the greenhouse effect is due to water vapour and a further quarter is due to clouds.

- Methane is produced when organic matter decomposes in an environment lacking oxygen.
  - natural sources include wetlands, termites and oceans
  - man-made sources include the mining and burning of **fossil fuels**, the digestive processes in animals such as cattle, rice paddies and the burying of waste in landfills.

D–C

The greenhouse effect

- The electromagnetic radiation from the Sun has a relatively short wavelength.

- This radiation is absorbed by and warms the Earth. The Earth then re-radiates the energy as infrared radiation with a longer wavelength.

- This longer wavelength radiation is absorbed by the greenhouse gases which warms the atmosphere.

B–A*

Dust warms, dust cools

- Dust in the atmosphere can have opposite effects:
  - the smoke from the factories reflects radiation from the town back to Earth. The temperature rises as a result
  - the ash cloud from a volcano reflects radiation from the Sun back into space. The temperature falls as a result.

D–C

Scientific data

- It is important that decisions about what to do about **global warming** are made on the basis of scientific evidence, not on the basis of unsubstantiated opinions.

- The vast majority of scientists agree that the evidence supports climate change. The average temperature of the Earth has increased steadily during the past 200 years.

- What scientists do not agree on is the extent to which human activity has contributed.

D–C

Remember!
There is disagreement among scientists about the seriousness of global warming. Most scientists agree that the global temperature is rising. They do not agree on more specific elements of the issue: How much will it warm up? What will happen if it does warm up? How far are humans responsible? What should we do to stop it?

EXAM TIP
Students working at B–A* will be expected to interpret data about global warming and climate change.

 Improve your grade

The greenhouse effect
Describe the greenhouse effect and explain how it contributes to global warming.
AO1 [4 marks]

Fuels for power

Measuring power

- Electrical appliances usually display a power rating in **watts** (W) or **kilowatts** (kW)

  power = voltage × current

  > What is the power rating of a kettle working at 9 amps (A) on the mains supply of 230 **volts** (V)?
  > power = **voltage** × **current**
  >   = 9 × 230
  >   = 2070 W

> **EXAM TIP**
> Students should be able to rearrange equations.

  energy supplied = power × time

- The unit of electrical **energy** used in the home is the **kilowatt-hour** (kWh).

  cost of electricity used = energy used × cost per kWh

  > Calculate the cost of using a 250 W television for 30 minutes if one kWh of electricity costs 10p?
  > energy used = power × time = 0.25 × 0.5 = 0.125 kWh
  > cost of electricity = energy used × cost per kWh = 0.125 × 10 = 1.25p

Cheaper electricity

- We pay less for electricity during the night when not as much is needed, but it still has to be produced.

Energy sources

- Some energy sources are more appropriate than others in a particular situation.

- The choice of energy sources depends on several factors:
  - availability
  - ease of extraction
  - effect on the environment
  - associated risks.

The National Grid

- The **National Grid** is a series of **transformers** and power lines that transport electricity from the power station to the consumer.

- In the National Grid, transformers are used to step up the voltage to as high as 400 000 V. The high voltage leads to:
  - reduced energy loss
  - reduced distribution costs
  - cheaper electricity for consumers.

- Transformers are then used again to step down the voltage to a more suitable level for the consumer.

Transmission losses

- When a current passes through a wire the wire gets hot. The greater the current, the hotter the wire:
  - when a transformer increases the voltage, the current is reduced which means there is less heating effect and therefore less energy lost to the environment.

EXAM TIP
Make sure you give enough detail in your answers. Be guided by the number of marks available for the question.

Improve your grade

Cost of electricity

Tracey's flat has electric storage heaters which heat up at night and release the heat slowly during the day. Why is this cheaper for Tracey? *AO1* [2 marks]

Nuclear radiations

Ionisation

- **Atoms** contain the same number of **protons** and electrons – this means they are neutral.
- **Ionisation** involves gaining or losing electrons:
  - when the atom gains electrons, it becomes negatively charged
  - when the atom loses electrons, it becomes positively charged.

- The formation of **ions** can cause chemical reactions.
  - Such reactions may disrupt the normal behaviour of molecules inside the body e.g. they may cause strands of **DNA** to break or change; protein molecules may change their shape and these effects are potentially harmful.

Properties of ionising radiations

- Alpha, beta and gamma **radiations** come from the **nucleus** of an atom.
- Alpha radiation causes most ionisation and gamma radiation the least.
- Alpha radiation is short ranged (a few centimetres) and is easily absorbed by a sheet of paper or card.
- Beta radiation has a range of about 1 m and is absorbed by a few millimetres of aluminium.
- Gamma radiation is much more penetrating and, although a few centimetres of **lead** will stop most of the radiation, some can pass through several metres of lead or concrete.

- Experiments can be done to identify each type of radiation from its penetrating power - when carrying out these experiments **background radiation** should always be taken into account.

Uses of radioactivity

- Smoke alarms contain a source of alpha radiation:
  - the radiation **ionises** the oxygen and nitrogen atoms in air which causes a very small electric **current** that is detected. When smoke fills the detector in the alarm during a fire, the air is not so ionised, the current is less and the alarm sounds.
- Thicknesses in a paper rolling mill can be controlled using a source of beta radiation and a detector:
  - the amount of radiation passing through the sheet is monitored and the pressure on the rollers adjusted accordingly.

- Gamma radiation kills **microbes** and **bacteria** so it can be used for sterilising medical instruments. It can also be used to check for leaks in pipes and welds.
- The passage of blood and other substances can be traced around the body using a beta or gamma source.

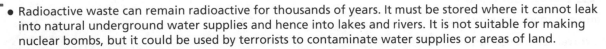

Nuclear waste

- **Plutonium** is a waste product from nuclear reactors which can be used to make nuclear bombs.
- Some low level **radioactive waste** can be buried in landfill sites. High level waste is encased in glass and buried deep underground or reprocessed.

- Radioactive waste can remain radioactive for thousands of years. It must be stored where it cannot leak into natural underground water supplies and hence into lakes and rivers. It is not suitable for making nuclear bombs, but it could be used by terrorists to contaminate water supplies or areas of land.

Advantages and disadvantages of nuclear power

- There are advantages in using nuclear power stations: fossil fuel reserves are not used and no **greenhouse gases** are discharged into the **atmosphere**.
- The disadvantages are its very high maintenance and decommissioning costs, and the risk of accidents similar to the one at Chernobyl.

 Improve your grade

Ionising radiation
Give two advantages and two disadvantages of nuclear power. *AO1* [4 marks]

Exploring our Solar System

Our Solar System

D–C

- **Comets** have very **elliptical orbits**:
  - they pass inside the orbit of Mercury and go out well beyond the orbit of Pluto.
- A **meteor** is made from grains of dust that burn up as they pass through the Earth's **atmosphere**:
  - they heat the air around them which glows and the streak is known as a 'shooting **star**'.

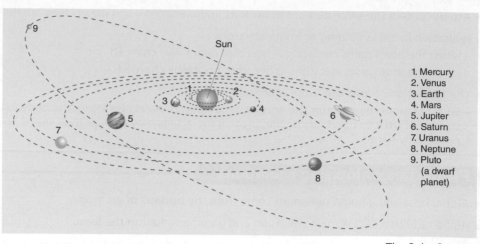

1. Mercury
2. Venus
3. Earth
4. Mars
5. Jupiter
6. Saturn
7. Uranus
8. Neptune
9. Pluto (a dwarf planet)

The Solar System

- **Black holes** are formed where large stars used to be:
  - you cannot see a black hole because light cannot escape from it
  - it has a very large mass but a very small size.

B–A*

- Moons orbit planets and planets orbit stars because **centripetal force** acts on them:
  - centripetal force acts towards the centre of the circular orbit
  - gravitational attraction is the source of the centripetal force.

Exploring the planets

D–C

- Unmanned **probes** can go where conditions are deadly for humans.
- Spacecraft carrying humans have to have large amounts of food, water and oxygen aboard.
- Astronauts can wear normal clothing in a pressurised spacecraft.
- Outside the spacecraft they need to wear special spacesuits:
  - a dark visor stops an astronaut being blinded
  - the suit is pressurised and has an oxygen supply for breathing
  - the surface of the suit facing towards the Sun can reach 120 °C
  - the surface of the suit facing away from the Sun may be as cold as −160 °C
- When travelling in space, astronauts are subjected to lower gravitational forces than on Earth.

B–A*

- Unmanned spacecraft cost less and do not put astronauts lives at risk:
  - they have to be very reliable because there is usually no way of repairing them when they break down.

A long way to go!

D–C

- Distances in space are very large.
- Light travels at 300 000 km/s:
  - light from the Sun takes about eight minutes to reach us on Earth
  - light from the next nearest star (Proxima Centauri) takes 4.22 years.

B–A*

- America plans a manned mission to Mars after 2020 at a cost of £400 billion.

Remember!
A **light-year** is the distance light travels in one year.

EXAM TIP

Sometimes a word is emphasised in the question to draw your attention to its importance. Be careful to answer accordingly.

Improve your grade

Unmanned space travel

Why do we send *unmanned* space probes to explore our Solar System? *AO1* [3 marks]

Threats to Earth

Asteroids

- **Asteroids** are mini-planets or planetoids orbiting the Sun:
  - most orbit between Mars and Jupiter
  - they are large rocks that were left over from the formation of the **Solar System**.

Asteroid or planet

- All bodies in space, including planets, were formed when clouds of gas and dust collapsed together due to gravitational forces of attraction.
- The mass of an object determines its gravitational force.
- Asteroids have relatively low masses compared to the mass of Jupiter.
- Jupiter's gravitational force prevents asteroids from joining together to form another planet.

Origin of the Moon

- Scientists believe our Moon was a result of the collision between two planets in the same orbit. The iron **core** of the other planet melted and joined with the Earth's core, less dense rocks began to orbit and they joined together to form our Moon.

- There is scientific evidence which supports this idea.
  - The average **density** of Earth is 5500 kg/m³ while that of the Moon is only 3300 kg/m³.
  - There is no iron in the Moon.
  - The Moon has exactly the same oxygen composition as the Earth, but rocks on Mars and meteorites from other parts of the Solar System have different oxygen compositions.

Evidence for asteroids

- Geologists examine evidence to support the theory that asteroids have collided with Earth:
  - near to a crater thought to have resulted from an asteroid impact, they found quantities of the metal iridium – a metal not normally found in the Earth's **crust** but common in meteorites
  - many fossils are found below the layer of iridium, but few fossils are found above it
  - **tsunamis** have disturbed the fossil layers, carrying some fossil fragments up to 300 km inland.

A comet's orbit

- Most **comets** pass inside the orbit of Mercury and well beyond the orbit of Pluto:
  - as the comet passes close to the Sun, the ice melts and solar winds blow the dust into the comet's tail which always points away from the Sun.
- Scientists are constantly monitoring and plotting the paths of comets and other **near-Earth objects** NEOs.

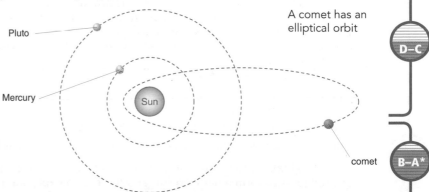

A comet has an elliptical orbit

- The speed of a comet increases as it approaches the Sun and decreases as it gets further away. This is because of the changing gravitational attraction.

NEOs

- If an NEO is on a collision course with Earth, it could be the end of life on Earth. To avoid this, one option would be to explode a rocket near to the NEO which could alter its course enough to miss Earth.

Improve your grade

Asteroids

What evidence have scientists gathered to show how the Moon was formed?
AO1 [3 marks]

The Big Bang

The expanding Universe

D–C

- Almost all of the galaxies are moving away from each other with the further galaxies moving fastest.

Models of the Universe

D–C

- With the help of the newly invented telescope, Galileo observed four moons orbiting Jupiter. This confirmed that not everything orbited the Earth and supported Copernicus' idea that planets orbit the Sun.

B–A*

- The Roman Catholic Church did not support Galileo's model as they believed that the Earth was at the centre of the **Universe** and it was a very long time before it was accepted.

- In the 17th century, Newton was working on his theory of universal gravitation which suggested that all bodies **attract** one another.

- Today, we believe that gravitational collapse is prevented because the Universe is constantly expanding as a result of the **Big Bang**.

Red shift

B–A*

- When a source of light is moving away from an observer, its **wavelength** appears to increase which shifts light towards the red end of the spectrum – **red shift**.

- When scientists look at light from the Sun, there is a pattern of lines across the spectrum. This same pattern is observed when they look at light from distant **stars** but it is closer to the red end of the spectrum.

- Scientists can use information from red shift to work out the age of the Universe.

A star's life history

B–A*

- The swirling cloud of gas and dust is a nebula:
  - nebula clouds are pulled together by **gravity** and, as the spinning ball of gas starts to get hot, it glows. This protostar cannot be seen because of the dust cloud
  - gravity causes the star to become smaller, hotter and brighter and after millions of years, the **core** temperature is hot enough for nuclear **fusion** to take place. As hydrogen nuclei join together to form **helium** nuclei, energy is released and the star continues to shine while there is enough hydrogen.

- Small stars shine for longer than large stars because they have less hydrogen but use it up at a slower rate and what happens at the end of a star's life depends on its size.

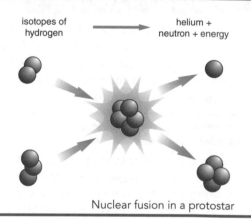

isotopes of hydrogen → helium + neutron + energy

Nuclear fusion in a protostar

The end of a star's life

D–C

- A medium-sized star, like the Sun, becomes a red giant: while the core contracts, the outer part cools, changes colour from yellow to red and expands:
  - gas shells, called planetary nebula, are thrown out
  - the core becomes a white dwarf shining very brightly but eventually cools to become a black dwarf

- Large stars become red supergiants: as the core contracts and the outer part expands and, suddenly, the core collapses to form a neutron star and there is an **explosion** called a supernova
  - neutron stars are very dense
  - remnants from a supernova can merge to form a new star
  - the core of the neutron star continues to collapse, becomes even more dense and could form a **black hole**.

B–A*

- a black hole has a very large mass concentrated in a small volume so it has a very large **density** and its large mass means it has a very strong gravitational pull.

Improve your grade

The Big Bang

What is red shift and how does it provide evidence for the Big Bang? *AO1* [4 marks]

P2 Summary

Energy from the Sun

Kinetic energy from moving air turns the blades on a wind turbine to produce electricity.

Some gases in the Earth's atmosphere trap heat from the Sun increasing global warming.

Scientists disagree about how much effect humans have on the increase in levels of these greenhouse gases.

Passive solar heating uses glass to help keep buildings warm.

Photocells do not need fuel or cables, need little maintenance and cause no pollution.

Electricity generation

A dynamo produces electricity when coils of wire rotate inside a magnetic field.

The size of the current depends on the number of turns, the strength of the field and the speed of rotation.

Transformers change the size of the voltage and current.

The National Grid transmits electricity around the country at high voltage and low current.

This reduces energy loss.

In power stations, fuels release energy as heat.

Water is heated to produce steam.

The steam drives turbines.

Turbines turn generators.

Generators produce electricity.

$$\text{efficiency} = \frac{\text{energy output}}{\text{energy input}}$$

Nuclear fuels are radioactive.

The radiation produced can cause cancer.

Waste products remain radioactive for a long time.

The main forms of ionising radiation are alpha, beta and gamma.

Their uses depend on their penetrative and ionisation properties.

Our Solar System

Planets, asteroids and comets orbit the Sun in our Solar System.

Centripetal forces keep bodies in orbit.

Medium-sized stars, such as our Sun, were formed from nebulae and will eventually become red giants, white dwarfs and finally black dwarfs.

When two planets collide, a new planet and a moon may be formed.

The Universe

The Universe consists of many galaxies.

Models of the Universe have changed over time and sometimes these changes take a long time to be accepted.

Most asteroids orbit between Mars and Jupiter but some pass closer to the Earth. They are constantly being monitored. An asteroid strike could cause climate change and species extinction.

The Universe is explored by telescopes on the Earth and in space.

Large distances mean that it takes a long time for information to be received and inter-galactic travel unlikely.

Scientists believe that the Universe started with the Big Bang.

The evidence is red shift.

Molecules of life

Cell structure

- The number of **mitochondria** in the cytoplasm of a cell depends on the activity of the cell. This is because **respiration** occurs in mitochondria. Cells such as liver or muscle cells have large numbers of mitochondria. This is because the liver carries out many functions and muscle cells need to contract. Both types of cell therefore need lots of energy.

- **Ribosomes** are smaller than mitochondria and are also found in the cytoplasm. They are too small to be seen with a light microscope and are the site of protein synthesis.

DNA and the genetic code

- The **nucleus** contains **genes**. Each gene:
  – is a section of a **chromosome** made of **DNA**
  – codes for a particular protein.

- DNA is made of two strands coiled to form a double helix, each strand containing chemicals called bases. There are four different types of bases, with cross links between the strands formed by pairs of bases. Each gene contains a different sequence of bases.

- Proteins are made in the cytoplasm but DNA cannot leave the nucleus. This means that a copy of the gene needs to be made that can leave the nucleus and carry the code to the cytoplasm.

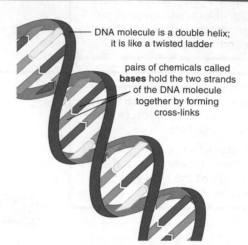

DNA molecule is a double helix; it is like a twisted ladder

pairs of chemicals called **bases** hold the two strands of the DNA molecule together by forming cross-links

The structure of DNA

- The four bases in DNA are called A, T, C and G. The cross links holding the two strands together are always between the same bases, A–T and G–C. This is called complementary base pairing.

- The **DNA base** code controls which protein is made. This is because the base sequence in the DNA codes for the **amino acid** sequence in the protein. Each amino acid is coded for by a sequence of three bases.

- The code needed to produce a protein is carried from the DNA to the ribosomes by a **molecule** called **messenger RNA**, or mRNA.

- Many of the proteins that are made are **enzymes**, which can control the activity of the cell.

Discovering the structure of DNA

- Watson and Crick built a model of DNA using data from other scientists. Two of the important pieces of data they used were:
  – photographs taken using **x-rays** which showed that DNA had two chains wound in a helix
  – data indicating that the bases occurred in pairs.

- Watson and Crick worked out the structure of DNA in 1953 and shared the Nobel prize for this discovery in 1962. There is often such a delay between a discovery and the award of prizes because other scientists need to check the discovery to make sure that it is correct.

How science works

You should be able to:

- describe examples of how scientists made a series of observations in order to develop new scientific explanations
- understand that unexpected observations or results can lead to new scientific developments in the understanding of science.

Improve your grade

Protein synthesis

Describe where and how proteins are coded for and made. *AO1* [6 marks]

Proteins and mutations

Grouping organisms

- All proteins are made of long chains of **amino acids** joined together.

- Proteins have different functions. Some examples are:
  - structural proteins used to build cells and tissues, e.g. **collagen**
  - **hormones**, which carry messages to control a reaction, e.g. **insulin** controls **blood sugar levels**
  - carrier proteins, e.g. **haemoglobin**, which carries oxygen
  - **enzymes**.

- Each protein has its own number and order of amino acids. This makes each type of protein **molecule** a different shape and gives it a different function.

Enzymes

- Enzymes speed up reactions in the body and so are called **biological catalysts**.

- They catalyse chemical reactions occurring in **respiration**, **photosynthesis** and protein synthesis of living cells.

- The **substrate** molecule fits into the **active site** of the enzyme like a key fitting into a lock:
  - This is why enzymes are described as working according to the 'lock and key mechanism'.
  - It also explains why each enzyme can only work on a particular substrate. This is called specificity and it happens because the substrate has to be the right shape.

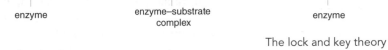

The lock and key theory

- Enzymes all work best at a particular temperature and pH. This is called the optimum. Any change away from the optimum will slow down the reaction.

- Enzyme activity is affected by **pH** and temperature:
  - At low temperatures molecules are moving more slowly and so the enzyme and substrate are less likely to collide.
  - At very high or low pH values and at high temperatures the enzyme active site changes shape. This is called denaturing. The substrate cannot fit, so cannot react so quickly.

- It is possible to work out how temperature alters the **rate of reaction** by calculating the **temperature coefficient**, called Q_{10}. This is done for a 10 °C change in temperature, using:

$$Q_{10} = \frac{\text{rate at higher temperature}}{\text{rate at lower temperature}}$$

Mutations

- **Mutations** may occur spontaneously but can be made to occur more often by **radiation** or chemicals.

- When they occur, mutations:
  - may lead to the production of different proteins
  - are often harmful but may have no effect
  - occasionally they might give the individual an advantage.

> ### EXAM TIP
> A mutation changes the *order* of the four bases. Do not say that different bases are involved.

- Although every cell in the body has the same **genes** it does not mean that the all the same proteins are made. This is because different genes are switched off in different cells. This allows different cells to perform different functions.

- Gene mutations alter or prevent the production of the protein that is normally made, because they change the base code of **DNA**, and so change the order of amino acids in the protein.

Improve your grade

Mutations and enzymes

A mutation can occur in a gene that codes for an enzyme. Explain how a mutation could lead to an enzyme failing to work properly. *AO1* [6 marks]

Respiration

Why is respiration important?

- **Respiration** releases energy from food and this energy is trapped in a **molecule** called **ATP**. ATP can then be used to provide the energy for many different processes in living organisms.

Aerobic respiration

- **Aerobic respiration** involves the use of oxygen. The symbol equation for aerobic respiration is:

$$C_6H_{12}O_6 + 6O_2 \rightarrow 6CO_2 + 6H_2O$$

EXAM TIP
When you write this particular equation, make sure that all the letters are capitals and that the numbers are the right size and in the right position.

Anaerobic respiration

- During exercise, despite an increase in breathing rate and heart rate, the muscles often do not receive sufficient oxygen. They start to use **anaerobic respiration** in addition to aerobic respiration.

- The word equation for anaerobic respiration is:

  glucose → lactic acid (+ energy)

- Anaerobic respiration has two main disadvantages over aerobic respiration.
  - The lactic acid that is made by anaerobic respiration builds up in muscles, causing pain and fatigue.
  - Anaerobic respiration releases much less energy per glucose molecule than aerobic respiration.

- The incomplete breakdown of glucose resulting in the build up of lactic acid is called the **oxygen debt.**

- During recovery the breathing rate and heart rate stay high so that:
  - rapid blood flow can carry lactic acid away to the liver
  - extra oxygen can be supplied, enabling the liver to break down the lactic acid.

An athlete running long distances tries to use only aerobic respiration

Measuring respiration rate

- It is possible to set up different experiments to measure the rate of respiration. Two ways to do this involve:
  - measuring how much oxygen is used up – the faster it is consumed, the faster the respiration rate
  - the rate at which **carbon dioxide** is made.

- Scientists can use these results to calculate the **respiratory quotient.** This is worked out using the formula:

$$RQ = \frac{\text{carbon dioxide produced}}{\text{oxygen used}}$$

Remember!
The RQ for glucose is 1, from the equation:
$$\frac{6CO_2}{6O_2} = 1$$

- The **metabolic rate** is described as the sum of all the reactions that are occurring in the body. If the metabolic rate is high, more oxygen is needed, as aerobic respiration is faster.

- Changes in temperature and **pH** can also change the respiration rate because they affect **enzymes**, and respiration is controlled by enzymes.

Improve your grade

Respiration and enzymes
An experiment was set up to measure the oxygen uptake of insects. The data shows that the maximum uptake was at about 35 °C. Above and below that temperature less oxygen was used.

Explain what the oxygen is used for and why it is used fastest at 35 °C. *AO1* [2 marks], *AO2* [2 marks]

Cell division

Becoming multicellular

- There are a number of advantages of being **multicellular**, as humans are. It allows an organism to become larger and more complex. It also allows different cells to take on different jobs. This is called **cell differentiation**.

- However, when an organism becomes multicellular, it needs to have systems that can:
  - allow communication between all the cells in the body
  - supply all the cells with enough nutrients
  - control exchanges with the environment such as heat and gases.

Mitosis

- The process that produces new cells for growth is called **mitosis**.

- The cells that are made by mitosis are genetically identical. Before cells divide, **DNA** replication must take place. This is so that each cell produced still has two copies of each **chromosome**.

- Body cells in **mammals** have two copies of each chromosome, so they are called **diploid** cells.

- Before mitosis happens, DNA is replicated. This involves:
  - the two strands of the DNA **molecule** 'unzipping' to form single strands
  - new double strands forming by **DNA bases** lining up in complementary pairings.

- Then mitosis occurs. The chromosomes line up along the centre of the cell and divide. The copies then move to opposite poles (ends) of the cell.

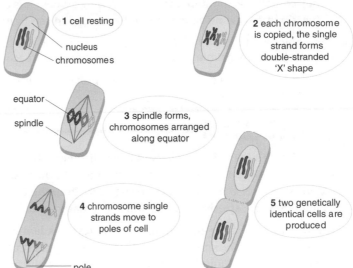

1 cell resting

nucleus
chromosomes

2 each chromosome is copied, the single strand forms double-stranded 'X' shape

equator
spindle

3 spindle forms, chromosomes arranged along equator

4 chromosome single strands move to poles of cell

pole

5 two genetically identical cells are produced

The process of mitosis

Meiosis

- The type of cell division that produces **gametes** is called **meiosis**.

- Gametes are **haploid** cells because they contain only one chromosome from each pair. This means that the zygote gets one copy of a **gene** from one parent and another copy from the other parent. This produces genetic **variation**.

- The structure of a sperm cell is adapted to its function. It has:
  - many **mitochondria** to provide the energy for swimming to the egg
  - an **acrosome** that releases **enzymes** to digest the egg membrane.

- In meiosis, there are two divisions. First, the single strands are copied to make X-shaped chromosomes and chromosomes with the same genes pair up. Then:
  - In the first division, one chromosome from each pair moves to opposite poles of the cell.
  - In the second division, the copies of each chromosome come apart and move to opposite poles of the cell.

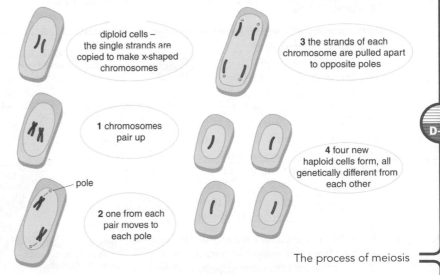

diploid cells – the single strands are copied to make x-shaped chromosomes

3 the strands of each chromosome are pulled apart to opposite poles

1 chromosomes pair up

4 four new haploid cells form, all genetically different from each other

pole

2 one from each pair moves to each pole

The process of meiosis

Improve your grade

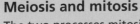

Meiosis and mitosis

The two processes mitosis and meiosis occur in the human body. Compare each process, writing about where they occur and any differences in the process. *AO1* [6 marks]

EXAM TIP

Be careful not to mix up mitosis and meiosis.

The circulatory system

Blood

- The liquid part of the blood called **plasma** carries a number of important substances around the body:
  - dissolved food substances such as glucose
  - **carbon dioxide** from the tissues to the lungs
  - **hormones** from the glands where they are made, to their target cells
  - plasma proteins such as antibodies
  - waste substances such as urea.

- **Red blood cells** are adapted to their function of carrying oxygen in a number of ways:
  - They are very small so that they can pass through the smallest blood vessels.
  - They are shaped like biconcave discs so that they have a large surface area to exchange oxygen quicker.
  - They contain **haemoglobin** to combine with oxygen. It is the haemoglobin that makes them appear red.
  - They don't have a **nucleus** so more haemoglobin can fit in.

- Haemoglobin in red blood cells reacts with oxygen in the lungs, forming oxyhaemoglobin. The reaction is reversible: when the oxyhaemoglobin reaches the tissues, the oxygen is released.

- The biconcave shape of the red blood cell provides a larger surface area to volume ratio to exchange oxygen more quickly.

Blood vessels

- The different types of blood vessels have different jobs:
  - **Arteries** transport blood away from the heart to the tissues.
  - **Veins** transport the blood back to the heart from the tissues.
  - **Capillaries** link arteries to veins and allow materials to pass between the blood and the tissues.

- The structures of arteries, veins and capillaries are adapted to carry out specific functions:
  - Arteries have a thick muscular and elastic wall to resist the high pressure.
  - Veins have large lumen and valves to try and keep the blood moving back to the heart because the pressure is low.
  - Capillaries have permeable walls so substances can be transferred between the blood and the tissues.

Remember!
The blood in the pulmonary vein is oxygenated and in the pulmonary artery it is deoxygenated, unlike in other veins and arteries.

The heart

- The different parts of the heart work together to circulate the blood.
  - The left and right atria receive blood from veins.
  - The left and right ventricles pump blood out into arteries.
  - The semilunar, tricuspid and bicuspid valves prevent any backflow of blood.
  - The pulmonary veins and the vena cava are the main veins carrying blood back to the heart.
  - The aorta and pulmonary arteries carry blood away from the heart.

- The left ventricle has a thicker muscle wall than the right ventricle because it has to pump blood all round the body rather than just to the lungs, which are close by.

RIGHT

pulmonary artery, takes deoxygenated blood to the lungs

aorta, takes oxygenated blood to the body

LEFT

semi-lunar valve

vena cava, brings deoxygenated blood from the body

right atrium

tricuspid valve

valve tendon

pulmonary vein, brings oxygenated blood from the lungs

left atrium

bicuspid valve

right ventricle, thinner wall as pumps blood a relatively short distance to the lungs

left ventricle, has thick muscular wall to pump blood at higher pressure all the way round the body

Blood flow in the heart

- The blood is pumped to the lungs and returns to the heart to be pumped to the body. This is called a **double circulatory system**. This means the blood is at a higher pressure and so flows to the tissues at a faster rate.

Improve your grade

Valves

Valves are found in veins, at the start of the arteries leaving the heart and in the heart.
Write about the importance of these different valves. *AO1/2* [2 marks]

Growth and development

Different types of cells

- Bacterial cells differ from plant and animal cells in that they lack a 'true' **nucleus, mitochondria** and chloroplasts.

- As bacterial cells do not have a true nucleus, **DNA** is found in the cytoplasm as a single circular strand or **chromosome**.

Measuring growth

- The diagram of a typical growth curve shows the main phases of growth.

- Two of these phases involve rapid growth; one is just after birth and the other in adolescence.

- Dry mass is the best measure of growth.

- Measuring growth by length is easy to do but only measures growth in one direction.

- Measuring wet mass is hard to do for some organisms, e.g. trees, but is easy for animals. However, the water content of organisms can vary with time.

- Dry mass can only be measured by killing the organism and driving off the water but it does measure the true growth of the whole organism.

- Different parts of an organism may grow at different rates compared to the whole organism. This is because different parts of the organism may be needed at different times during the life of the organism.

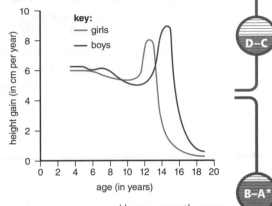

Human growth curve

Differentiation

- Cells called **stem cells** stay undifferentiated. They can develop into different types of cells.

- Stem cells can be obtained from embryos and could potentially be used to treat some medical conditions.

- However, there are issues arising from stem cell research in animals. Some people think that it is wrong because the embryos are destroyed. Others think that this is acceptable as it can treat life-threatening diseases.

- Stem cells can be found in the adult as well as in the embryo. Embryonic stem cells can form a greater range of cell types and are easier to find.

EXAM TIP

Make sure that you can give both sides of this argument, even if you have strong views.

Plant and animal growth

- Differences between plant and animal growth include the following.
  - Animals tend to only grow to a certain size but many plants can carry on growing.
  - Plant cell division only happens in areas called **meristems**, found at the tips of roots and shoots.
  - The main way that plants gain height is by cells enlarging rather than dividing.
  - Many plant cells keep the ability to differentiate, but most animal cells lose it at an early stage.

Improve your grade

A plant growth curve

Katie wants to plot a growth curve for a broad bean plant using dry mass.
Given 100 broad bean seeds, explain how you would collect the data to plot
the graph and explain why she wants to plot dry mass. *AO1* [5 marks]

New genes for old

Selective breeding

- There are possible problems with **selective breeding** programmes. They may lead to **inbreeding**, where two closely related individuals mate, and this can cause health problems within the **species**. Certain breeds of dog show these problems.

- Inbreeding can reduce the variety of **alleles** in the population (the **gene pool**). This can lead to:
  - an increased risk of harmful recessive characteristics showing up in offspring
  - a reduction in **variation**, so that **populations** cannot adapt to change so easily.

Genetic engineering

- **Genetic engineering** has advantages and risks:
  - One advantage is that organisms with desired features can be produced very quickly.
  - However, there is a risk that the inserted genes may have unexpected harmful side effects.

- Examples of organisms that have been made using genetic engineering include the following:
  - Rice that contains beta-carotene has been made by inserting the genes that control beta-carotene production from carrots. Humans can then convert the beta-carotene from rice into Vitamin A. This is important because in some parts of the world people rely on rice, which normally has very little Vitamin A.
  - Genetically engineered **bacteria** have been made that produce human **insulin**.
  - Crop plants have been made that are resistant to **herbicides**, frost damage or disease.

- A number of ethical issues are involved in genetic engineering:
  - Some people are worried about possible long-term side effects, e.g. that genetically engineered plants or animals will disturb natural ecosystems.
  - Other people think that it is morally wrong, whatever the intended benefits.

- To carry out genetic engineering, four steps are taken:
  - The desired characteristics are selected.
  - The genes responsible are identified and removed (isolation).
  - The genes are inserted into other organisms.
  - The organisms are allowed to reproduce (replication).

Remember!
It is the *gene* that is put into another organism, e.g. the gene for human insulin, not human insulin itself.

Gene therapy

- The process of using genetic engineering to change a person's genes and cure certain disorders is called **gene therapy**.

- Gene therapy could involve body cells or **gametes**. Changing the genes in gametes is much more controversial. This is because it is sometimes difficult to decide which genes parents should be allowed to change. It could lead to 'designer babies'.

 Improve your grade

Spider silk

Spider silk is very strong and could be very useful in industry. Goats have now been produced that make spider silk in their milk.

Describe how this could be done and suggest reasons why this method of production might be more useful. *AO2* [4 marks]

Cloning

Cloning animals

- Dolly the sheep was produced by a process called **nuclear transfer**. This involves removing the **nucleus** from a body cell and placing it into an egg cell that has had its nucleus removed.

- Animals could be **cloned** to:
  - mass-produce animals with desirable characteristics
  - produce animals that have been genetically engineered to provide human products
  - produce human embryos to supply **stem cells** for therapy.

- There are some ethical dilemmas concerning human cloning. Some people think that it is wrong to clone people as they will not be 'true individuals'.

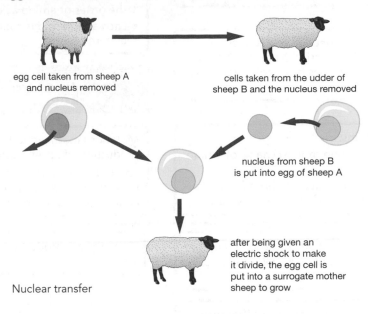

egg cell taken from sheep A and nucleus removed

cells taken from the udder of sheep B and the nucleus removed

nucleus from sheep B is put into egg of sheep A

Nuclear transfer

after being given an electric shock to make it divide, the egg cell is put into a surrogate mother sheep to grow

D–C

- When Dolly was produced by nuclear transfer, it involved a number of steps:
  - The donor egg cell had its nucleus removed.
  - The egg cell nucleus was replaced with the nucleus from an udder cell.
  - The egg cell was then given an electric shock to make it divide.
  - The embryo was implanted into a surrogate mother sheep.
  - The embryo then grew into a clone of the sheep from which the udder cell came.

- Cloning technology could be useful in a number of ways. Some of these may carry risks. For example, genetically modified animals could be cloned to supply replacement organs for humans. Some people are worried that this could lead to diseases being spread from animals to humans.

B–A*

Remember!
Cloning would not be very successful in recreating endangered or extinct animals, as all the clones would be the same sex and genetically identical.

Cloning plants

- Producing cloned plants has advantages and disadvantages.
  - Advantages: growers can be sure of the characteristics of each plant since all the plants will be genetically identical. It is also possible to mass-produce plants that may be difficult to grow from seed.
  - Disadvantages: if the plants become susceptible to disease or to change in the environmental conditions, then all the plants will be affected. There is a lack of genetic **variation** in the plants.

D–C

- Plants can be cloned by a process called **tissue culture**, as follows:
  - A plant is selected that has certain characteristics.
  - A large number of small pieces of tissue are then cut from the plant.
  - The small pieces of tissue are grown in test tubes or dishes containing a growth medium.
  - Aseptic technique is used at all stages to stop any **microbes** infecting the plants.

- Cloning plants is easier than cloning animals because many plant cells retain the ability to differentiate. Animal cells, however, usually lose this ability at an early stage.

B–A*

Improve your grade

Cloning plants

A garden centre wants to sell an attractively coloured geranium plant.
They decide to produce many clones of the plant using tissue culture.

(a) What are the possible disadvantages of this method of reproduction?
  AO1 [2 marks]

(b) Why is this method not possible for producing goldfish for garden ponds? *AO2* [2 marks]

B3 Summary

Protein synthesis occurs on ribosomes in the cytoplasm.

The code needed to produce a protein is carried from the DNA to the ribosomes by a molecule called mRNA.

Proteins:
- are made of long chains of amino acids
- can be structural, hormones, carrier molecules or enzymes.

Proteins are coded for by DNA.
- The base sequence codes for the order of amino acids.
- Each amino acid is coded for by three bases.

DNA, proteins and mutations

Mutations may lead to the production of different proteins. This is because a change in bases in DNA can change the amino acid sequence.

Chromosomes are long, coiled molecules of DNA, divided up into regions called genes.

Enzymes are specific and work by a 'lock and key' mechanism.

Enzyme activity is affected by pH and temperature. This is due to:
- lower collision rates at lower temperatures
- denaturing at extremes of pH and high temperatures.

In meiosis, the chromosome number is halved and each cell is genetically different.

There are a number of differences between plant growth and animal growth.

Growth can be measured by a change in wet mass, dry mass or length. Each method has advantages and disadvantages but dry mass is the best measure.

Gametes are produced by meiosis

Cell division and growth

Undifferentiated cells called stem cells can develop into different cells, tissues and organs.

New cells for growth are produced by mitosis.

The new cells made by mitosis are genetically identical.

Being multicellular allows organisms to:
- be larger
- use cell differentiation
- be more complex.

The symbol equation for aerobic respiration is

$C_6H_{12}O_6 + 6O_2 \rightarrow 6CO_2 + 6H_2O$
The energy released is stored in ATP.

Red blood cells carry oxygen around the body and are especially adapted for this function.

The heart has four chambers and is part of a double circulatory system.

Anaerobic respiration takes place during hard exercise when there is insufficient oxygen available.

Respiration and the circulation

- Arteries have thick elastic walls and carry blood away from the heart.
- Veins have large lumens and valves, and carry blood back to the heart.
- Capillaries are permeable and link arteries to veins.

Anaerobic respiration produces lactic acid, which:
- builds up in muscles causing pain and fatigue
- acts as an oxygen debt and has to be broken down in the liver after exercise.

Gene therapy involves changing a person's genes to try and cure disorders.

Genetic engineering can be used to produce useful products but raises some ethical issues.

A selective breeding programme can produce organisms with desired characteristics but may reduce the gene pool leading to problems of inbreeding.

Changing genes and cloning

Cloning plants is easier than cloning animals because many plant cells retain the ability to differentiate.

Dolly the sheep was produced by the process of nuclear transfer – this involves placing the nucleus of a body cell into an egg cell.

New cloning technology will:
- produce a number of benefits
- involve certain risks
- raise ethical issues.

Plants can be cloned by tissue culture, which provides a number of benefits.

Ecology in the local environment

Distribution of organisms

- An ecosystem, such as a garden, is made up of all the plants and animals living there and their surroundings. Where a plant or animal lives is its **habitat**.

- All the animals and plants living in the garden make up the **community**. The number of a particular plant or animal present in the community is called its **population**.

- Natural ecosystems, such as native woodlands and lakes, have a large variety of plants and animals living there – this means it has good **biodiversity**. Artificial ecosystems, such as forestry plantations and fish farms, have poor biodiversity.

- The distribution of organisms can be mapped using a **transect** line. A long length of string is laid across an area such as a path or sea shore. At regular intervals the organisms in a square frame called a quadrat can be counted (for animals) or assessed for percentage cover (for plants). The data can be displayed as a **kite diagram**.

Remember!
Artificial ecosystems are created by humans, for the benefit of humans

D–C

- In artificial ecosystems, humans deliberately keep and protect only one **species** (such as salmon in a fish farm) and remove any other organisms that would compete with it and lower the yield. This does not happen in a natural ecosystem.

- A transect line can show zonation in the distribution of organisms. Changes in abiotic (not biological) factors such as exposure on a sea shore or trampling near a footpath, cause zonation.

- Food chains and food webs show that plants and animals are interdependent, with energy being transferred from one organism to another. The exchange of gases in **photosynthesis** and **respiration** ensures an overall balance of these gases. An ecosystem Is therefore self-supporting in all factors apart from having to have the Sun as an energy source.

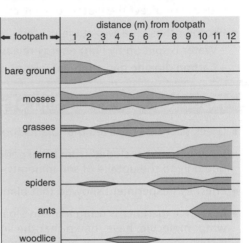

Kite diagram showing distribution of organisms near a path

B–A*

Population size

- Population size can be estimated by obtaining data from a small sample and scaling up. For example, Shabeena has a large lawn with some dandelions in it. She counts the number of dandelions in ten 1 m² quadrats and calculates the mean as 8 dandelions/1 m². Her lawn is 100 m². She calculates that there are 800 dandelions in her lawn.

- Shabeena also wants to estimate the woodlouse population in her garden. She uses a capture–recapture method.

$$\text{Population size} = \frac{\text{number in 1st sample} \times \text{number in 2nd sample}}{\text{number in 2nd sample previously marked}}$$

- Shabeena captures 36 woodlice in a pitfall trap, marks them with a white dot and releases them. Later she captures 48 woodlice, 6 with a white dot.

$$\text{Population size} = \frac{36 \times 48}{6} = 288 \text{ woodlice}$$

D–C

- Shabeena realises that if she had used a bigger quadrat and more samples, her estimation would have been more accurate.

- Using a capture–recapture method assumes:
  - There are no deaths or reproduction and no movement of animals into and out of the area.
  - Identical sampling methods are used for both samples.
  - The markings do not affect the survival of the woodlice.

B–A*

Improve your grade

Distribution of animals
Rick wants to find out about the zonation of different types of limpets down a sea shore. Explain:

(a) how he should do this
(b) what can cause this zonation.
AO1/2 [5 marks]

Photosynthesis

The chemistry of photosynthesis

- The **balanced symbol equation** for **photosynthesis** is:

$$6CO_2 + 6H_2O \xrightarrow[\text{chlorophyll}]{\text{light energy}} C_6H_{12}O_6 + 6O_2$$

EXAM TIP

It is important to write the correct chemical formula (e.g. CO_2, not CO2).

- The simple sugars such as glucose can be:
  - used in **respiration**, releasing energy
  - converted into cellulose to make cell walls
  - converted into proteins for growth and repair
  - converted into starch, fats and oils for storage.

- Starch is used for storage since it is insoluble and does not move from storage areas. Unlike glucose, it does not affect the water concentration of cells and cause **osmosis**.

- Photosynthesis is a two-stage process.
  - Water is split up by light energy releasing oxygen gas and hydrogen **ions**.
  - **Carbon dioxide** gas combines with the hydrogen ions producing glucose and water.

Historical understanding of photosynthesis

- Greek scientists believed that plants took **minerals** out of the soil to grow and gain mass.

- Van Helmont concluded from his experiment on growing a willow tree that plant growth could not be due only to the uptake of soil minerals – it must depend on something else.

- Priestley's experiment showed that plants produce oxygen.

- Modern experiments using a green alga called *Chlorella* and an **isotope** of oxygen, ^{18}O, as part of a water **molecule**, have shown that the light energy is used to split water, not carbon dioxide. The water is split up into oxygen gas and hydrogen ions. Isotopes are different forms of the same element.

The rate of photosynthesis

- The rate of photosynthesis can be increased by the plant having:
  - more carbon dioxide
  - more light
  - a higher temperature which increases **enzyme** action.

- Photosynthesis will only take place during daytime (in the light). However, since plants are living organisms they must also respire, releasing energy at all times.

- Plants respire at all times by taking in oxygen and releasing carbon dioxide.

  During the day, when it is light, they also carry out photosynthesis, taking in carbon dioxide and releasing oxygen: the same gas exchange as respiration but in reverse. The rate of gas exchange in photosynthesis is more than that of respiration in terms of quantities, so respiration can only be noticed at night (in darkness).

- Since photosynthesis depends on light, temperature and carbon dioxide, a lack of one of these factors will limit the rate of photosynthesis. They are called limiting factors.

> **Remember!**
> Plants respire at all times, not just at night.

Effect of light on the rate of photosynthesis – at B, the rate of photosynthesis is being limited either by carbon dioxide concentration or temperature

Improve your grade

Exchange of gases in plants
When gas exchange in plants is analysed, they seem to respire only at night. Explain why.
AO2 [4 marks]

Leaves and photosynthesis

Leaf structure

- A green leaf has many specialised cells, as shown in the diagram.

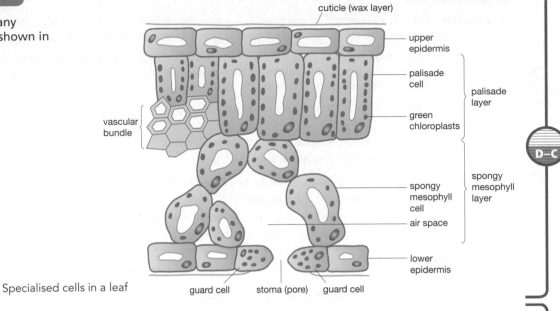

Specialised cells in a leaf

Labels: cuticle (wax layer), upper epidermis, palisade cell, palisade layer, green chloroplasts, vascular bundle, spongy mesophyll cell, spongy mesophyll layer, air space, lower epidermis, guard cell, stoma (pore), guard cell

D–C

- The cells are adapted for efficient **photosynthesis**.
  - The outer epidermis lacks chloroplasts and so is transparent; there are no barriers to the entry of light.
  - The upper **palisade** layer contains most of the leaf's chloroplasts, as they will receive most of the light.
  - The **spongy mesophyll cells** are loosely spaced so that **diffusion** of gases between cells and the outside atmosphere can take place.
  - The arrangement of mesophyll cells creates a large surface area/volume ratio so that large amounts of gases can enter and exit the cells.

B–A*

Leaf adaptations for photosynthesis

- Leaves are adapted so that photosynthesis is very efficient.
  - They are usually broad so that they have a large surface area to get as much light as possible.
  - They are usually thin so that gases can **diffuse** through easily and light can get to all cells.
  - They contain **chlorophyll** and other pigments so that they can use light from a broad range of the light spectrum.
  - They have a network of **vascular bundles (veins)** for support and transport of chemicals such as water and glucose.
  - They have specialised guard cells which control the opening and closing of **stomata** therefore regulating the flow of **carbon dioxide** and oxygen as well as water loss.

D–C

Remember!
The small holes in the lower epidermis are called stomata; they are surrounded by two guard cells.

- By having many pigments (chlorophyll a and b, **carotene** and **xanthophylls**) the plant cells can maximise the use of the Sun's energy. Each pigment absorbs light of different wavelengths.

B–A*

Improve your grade

Absorption of light
Look at the graph.
(a) Which parts of the light spectrum are i) reflected and ii) used by leaves? *AO2* [2 marks]
(b) Explain why a leaf has many different pigments. *AO1* [2 marks]

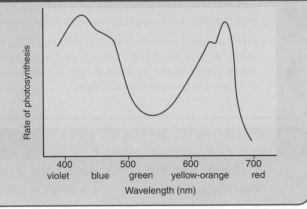

y-axis: Rate of photosynthesis
x-axis: Wavelength (nm) — 400 violet, 500 blue, green, 600 yellow-orange, 700 red

Diffusion and osmosis

Diffusion

- **Diffusion** is the net movement of particles in a gas or liquid from an area of high concentration to an area of low concentration, resulting from the **random** movement of the particles.
- This explains how **molecules** of water, oxygen and **carbon dioxide** can enter and leave cells through the **cell membrane**. If a plant cell is using up carbon dioxide, there is a lower concentration of it inside the cell, so carbon dioxide will enter by diffusion.
- Leaves are adapted to increase the rate of diffusion of carbon dioxide and oxygen by having:
  – (usually) a large surface area
  – specialised openings called **stomata**, which are spaced out
  – gaps between the **spongy mesophyll cells**.

- The rate of diffusion is not a fixed quantity. The rate can be increased by having:
  – a shorter distance for the molecules to travel
  – a steeper concentration gradient (a greater difference in concentration between the two areas)
  – a greater surface area for the molecules to **diffuse** from, or into.

Osmosis

- **Osmosis** is a type of diffusion; It depends on the presence of a **partially-permeable membrane** that allows the passage of water molecules but not large molecules like glucose.
- Osmosis is the movement of water across a partially-permeable membrane from an area of high water concentration (a dilute solution) to an area of low water concentration (a concentrated solution).

- Osmosis is a consequence of the random movement of water molecules, which is not restricted by a partially permeable membrane. The net movement of water molecules will be from an area where there are many to one where there are few.
- Knowing the different concentrations of water inside and outside cells makes it possible to predict the net movement of water molecules.

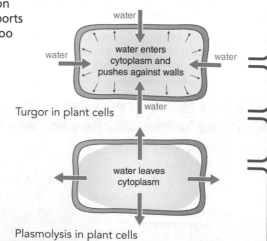

Osmosis in a plant cell

permeable cell wall
potato cell
cytoplasm
small water molecules
partially-permeable membrane
large molecules such as sugar cannot pass through the partially-permeable membrane
diffusion

Remember!
Osmosis requires the presence of a partially permeable cell membrane.

Water in cells

- The entry of water into plant cells increases the pressure pushing on the cell wall, which is rigid and not elastic. This **turgor pressure** supports the cell, stopping it, and the whole plant, from collapsing. When too much water leaves a cell, it loses this pressure and the plant wilts.

- A plant cell full of water is said to be **turgid**. When the cell loses water the cell contents shrink and become **plasmolysed** and the cell is called **flaccid**.

- Animal cells also react to intake and loss of water due to osmosis. They will also shrink and collapse when they lose too much water, and swell up when too much water enters.

- Since animal cells lack a supporting cell wall, when too much water enters, they will swell up and burst (**lysis**). When too much water leaves an animal cell, it shows **crenation** by shrinking into a scalloped shape.

water
water
water enters cytoplasm and pushes against walls
water
water
Turgor in plant cells

water leaves cytoplasm
Plasmolysis in plant cells

Improve your grade

Osmosis in plant and animal cells
Describe and explain the effects of water entry into plant and animal cells. *AO1/2* [5 marks]

Transport in plants

Xylem and phloem cells

- **Xylem** and **phloem** are made up of specialised plant cells. Both types of tissue are continuous from the root, through the stem and into the leaf.

- Xylem and phloem cells form **vascular bundles** in dicotyledonous (broad-leaved) plants.

- Xylem cells carry water and **minerals** from the roots to the leaves and are therefore involved in transpiration. Phloem cells carry food substances such as sugars up and down stems to growing and storage tissues. This transport of food substances is called translocation.

- The xylem cells are called vessels. They are dead cells, and the lack of living cytoplasm leaves a hollow lumen. Their cellulose walls have extra thickening of lignin, giving great strength and support.

- Phloem cells are living cells and are arranged in columns.

Transpiration

- Transpiration is the **evaporation** (changing from a liquid into a gas) and **diffusion** of water from inside leaves. This loss of water from leaves helps to create a continuous flow of water from the roots to the leaves in xylem cells.

- Root hairs are projections from root hair cells. They produce a large surface area for water uptake by **osmosis**.

- Transpiration ensures that plants have water for cooling by evaporation, **photosynthesis** and support from cells' **turgor pressure**, and for transport of minerals.

- The rate of transpiration is increased by an increase in light intensity, temperature and air movement and a decrease in humidity (the amount of water vapour in the atmosphere).

- The structure of a leaf is adapted to prevent too much water loss, which could cause wilting. Water loss is reduced by having a waxy cuticle covering the outer epidermal cells and by most **stomatal** openings being situated on the shaded lower surface.

- Plant leaves are adapted for efficient photosynthesis by having stomata for entry and exit of gases. The **spongy mesophyll cells** are also covered with a film of water in which the gases can dissolve. This water can therefore readily escape through the stomata.

- The rate of transpiration can be increased by:
  - an increase in light intensity, which results in stomata being open
  - an increase in temperature, causing an increase in the evaporation of water
  - an increase in air movement, blowing away air containing a lot of evaporated water
  - a decrease in humidity (the amount of water vapour in the atmosphere), allowing more water to evaporate.

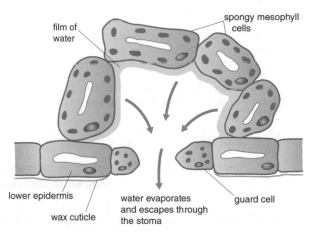

Movement of water through a stoma

- The structure of a leaf is adapted to reduce water loss. Its guard cells are able to change the size of the stomatal openings. The guard cells contain chloroplasts, so photosynthesis (in the presence of water and light) will produce sugars, increasing turgor pressure, causing the cells to swell. Due to differential thickness of their walls, the guard cells curve, opening the stoma.

- Further reduction in water loss is brought about by having fewer stomata, smaller stomata, the position of stomata (mainly in the lower epidermis), and their distribution.

Improve your grade

Transport in plants

Marram grass grows in exposed sand dunes. It has narrow leaves, stomata sunk in pits, many hairs on its leaves and leaves that can curl up into a tube. Explain how these adaptations help it to survive. *AO2* [4 marks]

Plants need minerals

Use of minerals

- Plants need **minerals**, such as:
  - nitrates, to make proteins, which plants use for **cell** growth
  - phosphates, which are involved in **respiration** and growth
  - potassium compounds, which are involved in respiration and **photosynthesis**
  - magnesium compounds, which are involved in photosynthesis.

- Elements from soil minerals are used to produce useful compounds.
  - Nitrogen (from nitrates) is used to produce **amino acids**, which combine to form a variety of proteins.
  - Phosphorus (from phosphates), is used to make **DNA**, which contains the plant's genetic code, and cell membranes.
  - Potassium (from potassium compounds) is used to help **enzyme** action in photosynthesis and respiration; enzymes speed up chemical reactions.
  - Magnesium (from magnesium compounds) is used to make **chlorophyll**, which is essential for photosynthesis.

Mineral deficiency

- The lack of certain minerals results in specific symptoms:
  - Lack of nitrate causes poor growth and yellow leaves.
  - Lack of phosphate causes poor root growth and discoloured leaves.
  - Lack of potassium causes poor flower and root growth and discoloured leaves.
  - Lack of magnesium causes yellow leaves.

Mineral uptake

- Minerals are usually present in soil in low concentrations.

- Minerals are taken up by root hair cells by active transport, rather than by **diffusion** or **osmosis**. A system of carriers transport selected minerals across the cell **membrane**.

- **Active transport** enables minerals, present in the soil only in low concentrations, to enter root hairs already containing higher amounts of minerals.

- This uptake of minerals against a concentration gradient requires energy from respiration.

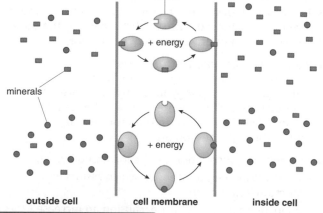

Active transport of minerals

Remember!
Mineral uptake involves active transport, rather than diffusion or osmosis.

EXAM TIP
Root *hairs* (rather than roots) absorb water.

Improve your grade

Mineral uptake
Look at the graph.

What conclusions can be made from this data?

AO3 [4 marks]

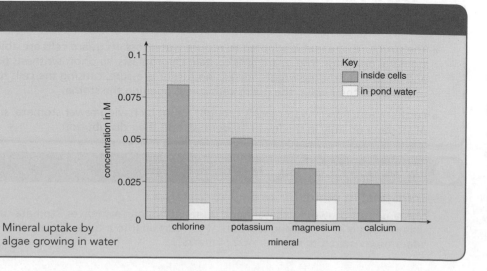

Mineral uptake by algae growing in water

Decay

Decay

- Earthworms, maggots and woodlice are called **detritivores** because they feed on dead and decaying material (**detritus**).

- Detritivores increase the rate of **decay** by breaking up the detritus and so increasing the surface area for further microbial breakdown.

- The rate of decay can be increased by increasing the temperature, amount of oxygen and water.

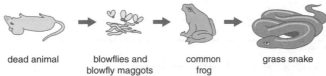

dead animal → blowflies and blowfly maggots → common frog → grass snake

Two food chains

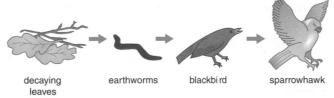

decaying leaves → earthworms → blackbird → sparrowhawk

- Increasing the temperature to an optimum of 37 °C for **bacteria** or 25 °C for fungi will increase their rate of **respiration**. Higher temperatures will denature enzymes.

- By increasing the amount of oxygen, bacteria will use **aerobic respiration** to grow and reproduce faster.

- Increasing the amount of water will allow for material to be digested and absorbed more efficiently and increase growth and reproduction of bacteria and fungi.

- A **saprophyte**, for example a fungus, is an organism that feeds on dead and decaying material.

- Fungi produce enzymes to digest food outside their cells and then reabsorb the simple **soluble** substances. This type of digestion is called extracellular digestion.

Food preservation

- Food preservation methods reduce the rate of decay.
  - In canning, foods are heated to kill bacteria and then sealed in a **vacuum** to prevent entry of oxygen and bacteria.
  - Cooling foods will slow down bacterial and fungal growth and reproduction.
  - Freezing foods will kill some bacteria and fungi and slow down their growth and reproduction.
  - Drying foods removes water so bacteria cannot feed and grow.
  - Adding salt or sugar will kill some bacteria and fungi, as the high osmotic **concentration** will remove water from them.
  - Adding vinegar will produce very acid conditions killing most bacteria and fungi.

> **EXAM TIP**
>
> Fungi and bacteria do not use photosynthesis to make food.

> **Remember!**
> Keeping food cool in a refrigerator only slows down the growth of bacteria. It does not stop it completely.

 Improve your grade

Growing mushrooms

Lynn wants to grow mushrooms (a fungus).

Explain what conditions she should provide for optimum growth. *AO1/2* [3 marks]

Farming

Pesticides

D–C

- The use of pesticides such as **insecticides**, **fungicides** and **herbicides** has disadvantages.
  - They can enter and accumulate in food chains causing a lethal dose to **predators**.
  - They can harm other organisms living nearby which are not pests.
  - Some are persistent (take a very long time to break down and become harmless).

Remember!
Both organic and intensive farming have advantages and disadvantages.

Organic farming

D–C

- Organic farming does not use artificial **fertilisers** or pesticides.

- It uses animal manure and **compost** (instead of artificial fertilisers), **crop rotation** (to avoid build-up of soil pests), nitrogen-fixing crops as part of the rotation, and varying seed planting times to get a longer crop time and avoid certain times of the life cycle of insect pests.

- It avoids expensive fertilisers and pesticides and their disadvantages. However, the crops are smaller and the produce more expensive. Many people believe that organic crops are healthier and tastier than other crops.

Biological control

D–C

- **Biological control** uses living organisms to control pests. Examples are using ladybirds and certain wasp species to eat aphids, which damage plants.

- Biological control can avoid the disadvantages of artificial insecticides and as they use living organisms, once introduced they usually do not need replacing.

- However, many attempts at biological control have caused other problems such as the introduced **species** eating other useful species and then showing a rapid increase in their **population** so they themselves become pests and then spread into other areas or countries, e.g. the use of cane toads in Australia.

- Introducing a species into a **habitat** to kill another species can affect the food sources of other organisms in a food web, causing unexpected results.

Hydroponics and intensive farming

D–C

- **Intensive farming**, which makes use of artificial pesticides and fertilisers, is very efficient in producing large crop yields cheaply. However, intensive farming methods raise concerns about animal cruelty, as animals are kept in small areas, and about the effects of extensive use of chemicals on soil structure and other organisms.

EXAM TIP
Hydroponics is a form of intensive farming.

- Plants can be grown without using soil using **hydroponics**. This system uses a regulated recycling flow of aerated water containing **minerals** and is usually done in glasshouses and polytunnels.

- Hydroponics is a type of intensive farming that is especially useful in areas of barren soil or low rainfall. Tomatoes are a common crop from hydroponics.

- Being a soil-free system, hydroponics has a better control over mineral levels and disease. Many plants can be grown in a small space. As there is no anchorage for plants when using water, artificial fertilisers are used.

B–A*

- Intensive farming improves the **efficiency** of energy transfer in food chains involving humans by reducing or removing competing organisms such as animal pests and weeds. Also by keeping animals inside sheds or barns (battery farming), they use less energy to keep warm and to move, and more energy on growth (cattle) or egg production (hens).

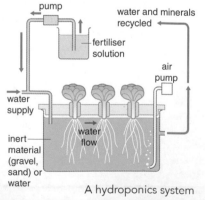

A hydroponics system

Improve your grade

Hydroponics

Look at the diagram of a hydroponics system. Explain how such a system is useful for growing plants such as lettuce in glasshouses. *AO1/2* [4 marks]

B4 Summary

An ecosystem, e.g. a garden, includes all living things and their surroundings.
Where an organism lives is called its habitat.

A community is made up of the organisms living there. A population is the number of a particular organism in a community.

Native ecosystems (woodland, lakes) have a wider biodiversity than artificial ecosystems (forestry plantations, fish farms).

Capture–recapture data can be used to calculate a population size estimate.

$$\text{Population size} = \frac{\text{number in 1st sample} \times \text{number in 2nd sample}}{\text{number in 2nd sample previously marked}}$$

Capture–recapture data assumes:
- no deaths, immigration, emigration
- identical sampling methods
- survival rates not affected by marking organisms.

Ecology in the local environment

Zonation is the gradual change in species distribution across a habitat.

The distribution of organisms can be mapped using a transect line and displayed as a kite diagram.

Photosynthesis is a two-stage process:
- Light energy splits water into oxygen gas and hydrogen ions.
- Carbon dioxide gas combines with the hydrogen to make glucose and water.

Isotope experiments showed that the oxygen comes from water.

Greek scientists (plants take minerals from the soil), Van Helmont (plant growth needs more than minerals) and Priestley (plants produce oxygen) all improved the understanding of photosynthesis.

The balanced symbol equation for photosynthesis is
$$6CO_2 + 6H_2O \xrightarrow[\text{chlorophyll}]{\text{light energy}} C_6H_{12}O_6 + 6O_2$$

Photosynthesis

Carbon dioxide, light and temperature are limiting factors in photosynthesis.

Leaves are adapted for efficient photosynthesis – large surface area, thin, contain pigments, have vascular bundles and guard cells.

Diffusion is the net movement of particles from a high concentration to a low concentration.

The rate of diffusion is increased by a shorter distance, a greater concentration gradient and a greater surface area.

Diffusion and osmosis

Transpiration is the evaporation and diffusion of water from leaves.

Osmosis is the net movement of water from a high to a low water concentration across a partially-permeable membrane.

Transpiration rate is increased by increased light intensity, temperature, air movement and a decrease in humidity.

Plants require nitrates for cell growth, phosphates for respiration and growth, potassium for respiration and magnesium for photosynthesis.

Organic farming uses animal manure, crop rotation, weeding and differing planting times.

Biological control has advantages (no chemical pesticides) and disadvantages (introduced predator may become a pest).

Minerals are absorbed by active transport which needs energy.

Minerals and farming

Hydroponics has advantages (better control of minerals and water) and disadvantages (lack of support).

Detritivores feed on dead and decaying material.

Saprophytic fungi use extracellular digestion.

Intensive food production improves the efficiency of energy transfer.

Rate of reaction (1)

Reaction rates

- The **rate of reaction** measures how much **product** is formed in a fixed period of time.

- Reactions are usually fast at the start and then slow down as the **reactants** are used up.

- The units commonly used when measuring the rates of reactions are:
  - g/s (grams per second) and g/min (grams per minute) when measuring the **mass** of a product formed. g/s is used for faster reactions and g/min for slower reactions.
  - cm³/s (centimetre cubed per second) and cm³/min (centimetre cubed per minute) when measuring the volume of gas produced. cm³/s is used for faster reactions and cm³/min for slower reactions.

> **EXAM TIP**
>
> Always include units when giving answers to questions involving numbers.

Calculating the rate of reaction

B–A*

- The rate of reaction can be worked out from the **gradient** of a graph. This can be calculated by drawing construction lines.
  - Choose part of the graph where there's a **straight line** (not a **curved line**).
  - Using the scales on the graph, measure the value of y and x.
  - Then divide y by x.

- The gradient of the graph is:

  $$\text{gradient} = \frac{y}{x}$$

- Add units, making sure these are the same as those used on the graph, cm³/s in this example.

Remember!
You need to be able to interpret data from tables and graphs to compare reaction rates.

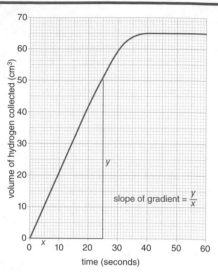

Calculating the gradient of a graph

Limiting reactants

D–C

- The **limiting reactant** is the reactant *not* in excess that gets used up by the end of the reaction.

- The amount of product formed in a reaction is *directly proportional* to the amount of the limiting reactant used. For example, if the amount of limiting reactant doubles, the amount of product formed doubles (as shown in the graph below).

- Reactions occur when particles collide together. Therefore, if the number of reacting particles of one reactant is limited, the number of collisions by particles of that reactant is limited.

- The mass of gas that would be collected at 15 seconds can be estimated by **interpolation**.

B–A*

- If the mass of magnesium was not a limiting factor, the volume of gas that would be collected after 40 seconds can be estimated by **extrapolation**, using the initial line.

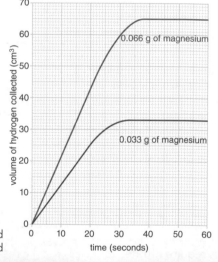

Doubling the limiting reactant used doubles the amount of product formed

How science works

You should be able to:

- choose the most appropriate format for presenting and processing data, using mathematical techniques such as statistics or calculating gradients from graphs. You should also be able to work out data within a range already given and suggest data outside the range.

Improve your grade

Change in rate of reaction

When limestone reacts with excess hydrochloric acid, carbon dioxide gas is made. Explain why the rate of reaction changes during the reaction. Use ideas about reacting particles. *AO2* [4 marks]

Rate of reaction (2)

Reacting particle model

- Chemical reactions take place when particles collide.

- The **rate of reaction** depends on the number of collisions between the reacting particles: the higher the number of collisions that take place, the faster the reaction.

- The rate of reaction can be increased by increasing the **concentration**, raising the *temperature* or the *pressure*.
  - As the concentration increases the particles become more crowded. This increases the number of collisions between reacting particles.
  - As the temperature increases, the particles gain **kinetic energy** and move around more quickly, so collisions between reacting particles are more successful, and also occur in greater number. (The size of the arrows on the diagram indicate the amount of kinetic energy the particles have.)
  - Increases in pressure forces particles closer together, increasing the rate of reaction.

- The rate of reaction depends on the **collision frequency**: this describes the number of *successful collisions* between reacting particles each second.

- For a successful collision to occur, each particle must have enough energy to react.
  - As the concentration increases, the number of collisions per second increases, and so the rate of reaction increases.
  - As the temperature increases, the **reactants** have more kinetic energy, so collisions are more energetic and frequent, so more likely to be successful.
  - Increasing the pressure in gases forces particles closer together, increasing collision frequency.

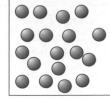

low concentration | high concentration

 reacting particle of substance **A**

 reacting particle of substance **B**

low temperature | high temperature

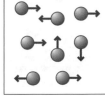 reacting particle of substance **A**

reacting particle of substance **B**

Particle diagrams representing changing conditions

 D–C

EXAM TIP

When asked to explain reaction rate changes, always use the idea of the rate of successful collisions of particles.

B–A*

Rate of reaction

- Both tables and graphs can show when a reaction has finished.
  - On a graph the line will be horizontal.
  - In a table the numbers will stop changing.

- Both tables and graphs can be used to compare the rate of reaction.
  - On graphs, the steeper the line, the faster the reaction.
  - In a table, the larger the change between readings, the faster the reaction.

- An essential skill is to be able to sketch graphs to show the affect of changing temperature, concentration or pressure on:
  - the rate of reaction
  - the amount of **product** formed in a reaction.

Remember!

Extrapolation means making an estimate beyond the range of results. To do this, assume that the trend shown by the existing data will continue. Interpolation means making an estimate between results in a range.

 D–C

Improve your grade

Temperature and rate of reaction

Explain, using the reacting particle model, why raising the temperature increases the rate of a chemical reaction. *AO1* [3 marks]

Rate of reaction (3)

Explosions and surface area

- Combustible powders can cause **explosions**.
  - Some powders reacts with oxygen to make large volumes of **carbon dioxide** and water vapour.
  - The owners of a factory using combustible powders such as flour, custard powder or sulfur must be very careful. They must ensure that the powders can't reach the open atmosphere and that the chances of a **spark** being produced near the powders is very small.
- Breaking up a block into smaller pieces increases the surface area.
- A powdered **reactant** has a much larger surface area than the same **mass** of reactant in block form.
- As the surface area of a solid is increased so is the **rate of reaction**.

small surface area   large surface area

Changing surface area

- In the 'small surface area' diagram:
  - The blue particles at the bottom represent a solid block of reactant.
  - The red particles represent the other reactant.
  - Most particles in the solid block cannot react as collisions can only occur at the surface.
- In the 'large surface area' diagram:
  - The blue particles from the block are now spread out.
  - More surfaces on these particles are exposed, so more collisions are possible, increasing the rate of reaction.

> **Remember!**
> Surface area increases when a block of solid is cut up into pieces.

- Reactions occur when reactant particles collide with sufficient energy.
- Powders can spread throughout a reaction mixture, increasing the **collision frequency** (collisions per second).
- In a block of material, collisions can only occur with the particles on the surface. Most of the particles in the block are trapped on the inside, so they are not available to react.

Catalysts and rates of reaction

- A **catalyst** changes the rate of reaction and is unchanged at the end of the reaction.
- Only a small quantity of catalyst is needed to catalyse a large mass of reactants; each catalyst is specific to a particular reaction.
- The data in the table shows the effect of adding a catalyst to a reaction.
  - The second reaction finishes at 20 seconds, whereas the first reaction finishes at 40 seconds.
  - The rate of reaction is highest in the first 10 seconds of the second reaction.

Time in seconds	10	20	30	40	50
volume of gas with no substance cm³	18	30	60	100	100
volume of gas with catalyst cm³	70	100	100	100	100

Improve your grade

Surface area and rate of reaction

Draw a sketch graph as on page 76. On this graph, sketch further lines to show the effect of changing the surface area of reactants on:

(a) the amount of product formed in a reaction

(b) the rate of reaction. *AO2* [3 marks]

C3 Chemical economics

Reacting masses

Relative formula mass

- The **relative atomic mass** of elements can be found on the **periodic table** (see page 248). Atomic mass (A_r) is the largest number shown for each **element**; these numbers can be found above the atomic symbol in the table.

- The relative atomic masses must be added up in the correct order, if there are brackets in the formula, in order to arrive at the correct relative formula mass (M_r).

To find the relative formula mass of calcium nitrate, $Ca(NO_3)_2$ from the relative atomic masses of the elements (A_r) Ca = 40, N = 14, O = 16:
1 Work out the masses inside of the bracket. $14 + (16 \times 3) = 62$
2 Multiply the total by the number outside the bracket. $62 \times 2 = 124$
3 Work out the remaining number. $40 = 40$
4 Add the totals from steps 2 and 3 to find the relative formula mass. $= 164$

Conservation of mass

- In any chemical equation, the total **mass** of the **reactants** equals the total mass of the **products**. This is called **conservation of mass**.
  - Mass is conserved because **atoms** cannot be created or destroyed, only rearranged into different **compounds**.

Remember!
The mass of product is directly proportional to the mass of the limiting reactant used.

In the reaction $KOH + HNO_3 \rightarrow KNO_3 + H_2O$
the formula masses of the reactants = 119
the formula masses of the products = 119

Calculating reacting masses

- In a chemical symbol equation, there are the same number and type of atoms on each side of the equation.

The symbol equation for the decomposition of copper carbonate is:
$CuCO_3 \rightarrow CuO + CO_2$
The relative atomic masses (A_r) are Cu = 64, C = 12, O = 16:
The relative formula masses are:
$64 + 12 + (16 \times 3) \rightarrow (64 + 16) + 12 + (16 \times 2)$
So: $124 \rightarrow 80 + 44$
So: 124 g $CuCO_3$ will give 80 g of CuO or 12.4 g $CuCO_3$ will give 8.0 g CuO.

Predicting the yield of a reaction

- Predictions can be made by using the symbol equation for a reaction.

To find the mass of copper carbonate needed to make 20 g of copper oxide.
1 Write a **balanced symbol equation**. $CuCO_3 \rightarrow CuO + CO_2$
2 Find the atomic masses of the elements. A_r (Cu = 64, C = 12, O = 16)
   Work out the relative formula mass for each formula needed. $CuCO_3 = 124$ CuO = 80
3 Add the units.   124 g $CuCO_3$ makes 80 g of CuO
4 Find the mass of $CuCO_3$ needed to make 1 g CuO. $\dfrac{80 \text{ g}}{124 \text{ g}}$ $CuCO_3$ makes 1 g CuO
5 Multiply up to the amount of $CuCO_3$ needed. $\dfrac{80 \text{ g} \times 20 \text{ g}}{124 \text{ g}} = 12.9 \text{ g}$

Improve your grade

Conservation of mass
Show that mass is conserved in this reaction.
$ZnCO_3 \rightarrow ZnO + CO_2$
A_r (Zn = 65, C = 12, O = 16)
AO2 [3 marks]

Percentage yield and atom economy

Calculating percentage yield and industrial costs

- Relative formula mass calculations can be used to find out how much **product** should be made (the predicted yield) in a chemical reaction.

- In practice, the actual yield produced is less, due to losses in the practical methods used.

- **Percentage yield** is calculated using the formula:

$$\text{Percentage yield} = \frac{\text{actual yield}}{\text{predicted yield}} \times 100$$

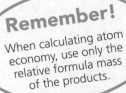

Remember!

When calculating atom economy, use only the relative formula mass of the products.

> A soap manufacturer calculates the yield required as 150 tonnes. However, when the soap is made, only 147.6 tonnes have been produced. The percentage yield can be calculated as follows:
>
> $\text{Percentage yield} = \dfrac{147.6}{150} \times 100 = 98.4$ tonnes

- Industrial processes need to have as high a percentage yield as possible so that they:
  - reduce the amount of **reactants** wasted, which is wasteful and costly
  - reduce their costs by ensuring that enough reactants are used, as too little reduces the amount of product.

Atom economy

- **Atom economy** can be found using the formula:

$$\text{Atom economy} = \frac{M_r \text{ of desired products}}{\text{Sum of } M_r \text{ of all products}} \times 100$$

EXAM TIP
You need to learn both of the formulae on this page, and be able to use them.

> A company makes potassium nitrate for use as a **fertiliser**, as a rocket propellant and in fireworks. H_2O is a waste product.
> The chemical can be made using this reaction:
>
> | | KOH | + | HNO_3 | → | KNO_3 | + | H_2O |
> |---|---|---|---|---|---|---|---|
> | M_r | 56 | + | 63 | → | 101 | + | 18 |
>
> $\text{Atom economy} = \dfrac{101}{119} \times 100 = 84.9\%$

- The atom economy can be found, using the same formula, for complex **balanced symbol equations**. The relative formula masses first need to be calculated from the atomic mass data before completing the calculation.

> Iron ore can be reduced to iron using the reaction:
> $Fe_2O_3 + 3CO \rightarrow 2Fe + 3CO_2$ [Fe = 56, O = 16, C = 12]
> Knowing that CO_2 is a waste product, the atom economy of this process can be calculated as follows.
> 1 Work out the relative formula mass for each product:
>   2Fe = 56 + 56 = 112 $3CO_2$ = 3(12 + 16 + 16) = 132
>
> 2 $\text{Atom economy} = \dfrac{112}{244} \times 100 = 45.9\%$

- Industry wants as high an atom economy as possible. The reasons for this are:
  - to reduce the production of unwanted products that will need to be disposed of (which often adds to overall costs)
  - to make the process more sustainable by making better use of the reactants, so conserving raw materials and avoiding the need to get rid of wasted products.

Improve your grade

Calculating atom economy
Zinc oxide can be made by this reaction:

$ZnCO_3 \rightarrow ZnO + CO_2$

CO_2 is a waste product. ZnO is the desired product. Calculate the atom economy for this reaction.
A_r (Zn = 65, C = 12, O = 16)
AO2 [2 marks]

Energy

Making and breaking bonds

- Bond breaking is an **endothermic** process.

- Bond making is an **exothermic** process.

Remember!
All chemical reactions involve bond making (exothermic) and bond breaking (endothermic).

- To decide if a chemical reaction is exothermic or endothermic, the amount of **energy** made and produced need to be compared.

- Diagrams can be used to represent bond energy changes. This diagram shows an overall exothermic reaction.

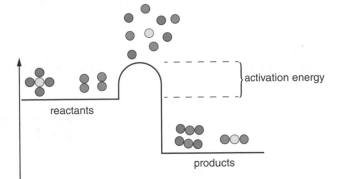

activation energy

reactants

products

Diagram representing
an exothermic reaction

- Stage 1: Energy is needed to break **reactants** into separate **atoms**.
- Stage 2: Atoms join to form new bonds, releasing energy.
- If more energy is released than needed, the reaction is *exothermic*.
- If more energy is needed than released, the reaction is *endothermic*.

Finding out the energy released by fuels

- The energy transferred by a fuel is calculated using the formula:

  Energy transferred (in J) = m × c × ΔT

  where m = **mass** in water heated (in g)
  c = **specific heat capacity** of water (4.2 J/g °C)
  ΔT = temperature change (in °C).

- To find out the energy released by 1 g of solid fuel:
  - Measure out the mass of 1 g of fuel.
  - Pour 100 g of water into a copper calorimeter (1 cm³ of water = 1 g).
  - Heat the water with the burning fuel.
  - Measure the temperature rise.
  - Repeat with different fuels to ensure fair testing. For example, the same distance from the flame to the calorimeter.
  - Ensure reliability by repeating results.

- Mass of fuel burnt can be found by measuring the mass of the burner and the fuel before and after heating.

- The formula energy = m × c × ΔT is used to calculate energy transferred.

- To calculate the energy in a gram of fuel the following formula is used:

  Energy per gram = $\dfrac{\text{energy released (in J)}}{\text{mass of fuel burnt (in g)}}$

EXAM TIP

When answering calculation questions, remember to show all the stages and include units in your answers.

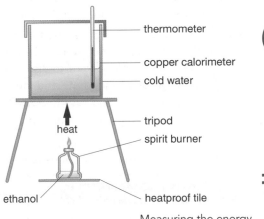

thermometer

copper calorimeter

cold water

tripod

heat

spirit burner

ethanol

heatproof tile

Measuring the energy transferred by a fuel

Improve your grade

Measuring energy transfer
Describe how you could compare the energy released when different liquid fuels burn.
AO1 [4 marks]

Batch or continuous?

Batch and continuous processing

- Chemicals needed in large quantities, such as ammonia and sulfuric acid, are made using continuous processes.
- Continuous processing:
  - makes large amounts of the **product** 24 hours a day, seven days a week
  - takes place in large chemical plants with good transport links; plants are highly automated so have minimal labour costs, making the product cheaper
  - takes less energy to maintain – as long as the process can be kept running – than to stop and start the process.
- Chemicals needed in small quantities, such as medicines, are made using **batch processing**.
- Batch processing:
  - makes a fixed amount
  - allows batches to be made and stored until needed
  - allows quantities to be made that can be sold within a given time as many drugs have a 'sell by' date
  - makes it easy to make a new batch when needed
  - makes it easy to change production to a different product.

- Although the advantages listed for continuous processing mean that the cost per tonne is very small, the disadvantages are that:
  - the process is inefficient if not in constant use
  - there is a very high initial building and set-up cost for these chemical plants.
- Batch processing has the advantage of flexibility for the reasons given above. However, there are also a number of disadvantages with this method of production.
  - Each batch has to be supervised so it is labour intensive and very costly.
  - Time is needed for cleaning if the product line is changed.
  - It is inefficient as the production is not in use all the time.

> **EXAM TIP**
>
> If asked to state advantages and disadvantages, use a different idea for each point.

Why are medicines so expensive?

- Drug prices are set to take account of the development costs.
  - It can take about 10 years to develop and test a new drug and each country has strict safety laws that **pharmaceutical** companies must follow.
  - Many **compounds** need to be made before one may be useful to develop.
  - Raw materials are often rare and costly.
  - Many raw materials are found in plants, and are difficult to extract.
- Extracting a chemical from a plant involves:
  - crushing to disrupt and break cell walls
  - boiling in a suitable solvent to dissolve the compounds
  - **chromatography** to separate and identify individual compounds
  - isolating, purifying and testing potentially useful compounds.
- Pure compounds have definite **melting points** and **boiling points**.
- Thin layer chromatography is used to test the purity of a compound by comparing the speed of movement against a known pure sample.

Drug development

- It is difficult and costly to get a licence for a new drug because:
  - thousands of compounds often need to be tested to find effective ones
  - likely compounds need to be tested on living tissue to ensure safety
  - long-term **trials** on humans are needed to identify possible side effects
  - many similar compounds need to be developed to try to reduce side effects
  - recommended doses need to be shown to be effective
  - the research needs to be independently verified
  - patents expire before costs are recouped and others can make a version.

Improve your grade

Drug development

Explain why it is often expensive to develop a new drug. *AO1* [4 marks]

Allotropes of carbon and nanochemistry

Allotropes

- Diamond, **graphite** and **fullerenes** are all **allotropes** of **carbon**. Allotropes are different structures of the same **element**.

- Fullerenes are carbon structures that form spheres or tubes and can be used:
  - to carry and deliver drug **molecules** around the body
  - to trap dangerous substances in the body and remove them.

part of diamond structure

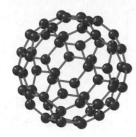
complete particle of buckminsterfullerene

part of graphite structure

Structure of diamond, buckminsterfullerene and and graphite

- **Buckminsterfullerene** contains 60 carbon **atoms** in a sphere. Its formula is written as C_{60}. Each sphere is so small it is measured in **nanometres**. A nanometre is 10^{-9} metres long.

D–C

Why diamonds and graphite are useful

- Diamonds and graphite are both giant covalent structures of carbon atoms. Every carbon atom in diamond makes strong **covalent bonds** in different directions and in graphite the carbon atoms bond to make layers.

- Diamond is the hardest natural substance known and has a very high **melting point** and **boiling point**. The imperfections that naturally occur in diamonds form cleaving plates which allow them to be shaped. These properties make them ideal for use in cutting tools, and as jewellery, as the facets reflect light.

- Graphite has a high melting and boiling point, but the layers can slide over each other. It is used in pencils, and as a high temperature lubricant.

D–C

Explaining the properties of diamond and graphite

- Giant covalent bonding involves **electron** sharing. Every carbon atom in diamond is covalently bonded to four others in a three-dimensional tetrahedral lattice with all the outer shell electrons being shared.

- The structure of diamond gives it specific properties:
  - strong covalent bonds in all directions make diamond hard. Large amounts of energy are required to break the bonds, giving a high melting point of 3350 °C
  - an absence of free electrons, meaning diamond does not conduct electricity.

- Every carbon atom in graphite is covalently bonded to three others in flat hexagonal layers. This formation leaves an unshared outer shell electron. This delocalised electron is free to move along the layer. The separate layers are only weakly **attracted** to each other.

- This structure gives graphite specific properties.
  - The strong covalent bonds give graphite a melting point similar to diamond.
  - The delocalised electrons make it a good **electrical conductor**.
  - When **force** is applied, the weak forces between the layers slide over each other; this slippery nature makes it an ideal high-temperature lubricant.

B–A*

More on giant molecular structures

- Giant covalent bonding forms **compounds** with millions of very strong fixed bonds.
  - Any compound with this structure will have a high melting point.
  - If the bonds are formed in different directions, the substance is hard.
  - If the bonds form in layers, it will be easy to cut in slices.
  - If there are no free electrons, the substance will not conduct electricity.

- **Nanotubes** can be used in **catalyst** systems. Atoms of the catalyst can be attached to the large surface areas on the nanotubes.

B–A*

Improve your grade

Structure of allotropes

Diamond and graphite are both giant carbon structures. Explain why graphite layers slide over each other, while diamond is hard in every direction. *AO1* [4 marks]

C3 Summary

Rates of reaction

Rate of reaction is how much product is made in a fixed time.

The higher the temperature the more successful the collisions between particles.

The higher the concentration, the higher the collision frequency of reacting particles.

Catalysts change the rate of reaction but remain unchanged at the end.

Fine powders are more combustible as they have a greater surface area exposed.

Increasing pressure increases the rate of reaction as there are higher numbers of successful collisions.

Rates of reaction can be calculated from the gradient of a graph of volume of product against time.

Reacting masses

Mass is conserved in a chemical reaction as the reactants become products.

In the reaction:
$2KOH + H_2SO_4 \rightarrow K_2SO_4 + 2H_2O$
to get 174 g of K_2SO_4 you need 112 g of KOH.

$$\text{Percentage yield} = \frac{\text{actual yield}}{\text{predicted yield}} \times 100$$

The atom economy needs to be as high as possible to reduce unwanted products.

The percentage yield needs to be as high as possible to reduce reactant waste and cost.

$$\text{Atom economy is} = \frac{M_r \text{ of desired products}}{\text{Sum of } M_r \text{ of all products}} \times 100$$

Energy

Bond breaking is an endothermic process.

Bond making is an exothermic process.

If the energy transferred during bond breaking is less than during bond making the process is exothermic overall.

Energy transferred (in J) = $m \times c \times \Delta T$
where m = mass in water heated (in g)
         c = specific heat capacity of water (4.2 J/g °C)
        ΔT = temperature change (in °C).

$$\text{Energy per gram} = \frac{\text{energy released (in J)}}{\text{mass of fuel burnt (in g)}}$$

Industrial processes

Continuous processes are used to make chemicals such as ammonia.

Batch processes are used to make pharmaceutical drugs.

Chemicals are extracted from plants by crushing, boiling, dissolving and chromatography.

Allotropes

Diamond, graphite and fullerenes are allotropes of carbon.

Diamond does not conduct electricity but graphite does, as it has electrons that can move between layers.

Atomic structure

Atoms

- The **nucleus** of an **atom** is made up of **protons** and **neutrons**. The relative **charge** and mass of **electrons**, protons and neutrons are given in the table (right).

	Relative charge	Relative mass
electron	−1	0.0005 (zero)
proton	+1	1
neutron	0	1

- If a **neutral** atom has an **atomic number** of 11 and a **mass number** of 23 it will be a sodium atom and have:
  - 11 protons (because of its atomic number) and 11 electrons
  - 12 neutrons, making a mass number of 23 (11 protons + 12 neutrons).
- You can deduce the numbers of protons, electrons and neutrons from symbols.

	Atomic number	Mass number	Number of protons	Number of electrons	Number of neutrons
sodium	11	23	11	11	12
$^{19}_{9}$F	9	19	9	9	10
carbon-12	6	12	6	6	6

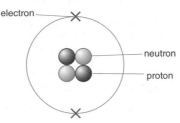

A helium atom has two protons and two electrons, so the atom is neutral because it has the same number of protons as electrons; the charge of the neutron is 0

- An atom is *neutral* because it has an equal number of electrons and protons. The *positive charges* balance out the *negative charges*.

If a charged particle has an atomic number of 3 and a mass number of 7 it must be a lithium particle. If it is not a neutral atom but has a charge of +1 it will have one less electron than protons, i.e. it will only have two electrons.

Charged particle	Number of protons	Number of electrons	Number of neutrons
Li$^+$	3	2	4

Remember!
An atom has a radius of about 1×10^{-10} m and a mass of about 1×10^{-23} g.

Isotopes

- **Elements** with the *same* atomic number but *different* mass numbers are **isotopes**.
- $^{12}_{6}$C has six protons and six neutrons. $^{14}_{6}$C has six protons and a mass of 14; it has eight neutrons (14 − 6). These are both isotopes. $^{14}_{6}$C can also be written as **carbon-14**.

Isotope	Electrons	Protons	Neutrons
$^{1}_{1}$H	1	1	0
$^{2}_{1}$H	1	1	1
$^{3}_{1}$H	1	1	2

- The different numbers of neutrons in isotopes can be deduced, as with these hydrogen isotopes.

Arrangement of electrons in atoms

- The elements in the **periodic table** are arranged in order of increasing atomic number.
  - An element with an **electronic structure** of 2.8.6 has three **electron shells**, so is in the third row.
  - The electronic structure 2.8.6 means the atomic number is 16, so the element is sulfur.
- Elements in the same group (with the same number of electrons in the outer shell) are arranged vertically.
- Elements in the same **period** (in order of how many shells the electrons occupy) are arranged horizontally.
- The electronic structure of each of the first 20 elements can be worked out using the atomic number of the element and the maximum number of electrons in each shell.

The development of atomic theory

- An early theory of atoms was developed by John Dalton. His explanation was *provisional*, but later was confirmed by better *evidence*. When J.J. Thompson, Rutherford and Bohr found new evidence, their explanations changed the model of the atom. With later evidence their *predictions* were confirmed.
- Ideas can also change rapidly following unexpected results. Geiger and Marsden had some unexpected results which made significant contributions to Rutherford and Bohr's ideas.

Improve your grade

Arrangement of electrons

How many electrons does a potassium atom have and why is it in the fourth row of the periodic table? (Use page 248). *AO2* [2 marks]

Ionic bonding

Why do atoms form bonds?

- **Atoms** with an outer shell of eight **electrons** have a **stable electronic structure**.

- Atoms can be made stable by transferring electrons. This called **ionic bonding**.

- **Metal** atoms lose electrons to get a stable electronic structure. If an atom loses electrons, a **positive ion** is formed, as there are fewer negatively **charged** electrons than the number of positively charged **protons** in the **nucleus**.

- **Non-metal** atoms gain electrons to get a stable electronic structure. If an atom gains electrons, a **negative ion** is formed. This is because there will be more negatively charged electrons than the number of positively charged protons in the nucleus.

- During ionic bonding, the metal atom becomes a positive ion and the non-metal atom becomes a negative ion. The positive ion and the negative ion then **attract** one another.

- We use the **'dot and cross' model** to describe ionic bonding.

- During the bonding of magnesium oxide:
  - magnesium atoms lose two electrons to form positive magnesium ions.
  - oxygen atoms gain two electrons to form negative oxide ions.
  - these are represented in the 'dot and cross' model showing each of the two ions. The model only shows the outer shell electrons.

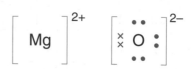

'Dot and cross' model of the ions in magnesium oxide

- Positive ions and negative ions are held together by attraction.

- During the bonding of sodium oxide:
  - a sodium atom only has one electron to lose to make a sodium ion, but an oxygen atom needs to gain two electrons to make an oxide ion.
  - two sodium atoms are needed to provide the two electrons needed, so two sodium ions are formed for every one oxide ion. This is represented in the diagram of atoms showing electron transfer. The ions formed are shown in the dot and cross diagram below, which show the outer electron shell only.

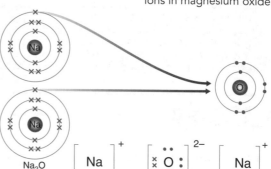

Sodium and oxygen ion transfer

Conducting electricity

- The structure of sodium chloride or magnesium oxide is a **giant ionic lattice**, in which positive ions have strong **electrostatic attraction** to negative ions. They always exist as solids.

- Sodium chloride **solution** can conduct electricity.

- Sodium chloride and magnesium oxide conduct electricity when they are **molten**.

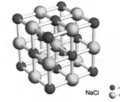

The structure of sodium chloride

More on ionic bonding

- These **physical properties** of sodium chloride and magnesium oxide mean:
  - they have high **melting points** as there are strong attractions between positive and negative ions.
  - they do not conduct electricity when solid because the ions cannot move.
  - they conduct electricity when in solution or as a molten liquid as the ions are free to move.

- The melting point of MgO is higher than NaCl because:
  - there are Mg^{2+}, not Na^+ ions and O^{2-}, not Cl^- ions, so there are stronger electrostatic attractions between positive and negative ions.
  - each magnesium atom donates *two* electrons to the oxygen atom, which makes a stronger bond when sodium atoms transfer one electron to chlorine atoms.
  - magnesium ions are very small in radius, so magnesium can get much closer to oxygen, which makes the bond stronger. More energy is needed to separate these ions.

Improve your grade

Structure and bonding

Explain why the melting point of magnesium oxide is so high. *AO1* [2 marks]

The periodic table and covalent bonding

Covalent bonding

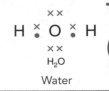

- **Non-metals** can share **electron** pairs between **atoms**. This is known as **covalent bonding**.

- The formation of simple **molecules** that contain **single** and **double** covalent **bonds** can be represented by **'dot and cross' models** which only show the outer shell electrons.

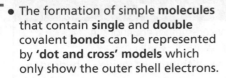 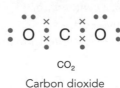

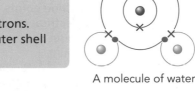

Chlorine $\quad$ Carbon dioxide $\quad$ Water

Predicting chemical properties

- The attraction between **carbon dioxide** molecules is called an **intermolecular force**. The attraction between water molecules is also an intermolecular force.

- The diagrams of water and carbon dioxide molecules represent how all the atoms bond to make the molecules, this time showing all the electrons in the atoms.

A molecule of water is made up two hydrogen atoms and one oxygen atom.
– An oxygen atom has six electrons in its outer shell, so needs two more electrons.
– Hydrogen atoms each have one electron in their only shell. The electron outer shell of the oxygen atom is shared with each electron of two hydrogen atoms.

A molecule of water

A molecule of carbon dioxide is also made up of three atoms, two oxygen atoms and one **carbon** atom.
– Carbon has four electrons in its outer shell, so it needs four more electrons to be complete.
– Oxygen atoms each have six electrons in their outer shell. So they each need two more electrons to be complete.
– So each oxygen outer shell is shared with two of the electrons of the carbon outer shell.
– In this way each of the oxygen atoms has a share of two more electrons so the shell is full.

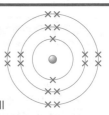

A molecule of carbon dioxide

- The weak intermolecular forces in substances that have a simple molecular structure, such as carbon dioxide and water, mean the molecules are easy to break apart so the substances have low **melting points**.

- As there are no free electrons, carbon dioxide and water do not conduct electricity.

Remember!
You should be able to predict the properties of other substances that have a simple molecular structure.

Group numbers and elements in periods

- The group number is the same as the number of electrons in the outer shell.

- If an atom of an **element** has electrons in only one occupied shell it will be found in the first **period**; if it has electrons in two occupied shells it will be in the second period; if it has electrons in three occupied shells, it will be found in the third, and so on.

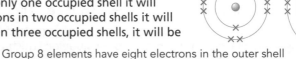

Group 8 elements have eight electrons in the outer shell

The development of the periodic table

- In 1865 Newlands put 56 elements into groups and saw that every eighth element behaved similarly. This was not accepted for 50 years until other scientists discovered more evidence.

- In 1869 Mendeleev arranged the elements in order in a table. He noticed periodic changes in properties. He saw there were gaps in his pattern and predicted new elements would be found.

Element	Electron pattern	Period
H	1	1
Li	2.1	2
Na	2.8.1	3

- In 1891 Mendeleev's table still did not contain the noble gases. Later investigations confirmed his idea of periodicity and his predictions on the discovery of missing elements.

Improve your grade

Electrical conductors
Explain why carbon dioxide does not conduct electricity. *AO1* [2 marks]

The group 1 elements

Properties of alkali metals

D–C

- Caesium and rubidium are both **group 1 metals** and so have similar properties to lithium, sodium and potassium:
  - They react vigorously with water.
  - Hydrogen gas is given off.
  - The **metal** reacts with water to form an **alkali** – the hydroxide of the metal.
- It is possible to predict how an **alkali metal**, such as caesium and rubidium, will behave by looking at the pattern of reactivity in other alkali metals.
  - Potassium reacts more vigorously than sodium.
  - Sodium reacts more vigorously than lithium.
- The **balanced symbol equation** for lithium and water is: $2Li + 2H_2O \rightarrow 2LiOH + H_2$

> ### EXAM TIP
> Construct balanced symbol equations in steps.
> **Step 1:** Write a word equation.
> lithium + water → lithium hydroxide + hydrogen
> **Step 2:** Put in the symbols and formulae.
> $Li + H_2O \rightarrow LiOH + H_2$
> **Step 3:** Balance up the number of atoms on either side.
> $2Li + 2H_2O \rightarrow 2LiOH + H_2$

Predicting the properties of alkali metals

B–A*

- If we know the **melting points, boiling points,** appearance and **electrical conductivity** of lithium, sodium and potassium, we can predict these properties in rubidium and caesium.

Reactivity of the alkali metals with water increases down group 1

reactivity increases down the group

	Melting point in °C	Boiling point in °C
$_3$Li	179	1317
$_{11}$Na	98	892
$_{19}$K	64	774

Explaining reactivity patterns

D–C

- Atoms of group 1 alkali metals have similar properties because they have one **electron** in their outer shell.

B–A*

- Atoms of alkali metals react, losing one electron and forming:
  - a full outer shell, to create a **stable electronic structure**
  - an **ion**, which has one more positive **charge** in its **nucleus** than negative electrons in the shells, so it becomes a **positive ion**.
- This can be represented by an equation: $Na - e^- \rightarrow Na^+$
- The easier it is for an **atom** of an alkali metal to lose one electron, the more reactive it is.

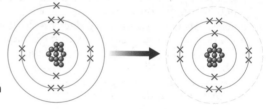

Making a positive ion

electron in outer shell is further from the nucleus, there is less attractive force, so the electron is more easily lost

electron in outer shell is closer to the nucleus, there is more attractive force, so the electron is less easily lost

Electronic structure and reactivity

Oxidation

B–A*

- If electrons are lost, the process is called **oxidation**. $K - e^- \rightarrow K^+$
- You can see why this process is oxidation from its **ionic equation**. An atom of potassium loses one electron. It becomes a potassium ion which is a positive ion. This is an example of oxidation.

Remember!
OIL – Oxidation Is Loss.

Flame tests

D–C

- A **flame test** can be used to find out if lithium, sodium or potassium are present in a **compound**.
- The test is carried out while wearing safety glasses.
  - A flame-test wire is moistened with dilute hydrochloric acid.
  - The flame-test wire is dipped into the solid chemical.
  - The flame test-wire is put in a blue Bunsen burner flame.
  - The colours of the flames are recorded.

Alkali metal in the compound	Colour of flame
lithium	red
sodium	yellow
potassium	lilac

Improve your grade

Reactivity in alkali metals

Explain why the reaction of rubidium with water is more vigorous than the reaction of sodium with water. *AO1* [2 marks]

The group 7 elements

Halogens, trends and predictions

- At room temperature, *chlorine* is a green gas, *bromine* is an orange liquid and *iodine* is a grey solid.

- **Group 7 elements** have similar properties as their atoms have seven **electrons** in their outer shell.

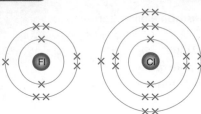

fluorine 2. 7   chlorine 2. 8. 7

bromine (outer shell only shown) 7   iodine (outer shell only shown) 7

Electronic structures of the halogens

- It is possible to predict the properties of **halogens**, such as fluorine or astatine, knowing the properties of the other halogens. This is because the properties follow trends, as shown in the table.

- Halogens have similar properties because when they react each **atom** gains one electron to form a **negative ion** with a **stable electronic structure**.

- The *nearer* the outer shell is to the **nucleus**, the easier it is for an atom to gain one electron. The *easier* it is to gain the electron, the *more reactive* the halogen.

Halogen	Melting point in °C	Boiling point in °C	State
$_9$F			
$_{17}$Cl	−101	−35	gas
$_{35}$Br	−7	59	liquid
$_{53}$I	114	184	solid
$_{85}$At			

D–C

B–A*

The reactions of halogens

- When halogens react with **alkali metals**, a **metal halide** is made. For example, when potassium reacts with iodine, the metal halide made is potassium iodide.

- The word equation for a reaction of a group 1 **element** and a group 7 element can be worked out from the two **reactants**:

  potassium + iodine → potassium iodide

- The balanced symbol equation for this reaction is:

  $2K + I_2 \rightarrow 2KI$

Remember!
The '2' in Cl_2 means there are two bonded atoms in the molecule of chlorine. This number cannot be changed!

D–C

- It is possible to construct a **balanced symbol equation** for the reaction of an **alkali metal** with a halogen without being given any of the formulae. For example:

  lithium + bromine → lithium bromide $2Li + Br_2 \rightarrow 2LiBr$

B–A*

Displacement reactions of halogens

- If halogens are bubbled through **solutions** of metal halides there are two possible outcomes: no reaction or a **displacement reaction**. A displacement reaction is shown in the example below.
  - Chlorine displaces the bromide to form **bromine** solution.

  chlorine + potassium bromide → potassium chloride + bromine (orange solution)
  $Cl_2 + 2KBr \rightarrow 2KCl + Br_2$ (balanced)
  - Bromine displaces iodides from solutions.

  $Br_2 + 2KI \rightarrow 2KBr + I_2$ (red–brown solution)

D–C

- The trend in reactivity is shown by displacement reactions. The reactivity of the halogens increases further up the group. Knowing this trend allows the reactions between halogens and metal halides to be predicted.

B–A*

Reduction

- If electrons are gained, the process is called **reduction**, as shown in this **ionic equation**.
  $Br_2 + 2e^- \rightarrow 2Br^-$
  - A **molecule** of bromine gains two electrons (one for each atom).
  - It now becomes two bromide **ions**. The bromide ion is a negative ion.
  - This is reduction.

Remember!
RIG – Reduction Is Gain.

B–A*

Improve your grade

Displacement reactions of halogens

Explain why a red–brown colour is seen if chlorine is bubbled into a solution of potassium iodide. *AO1* [2 marks]

Transition elements

Coloured compounds

- Compounds that contain a **transition element** are often coloured.
  - Copper compounds are often blue.
  - Iron(II) compounds are often pale green.
  - Iron(III) compounds are often orange/brown.

45 Sc scandium 21	48 Ti titanium 22	51 V vanadium 23	52 Cr chromium 24	55 Mn manganese 25	56 Fe iron 26	59 Co cobalt 27	59 Ni nickel 28	63.5 Cu copper 29	65 Zn zinc 30
89 Y yttrium 39	91 Zr zirconium 40	93 Nb niobium 41	96 Mo molybdenum 42	[98] Tc technetium 43	101 Ru ruthenium 44	103 Rh rhodium 45	106 Pd palladium 46	108 Ag silver 47	112 Cd cadmium 48
139 La* lanthanum 57	178 Hf hafnium 72	181 Ta tantalum 73	184 W tungsten 74	186 Re rhenium 75	190 Os osmium 76	192 Ir iridium 77	195 Pt platinum 78	197 Au gold 79	201 Hg mercury 80

The transition elements

Catalysts

- Transition elements and their compounds are often used as **catalysts**.
  - Iron is used in the **Haber process** to make ammonia, which is used in **fertilisers**.
  - Nickel is used in the manufacture of margarine to harden the oils.

> **Remember!**
> A catalyst is a chemical that speeds up a reaction but is not changed or used up by the reaction.

Thermal decomposition of metal carbonates

- If a transition **metal** carbonate is heated it undergoes **thermal decomposition** to form a metal oxide and **carbon dioxide**.

- On heating:
  - $FeCO_3$ decomposes forming iron oxide and carbon dioxide.
  - $CuCO_3$ decomposes forming copper oxide and carbon dioxide.
  - $MnCO_3$ decomposes forming manganese oxide and carbon dioxide.
  - $ZnCO_3$ decomposes forming zinc oxide and carbon dioxide.

- The reaction can be expressed as a word equation. For example, the word equation for the decomposition of $CuCO_3$ is:

copper carbonate → copper oxide + carbon dioxide

- The metal carbonates change colour during the reaction.

- To write a **balanced symbol equation** to describe the thermal decomposition, follow these steps:
  - Write a word equation to establish the **products** of the reaction.
  - Assign the formulae to the reactants and products. For example:

$$CuCO_3 \rightarrow CuO + CO_2$$
$$FeCO_3 \rightarrow FeO + CO_2$$
$$MnCO_3 \rightarrow MnO + CO_2$$
$$ZnCO_3 \rightarrow ZnO + CO_2$$

  - These are balanced.

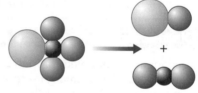

Model of the reaction

Precipitation reactions with sodium hydroxide solution

- Sodium hydroxide **solution** reacts with compounds of each transition metal to make a solid of a particular colour. The addition of sodium hydroxide solution identifies the presence of the transition metal **ions** in solution.
  - Cu^{2+} ions form a *blue* solid.
  - Fe^{2+} ions form a *grey/green* solid.
  - Fe^{3+} ions form an *orange/brown gelatinous* solid.

- These solids are metal hydroxide **precipitates**.

Writing a balanced symbol equation

- Sometimes equations of the **precipitation reactions** need to be balanced. For example:

$$Cu^{2+} + 2OH^- \rightarrow Cu(OH)_2 \qquad Fe^{2+} + 2OH^- \rightarrow Fe(OH)_2 \qquad Fe^{3+} + 3OH^- \rightarrow Fe(OH)_3$$

Improve your grade

Precipitation reactions

Explain how you can show that a solution contains copper ions. *AO1* [2 marks]

Metal structure and properties

Properties of metals

- **Metals** have specific properties that make them suitable for different uses.
- A property can be either *physical* or *chemical*.
  - An example of a **physical property** is the high *thermal conductivity* of copper. Saucepan bases need to be good conductors of heat. Copper can be used for the base or the whole of the pan.
  - An example of a **chemical property** is the *resistance to attack* by oxygen or acids shown by gold. Copper is also resistant, which is another reason why it is used for saucepans.
  - Other physical properties of metals include being lustrous and shiny, **malleable** or ductile.
  - Aluminium has a low **density** and is used where this property is important, such as in the aircraft industry and also in modern cars.

Metallic bonding

- Metals have high **melting points** and high **boiling points**, due to their strong **metallic bonds**. The bonds between these **atoms** are very hard to break. A lot of energy is needed to break the atoms apart.
- When metals conduct electricity, **electrons** in the metal move.
  - Copper, silver and gold conduct electricity very well.
- A metallic bond is a strong electrostatic **force** of **attraction** between close-packed positive metal **ions** and a 'sea' of **delocalised electrons**.
- Metals often have high melting points and boiling points, because a lot of energy is needed to overcome strong attractions between delocalised electrons and the **close-packed** positive metal ions.
- A metal conducts electricity because delocalised electrons within its structure can move easily.

electrons from outer shells of metal atoms are free to migrate

The structure of a metal   metal ions

Superconductors

- **Superconductors** are materials that conduct electricity with little or no **resistance**.
  - Copper, silver and gold conduct electricity well, but surprisingly do not become superconductors.
- The electrical resistance of mercury suddenly drops to zero at −268.8 °C. This phenomenon is called superconductivity.
- When a substance goes from its normal state to a superconducting state, it no longer has any magnetic fields inside it.
  - If a small magnet is brought near the superconductor, it is **repelled**.
  - If a small permanent magnet is placed above the superconductor, it levitates.
- The potential benefits of superconductors are:
  - loss-free **power transmission**
  - super-fast electronic circuits
  - powerful **electromagnets**.

Difficulties of superconductors

- There needs to be a good deal of development work before the true potential of superconductors is realised.
  - They work only at very low temperatures; this limits their use.
  - Superconductors that function at 20 °C need to be developed.

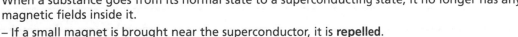

 Improve your grade

Electrical conductivity of metals

Explain how metals conduct electricity. *AO1* [2 marks]

Purifying and testing water

Water purification

- The water in a river is cloudy and often not fit to drink.

- To turn it into the clean water in taps it is passed through a water purification works.

- Some **pollutants**, such as nitrates from **fertilisers** and pesticides from crop spraying, get into the water before purification, and some, such as **lead** from old water pipes, get in after treatment.

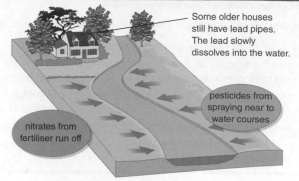

Some older houses still have lead pipes. The lead slowly dissolves into the water.

pesticides from spraying near to water courses

nitrates from fertiliser run off

The pollutants that can get into drinking water

- There are three main stages in water purification:
  - **sedimentation** – chemicals are added to make solid particles and **bacteria** settle out
  - **filtration** of very fine particles – a layer of sand on gravel filters out the remaining fine particles; some types of sand filter also remove microbes
  - **chlorination** – chlorine is added to kill **microbes**.

- These steps cost money so water must be conserved to provide clean drinking water.
  - Water is a renewable resource, but the supply is not endless. If there is not enough rain in the winter, **reservoirs** do not fill up.
  - In the UK more homes are being built, increasing the demand for water.
  - It takes energy to pump and purify, which increases **global warming**.

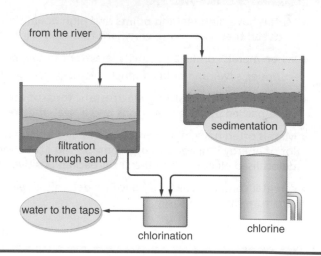

from the river

sedimentation

filtration through sand

water to the taps

chlorination

chlorine

The steps in the purification of water

More on water purification

- Some **soluble** substances such as pesticides and nitrates are not removed by the standard purification method (as they are dissolved in water) and extra processes are needed to remove them.

- **Sea water** has so many substances dissolved in it that it is undrinkable. Techniques such as **distillation** must be used to remove the dissolved substances. Distillation takes huge amounts of energy, so it is very expensive and only used when there is no fresh water.

Water tests

- Water can be tested with **precipitation reactions** using aqueous **silver nitrate** and **barium chloride** solutions.

- In a precipitation reaction, two **solutions** react to form a chemical that does not dissolve. This chemical suddenly appears in the liquid as a solid – a **precipitate**.

  barium chloride + magnesium sulfate → barium sulfate (white precipitate) + magnesium chloride
  silver nitrate + sodium bromide → silver bromide (cream precipitate) + sodium nitrate
  silver nitrate + sodium iodide → silver iodide (yellow precipitate) + sodium nitrate
  silver nitrate + sodium chloride → silver chloride (white precipitate) + sodium nitrate

- The **balanced symbol equations** for these precipitation reactions are:

  $BaCl_2 + MgSO_4 \rightarrow BaSO_4 + MgCl_2$

  $AgNO_3 + NaBr \rightarrow AgBr + NaNO_3$

  $AgNO_3 + NaI \rightarrow AgI + NaNO_3$

  $AgNO_3 + NaCl \rightarrow AgCl + NaNO_3$

Remember!
SO_4^{2-} ions are doubly charged, as are Mg^{2+} ions.

Improve your grade

Testing water

Explain, with the aid of a balanced equation, how you would test whether magnesium chloride is present in a sample of water. *AO1* [3 marks]

C4 Summary

The nucleus is made up of protons and neutrons.

The electron has a charge of −1, the proton has a charge of +1, the neutron has a charge of 0.

Metals form positive ions as they lose electrons.

Non-metals form negative ions as they gain electrons.

Isotopes are elements that have atoms of the same atomic number but different mass numbers.

Atomic structure and atomic bonding

Ionic bonding happens because of the transfer of electrons from metals to non-metals.

Sodium chloride and magnesium oxide form giant ionic lattices as their ions attract.

'Dot and cross' models can be used to represent the ionic bonding in compounds.

Non-metals combine by sharing electrons.

Carbon dioxide and water are simple molecules with weak intermolecular forces between the molecules.

The periodic table and covalent bonding

Carbon dioxide and water have low melting points as the intermolecular forces are weak.

The period to which an element belongs is the same as the number of occupied electron shells.

The group number is the same as the number of electrons in the outside shell.

Rubidium and caesium are group 1 metals which react violently with water to give off hydrogen and make an alkaline solution.

Group 1 metals have similar properties as they all need to lose one electron from their outside shell.

Groups

Flame tests can be used to identify the presence of lithium, sodium and potassium.

Chlorine is a green gas, bromine is a brown liquid and iodine is a grey solid.

If a group 1 metal reacts with a group 7 non-metal the word equation for the formation of a metal halide can be constructed.

Copper compounds are often blue, iron(II) compounds are often light green and iron(III) compounds are often orange/ brown.

The thermal decomposition of transition metal carbonates results in the metal oxide and carbon dioxide being made.

Cu^{2+} ions react with sodium hydroxide to make a blue solid in a precipitation reaction.

Metals and water

Superconductors conduct electricity with little or no resistance.

Drinking water is purified by filtration, sedimentation and chlorination.

Sulfates in water can be tested using barium chloride, halides can be tested using silver nitrate.

Metals have high melting points due their strong metallic bonding.

Speed

Measuring speed

- The formula for **speed** is:

  $$\text{average speed} = \frac{\text{distance}}{\text{time}}$$

- We write **'average speed'** because the speed of a car changes during a journey.

> **Remember!**
> There are 3600 seconds in one hour.

> An aircraft travels 1800 km in 2 hours.
> Average speed = $\frac{1800}{2}$ = 900 km/h = $900 \times \frac{1000}{3600}$ = 250 m/s

- The speed of a car at a certain point in time is called **instantaneous speed**.

  distance travelled = average speed × time
  $$= \frac{(u + v)t}{2}$$

> **Remember!**
> You may be asked to rearrange equations. Practise doing this.

- Increasing the speed means increasing the distance travelled in the same time. Increasing the speed reduces the time needed to cover the same distance.

Distance–time graphs

- **Distance–time graphs** allow a collection of data to be shown. It is easier to interpret data when they are plotted on a graph than when they are listed in a results table.

- The **gradient** of a distance–time graph tells you about the speed of the object. A higher speed means a steeper gradient.

- In graph **a**, the distance travelled by the object each second is the same. The gradient is constant, so the speed is constant. In graph **b**, the distance travelled by the object each second increases as the time increases. The gradient increases, so there is an increase in the speed of the object.

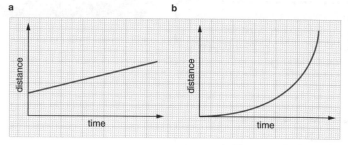

Distance–time graphs

- Speed is equal to the gradient of a distance–time graph; the higher the speed the steeper the gradient.

- A **straight line** indicates that the speed is constant.

- A **curved line** shows that the speed is changing.

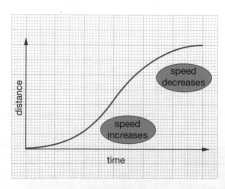

A curved line shows that the speed is not constant

The gradient of the graph is

$$\text{gradient} = \frac{AC}{BC} = \frac{(20 - 10)}{(5 - 0)} = \frac{10}{5} = 2$$

This means speed = 2 m/s

The higher the speed, the steeper the gradient

Improve your grade

Calculating time

How many hours will it take to travel 560 km at an average speed of 25 m/s?
AO1 [1 mark] *AO2* [2 marks]

Changing speed

Speed–time graphs

- A change of **speed** per unit time is called **acceleration**.
- If the speed is increasing, the object is accelerating. If the speed is decreasing, the object is decelerating.
- The area under a speed–time graph is equal to the **distance** travelled.
  - The speed of car B in the graph is increasing more rapidly than the speed of car A, so car B is travelling further than car A in the same time.
  - The area under line B is greater than the area under line A for the same time.
  - The speed of car D is decreasing more rapidly than the speed of car C, so car D isn't travelling as far as car C in the same time.
  - The area under line D is smaller than the area under line C for the same time.

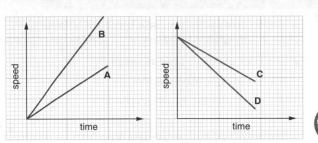

Speed–time graphs for 4 cars – A, B, C and D

EXAM TIP

Don't confuse distance–time and speed–time graphs. Always look at the axes carefully.

Acceleration

- The formula for measuring acceleration is:

$$\text{acceleration} = \frac{\text{change in speed (or velocity)}}{\text{time taken}}$$

- Acceleration is measured in metres per second squared (m/s^2).
- A negative acceleration shows the car is decelerating.

> A new car boasts a rapid acceleration of 0 to 108 km/h in 6 seconds.
>
> A speed of 108 km/h is $\dfrac{108 \times 1000}{60 \times 60} = 30$ m/s
>
> $\text{Acceleration} = \dfrac{\text{change in speed}}{\text{time taken}} = \dfrac{(30 - 0)}{6} = 5$ m/s^2
>
> This means the speed of the car increases by 5 m/s every second.

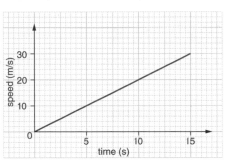

Acceleration is the gradient of a speed—time graph.

Relative velocities

- **Velocity** is a vector – it has both size (speed) and direction.
- When two cars are moving past each other, their **relative velocity** is:
  - the sum of their individual velocities if they are going in opposite directions
  - the difference of their individual velocities if they are going in the same direction.

Circular motion

- A vehicle may go around a roundabout at a constant speed but it is accelerating. This is because its direction of travel is changing; it's not going in a **straight line**. The driver needs to apply a **force** towards the centre of the roundabout to change direction. This gives the vehicle an acceleration directed towards the centre of the roundabout.
- Any object moving along a circular path moves at a tangent to the circle, or arc of a circle.

 Improve your grade

Calculating acceleration

A car is travelling at 10 m/s. It accelerates at 4 m/s^2 for 8 s. How fast is it then going?
AO1 [1 mark] *AO2* [2 marks]

Forces and motion

Force, mass and acceleration

- If the **forces** acting on an object are balanced, it's at rest or has a constant speed. If the forces acting on an object are unbalanced, it speeds up or slows down.

- The unit of force is the **newton (N)**.

  force = mass × **acceleration**

  Where F = **unbalanced** force in N, m = mass in kg and a = acceleration in m/s^2.

  > Marie pulls a sledge of mass 5 kg with an acceleration of 2 m/s^2 in the snow.
  > The force needed to do this is: F = ma = 5 × 2 = 10 N

- The equation F = ma is used to find mass or acceleration if the **resultant force** is known.

  > Professional golfers hit a golf ball with a force of approximately 9000 N. If the mass of the ball is 45 g, the acceleration during the very short time (about 0.005 milliseconds) of impact can be calculated.
  >
  > $$a = \frac{F}{m} = \frac{9000}{0.045} = 200\,000 \text{ m/s}^2$$

Car safety

- **Reaction time**, and therefore thinking distance, may increase if a driver is:
  - tired
  - under the influence of alcohol or other drugs
  - travelling faster
  - distracted or lacks concentration.

- **Braking distance** may increase if:
  - the road conditions are poor e.g. icy
  - the car has not been properly maintained e.g. worn brakes
  - the speed is increased.

- For safe driving, it is important to be able to stop safely:
  - Keep an appropriate distance from the car in front.
  - Have different speed limits for different types of road and locations.
  - Slow down when road conditions are poor.

- Factors affecting braking distance are:
  - The greater the mass of a vehicle the greater its braking distance.
  - The greater the speed of a vehicle the greater its braking distance.
  - When the brakes are applied the brake pads are pushed against the disc. This creates a large **friction** force that slows the car down. Worn brakes reduce the friction force, increasing the braking distance.
  - Worn tyres with very little **tread** reduce the grip of the wheels on a slippery road, leading to skidding and an increase in braking distance.
  - Increased braking force reduces the **stopping distance**.
- **Thinking distance** increases **linearly** with speed.
- Braking distance increases as a squared relationship e.g. the braking distance at 60 mph is nine times the braking distance at 20 mph.

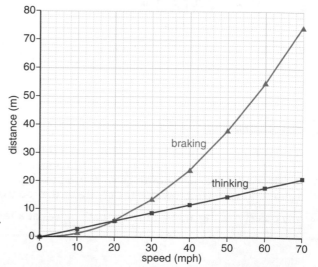

Thinking distance and speed is proportional, as is braking distance and speed

Improve your grade

Stopping distance

Explain why brakes and tyres are checked when a car has its annual MOT test. *AO2* [2 marks]

Work and power

Work

- **Work** is done when a **force** moves an object in the direction in which the force acts.
- The formula for **work done** is:

work done = force × distance moved (in the direction of the force)

> If a person weighs 700 N, the work he does against **gravity** when he jumps 0.8 m is:
> work done = force × distance moved = 700 × 0.8 = 560 J

Weight

- **Weight** is a measure of the gravitational attraction on a body acting towards the centre of the Earth.
- The formula for weight is:

weight = mass × **gravitational field strength**

- A mass of 1 kg has a weight of about 10 N on Earth.

Power

- **Power** is the rate at which work is done.
- The formula for power is:

$$\text{power} = \frac{\text{work done}}{\text{time taken}}$$

- The equation for power can be rearranged to work out:

work done = power × time

$$\text{time taken} = \frac{\text{work done}}{\text{power}}$$

> When the Eurostar travels at maximum **speed**, its power is 2 MW. The amount of work done, or energy transferred, in 2 hours is calculated by:
>
> work done = power × time = 2 000 000 × (2 × 60 × 60) = 14 400 000 000 J
>
> At maximum power, 14 400 MJ of energy would be transferred to other forms during a 2 hour journey. The Eurostar operates at maximum power for only a short part of the journey.

- A person's power is greater when they run than when they walk.
- Some cars are more powerful than others. They travel faster and cover the same distance in a shorter time and require more fuel. The power rating of a car depends on its engine size. More powerful cars have greater **fuel consumption**.
- Fuel is expensive and a car with high fuel consumption is expensive to run.
- Fuel **pollutes** the environment.
  – Car **exhaust gases**, especially **carbon dioxide**, are harmful.
  – Carbon dioxide is also a major source of **greenhouse gases**, which contribute to climate change.
- Power is also related to force and speed.

$$\text{power} = \frac{\text{work done}}{\text{time taken}} = \frac{\text{force} \times \text{distance}}{\text{time taken}} = \text{force} \times \frac{\text{distance}}{\text{time}} = \text{force} \times \text{speed}$$

Improve your grade

Calculating driving force

A Ford Focus has a power rating of 104 kW.

(a) Calculate the resultant force acting when the car is travelling at 90 km/h.

(b) Explain how this force compares with the driving force of the engine.
AO1 [3 marks] *AO2* [1 mark]

Energy on the move

Kinetic energy

D–C

- The **braking distance** of a car increases with increasing **speed**, but not proportionally.
- The formula for **kinetic energy** is:

kinetic energy $= \frac{1}{2} mv^2$, where m = mass in kg, v = **velocity** in m/s

> If a car has a mass of 1000 kg, its kinetic energy:
> – at 20 m/s is $\frac{1}{2} mv^2 = \frac{1}{2} \times 1000 \times (20)^2 = 200\,000$ J
> – at 40 m/s is $\frac{1}{2} mv^2 = \frac{1}{2} \times 1000 \times (40)^2 = 800\,000$ J

B–A*

- When a car stops, its kinetic energy changes into heat in the brakes, tyres and road.
- This can be shown by the formula:

work done by brakes = loss in kinetic energy

- The change in kinetic energy can be shown in the formula:

braking force × braking distance = loss in kinetic energy

- When the speed of the car doubles, the kinetic energy and the braking distance quadruple.
  – This is why there are speed limits and penalties for drivers who exceed them.

Fuel

D–C

- **Fuel consumption** data are based on ideal road conditions for a car driven at a steady speed in urban and non-urban conditions.

car	fuel	engine in litres	miles per gallon (mpg)	
			urban	non-urban
Renault Megane	**petrol**	2.0	25	32
Land Rover	petrol	4.2	14	24

> **EXAM TIP**
>
> Make sure you can interpret fuel consumption tables.

B–A*

- Factors that affect the fuel consumption of a car are:
  – the amount of **energy** required to increase its kinetic energy
  – the amount of energy required for it to do work against **friction**
  – its speed
  – how it is driven, such as excessive **acceleration** and **deceleration**, constant braking and speed changes
  – road conditions, such as a rough surface.

Electrically powered cars

D–C

- Electric cars are **battery** driven or **solar-powered**.
  – The battery takes up a lot of room.
  – They have limited range before **recharging**.
  – They are expensive to buy but the cost of recharging is low.
  – Solar-powered cars rely on the Sun shining and need backup batteries.
- Exhaust fumes from petrol-fuelled and diesel-fuelled cars cause serious **pollution** in towns and cities.
- Battery-driven cars do not **pollute** the local environment, but their batteries need to be recharged. Recharging uses electricity from a **power station**. Power stations pollute the local atmosphere and cause acid rain.

B–A*

- Solar-powered cars do not produce any **carbon dioxide** emissions.
- **Biofuels** may reduce carbon dioxide emissions but this is not certain because **deforestation** leads to an increase in carbon dioxide levels.

Improve your grade

Carbon dioxide emissions

Some scientists suggest that carbon dioxide emissions from burning biofuels may be at least 20% lower than those from fossil fuels. Some scientists argue that overall the emissions may be higher than from fossil fuels. Suggest why emissions may be higher. *AO3* [3 marks]

Crumple zones

Momentum and force

- The formula for **momentum** is:

  momentum = mass × velocity and the units are kgm/s

- To reduce injuries in a collision, **forces** should be as small as possible.

  $$\text{force} = \frac{\text{change in momentum}}{\text{time}}$$

- Spreading the momentum change over a longer time reduces the force.
- To minimise injury, forces acting on the people in a car during an accident must be minimised.

  force = mass × **acceleration**

- Force can be reduced by reducing the acceleration. This is done by:
  - increasing stopping or collision time
  - increasing stopping or collision distance.
- Safety features that do this include:
  - crumple zones – seat belts – **air bags** – crash barriers – **escape lanes**.

Car safety features

- Modern cars have safety features that absorb **energy** when a vehicle stops suddenly. These are:
  - brakes that get hot
  - **crumple zones** that change shape
  - **seat belts** that stretch a little
  - air bags that inflate and squash.
- On impact:
  - Crumple zones at the front and rear of the car absorb some of its energy by changing shape or 'crumpling'.
  - Seat belts stretch a little so that some of the person's **kinetic energy** is converted to elastic energy.
  - Air bags absorb some of the person's kinetic energy by squashing up around them.
- All these safety features:
  - absorb energy
  - change shape
  - reduce injuries
  - reduce momentum to zero slowly, therefore reducing the force on the occupants.

- Some people do not like wearing seat belts because:
  - there is a risk of chest injury
  - they may be trapped in a fire
  - drivers may be encouraged to drive less carefully because they know they have the protection from a seat belt.
- Despite **computer modelling**, crash tests using real vehicles and **dummies** provide more safety information.
- **ABS brakes** are a **primary safety feature** which helps to prevent a crash.
- In the ABS system, wheel-speed **sensors** send information to a computer about the rotational speed of the wheels. The computer controls the pressure to the brakes, via a pump, to prevent the wheels locking up. This increases the braking force (F) just before the wheels start to skid.

  kinetic energy lost = **work done** by the brakes

  $\frac{1}{2}mv^2 = Fd$ where m = mass of car, v = speed of car before braking, d = **braking distance**.
  - If F increases, the braking distance (d) decreases.
- Other primary safety features include:
  - **cruise control** which stops a driver becoming tired on a long journey and pressing harder on the accelerator
  - electric windows and **paddle shift controls** which allow the driver to concentrate on the road.

Improve your grade

Seat belts

Some people think that wearing a seat belt should be up to the individual, not the law. Explain how the wearing of seat belts can help to avoid injury but may not always do so. *AO1* [3 marks]

Falling safely

Falling objects

D–C

- All objects fall with the same **acceleration** due to **gravity** as long as the effect of **air resistance** is very small.

- The size of the air resistance **force** on a falling object depends on:
  - its **cross-sectional area** – the larger the area the greater the air resistance
  - its **speed** – the faster it falls the greater the air resistance.

- Air resistance has a significant effect on motion only when it is large compared to the **weight** of the falling object.

- The speed of a **free-fall** parachutist changes as he falls to Earth.

- In picture 1, the weight of the parachutist is greater than air resistance. He accelerates.

- In picture 2, the weight of the parachutist and air resistance are equal. The parachutist has reached **terminal speed** because the forces acting on him are balanced.

- In picture 3, the air resistance is larger than the weight of the parachutist. He slows down and air resistance decreases.

- In picture 4, the air resistance and weight of the parachutist are the same. He reaches a new, slower terminal speed.

Terminal speed

B–A*

- In picture 1, the parachutist accelerates, displacing more air molecules every second. The air resistance force increases. This reduces his acceleration. So, the higher the speed, the more air resistance.

- In picture 2, the parachutist's weight is equal to the air resistance; the forces on him are balanced. He travels at a constant speed – terminal speed.

- In picture 3, when the parachute opens, the upward force increases suddenly as there is a much larger surface area, displacing more air molecules every second. So, the larger the area, the more air resistance. The parachutist decelerates, displacing fewer air molecules each second, so the air resistance force decreases.

- In picture 4, the parachutist reaches a new slower terminal speed when his weight is equal to the air resistance once more, so he lands safely.

- **Drag** racers and the Space Shuttle use parachutes to slow them down rapidly.

Gravitational field strength – g

B–A*

- The force on each kilogram of mass due to gravity, g = 10 N/kg approximately on Earth.
  - g is also known as acceleration due to gravity, g = 10 m/s^2 approximately.

- The **gravitational field strength**:
  - is unaffected by atmospheric conditions
  - varies with position on the Earth's surface (9.78 N/kg at the equator and 9.83 N/kg at the poles)
  - varies with height above or depth below the Earth's surface.

Improve your grade

Gravitational field strength

(a) Explain why a 1 kg ball dropped from a height of 10m above the ground at the North Pole, takes less time to reach the ground than an identical 1 kg ball dropped from 10 m above the ground at the equator.

(b) How would the time taken be different if the ball was taken to the top of a mountain and dropped from a height of 10 m?
AO1 [2 marks] AO2 [1 mark]

The energy of games and theme rides

Gravitational potential energy

- An object held above the ground has **gravitational potential energy**.
- This is shown by the formula:

  GPE = mgh where

  m = mass, h = vertical height moved,
  g = **gravitational field strength** (10 N/kg)

- GPE is measured in **joules** (J).

Remember!
You will be expected to rearrange equations. Practise doing this.

D–C

Energy transfers

- A bouncing ball converts gravitational potential energy to **kinetic energy** and back to gravitational potential energy. It does not return to its original height because **energy** is transferred.

B the ball has *just* reached the ground so it has kinetic energy but no gravitational potential energy

D the ball has gravitational potential energy but no kinetic energy at the top of the bounce

2.00 m

1.20 m

A   B   C   D

Stages of energy transfer when a ball is dropped

A the ball has gravitational potential energy but no kinetic energy

C the ball squashes and has **elastic potential energy** which is converted to kinetic energy as it leaves the ground

D–C

- When skydivers reach **terminal speed**, their kinetic energy ($\frac{1}{2}$ mv²) has a maximum value and remains constant. The gravitational potential energy lost as they fall is used to do **work** against **friction (air resistance)**.
  - When terminal speed is reached it can be shown as:

    change in GPE = work done against friction

B–A*

How a roller coaster works

- A roller coaster uses a motor to haul a train up in the air. The riders at the top of a roller coaster ride have a lot of gravitational potential energy.
- When the train is released it converts gravitational potential energy to kinetic energy as it falls. This is shown by the formula:

  loss of gravitational potential energy (GPE) = gain in kinetic energy (KE)

D–C

- Ignoring friction, as the train falls:

  mgh = $\frac{1}{2}$ mv²

  h = $\frac{mv^2}{2\,mg}$ = $\frac{v^2}{2\,g}$

  - so this is independent of the mass of the falling object.

B–A*

- Each peak is lower than the one before because some energy is transferred to heat and sound due to friction and air resistance.
- This is shown by the formula:

  GPE at top = KE at bottom + energy transferred (to heat and sound) due to friction

- If speed doubles, KE quadruples (KE ∝ v²).
- If mass doubles, KE doubles (KE ∝ m).

D–C

Improve your grade

Energy changes during free-fall

Mel is a free-fall parachutist. During her time in free fall she reaches terminal speed.
Explain how her gravitational potential energy and kinetic energy change during her descent.
AO1 [3 marks]

P3 Summary

Distance–time graphs

The gradient of a distance–time graph is the speed.

The steeper the gradient, the higher the speed.

Speed and acceleration

$$distance = average\ speed \times time$$

$$acceleration = \frac{change\ in\ speed}{time}$$

Relative velocity is the difference between the **velocities** of two bodies having regard to their direction.

Speed–time graphs

The area under a speed–time graph is the distance travelled.

The gradient of a speed–time graph is the acceleration.

The steeper the gradient, the greater the acceleration.

A negative gradient means the body is slowing down.

Pairs of forces

Forces always occur in pairs:

- that are the same size
- act in opposite directions
- act on different objects.

Forces and motion

Forces can make things speed up or slow down.

force = mass × acceleration

$$weight = mass \times \frac{gravitational}{field\ strength}$$

When a car stops the

$$\frac{total}{stopping\ distance} = \frac{thinking}{distance} + \frac{braking}{distance}$$

Thinking distance depends on the state of the driver and the speed of the vehicle (linear).

Braking distance depends on the state of the road, vehicle and speed (squared).

Moving objects possess kinetic energy.

The faster they travel, the more kinetic energy they possess.

The greater their mass, the more KE they possess.

$$KE = \frac{1}{2}mv^2$$

Fuel consumption depends on the energy required to increase kinetic energy.

Roller coasters use gravitational potential energy as the source of movement.

GPE = mgh

During a roller coaster ride, GPE is transferred into KE and back again.

Work, energy and power

Work (J) is done when a force moves through a distance.

Energy (J) is needed to do work.

Power (W) is a measure of how quickly work is done.

work done = force × distance

$$power = \frac{work\ done}{time}$$

power = force × speed

Modern cars have lots of safety features that absorb energy when the cars stop:

- crumple zones
- seat belts
- air bags.

$$force = \frac{change\ in\ momentum}{time}$$

Spreading momentum change over a longer time reduces injury.

ABS brakes mean the driver can keep control when braking without skidding.

Terminal speed is the maximum speed reached by a falling object. This happens when the forces acting on the object are balanced.

Falling safely

Falling objects get faster as they fall.

They are pulled towards the centre of the Earth by their weight.

The acceleration of all objects is the same at any point on the Earth's surface.

Gravitational field strength varies across the Earth's surface and with height above the surface.

The forces on a falling object depend on:

- the speed of the object (higher speed means more drag)
- the area of the object (larger area means more drag).

Sparks

Electrons

- An **atom** consists of a small positively charged **nucleus** surrounded by an equal number of negatively charged **electrons**.

- In a stable, neutral atom there are the same amounts of positive and negative **charges**.

- All electrostatic effects are due to the movement of electrons.

- The law of electric charge states that: like charges **repel**, unlike charges **attract**

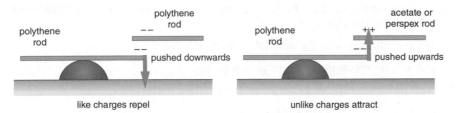

Like charges repel and unlike charges attract

- When a polythene rod is rubbed with a duster, electrons are transferred from the duster to the polythene, making the polythene rod negatively charged.

- When an acetate rod is rubbed with a duster, electrons are transferred from the acetate to the duster, leaving the acetate rod positively charged.

- In general an object has:
  - a negative charge due to an excess of electrons
  - a positive charge due to a lack of electrons.

- atoms or molecules that have become charged are ions.

Remember!
Only the electrons can be transferred.

Electrostatic shocks

- When inflammable gases or vapours are present, or there is a high concentration of oxygen, a **spark** from **static** electricity could ignite the gases or vapours and cause an **explosion**.

- If a person touches something at a high **voltage**, large amounts of electric charge may flow through their body to earth.

- Even small amounts of charge flowing through the body can be fatal.

- Static electricity can be a nuisance but not dangerous.
  - Dust and dirt are attracted to insulators, such as television screens.
  - Clothes made from synthetic materials often 'cling' to each other and to the body.

- Electric **shocks** can be avoided in the following ways:
  - if an object that is likely to become charged is connected to earth, any build-up of charge would immediately flow down the **earth wire**
  - in a factory where machinery is at risk of becoming charged, the operator stands on an insulating rubber mat so that charge cannot flow through them to earth
  - shoes with insulating soles are worn by workers if there is a risk of charge building up so that charge cannot flow through them to earth
  - fuel tankers are connected to an aircraft by a conducting cable during refuelling.

- Anti-static sprays, liquids and cloths made from conducting materials carry away electric charge. This prevents a build-up of charge.

 Improve your grade

Static charge
Connor is in the library walking on a nylon carpet. He touches a metal bookshelf and receives an electric shock. Explain how he became charged and why he received a shock.
AO2 [3 marks]

Uses of electrostatics

Dust precipitators

- A dust precipitator removes harmful particles from the chimneys of factories and power stations that **pollute** the atmosphere.
- A metal grid (or wires) is placed in the chimney and given a large **charge** from a high-**voltage** supply.
- Plates inside the chimney are **earthed** and gain the opposite charge to the grid.
- As the dust particles pass close to the grid, they become charged with the same charge as the grid.
- Like charges **repel**, so the dust particles are repelled away from the wires. They are **attracted** to the oppositely charged plates and stick to them.
- At intervals the plates are vibrated and the dust falls down to a collector.

- The dust particles gain or lose **electrons** to become charged.
- The charge on the dust particles induces a charge on the earthed metal plate.
- Opposite charges attract so the dust is attracted to the plate.

Paint spraying

- Static electricity is used in paint spraying.
  - The spray gun is charged.
  - All the paint particles become charged with the same charge.
  - Like charges repel, so the paint particles spread out giving a fine spray.
  - The object to be painted is given the opposite charge to the paint.
  - Opposite charges attract, so the paint is attracted to the object and sticks to it.
  - The object gets an even coat, with limited paint wasted.

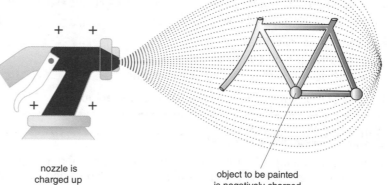

nozzle is charged up positively

object to be painted is negatively charged

Electrostatic principles in paint spraying

- If the object to be painted is not charged, the paint moves onto it but:
  - the object becomes charged from the paint, gaining the same charge
  - further paint droplets are repelled away from the object.
- Therefore the object to be painted is given the opposite charge to the paint. If the paint is negatively charged, having gained electrons, the object should be positively charged, by losing electrons.

Defibrillators

- **Defibrillation** is a procedure to restore a regular heart rhythm by delivering an electric **shock** through the chest wall to the heart.
  - Two **paddles** are charged from a high-voltage supply.
  - They are then placed firmly on the patient's chest to ensure good electrical contact.
  - Electric charge is passed through the patient to make their heart contract.
  - Great care is taken to ensure that the operator does not receive an electric shock.

- If a defibrillator is switched on for 5 milliseconds (0.005 s), the power can be calculated from:

$$\text{power} = \frac{\text{energy}}{\text{time}} = \frac{400}{0.005} = 80\,000 \text{ W}$$

Improve your grade

Spray painting

Static electricity is useful in spray-painting cars.

Explain how by writing about:
- electrostatic charge
- electrostatic force
- why it is used.

AO1 [3 marks]

Safe electricals

Resistance

- A **variable resistor,** or **rheostat**, changes the **resistance**. Longer lengths of wire have more resistance; thinner wires have more resistance.

- **Voltage** (**potential difference**) is measured in **volts** (V) using a **voltmeter** connected in parallel.
  - For a fixed resistor, as the voltage across it increases, the **current** increases.
  - For a fixed power supply, as the resistance increases, the current decreases.

- The formula for resistance is:

$$\text{resistance} = \frac{\text{voltage}}{\text{current}} \qquad R = \frac{V}{I}$$

- Resistance is measured in **ohms** (Ω).

- The formula for resistance can be rearranged to find out:

$$\text{voltage } V = IR \quad \text{or} \quad \text{current } I = \frac{V}{R}$$

Remember!
Always remember to include the correct unit with your answer.

Live, neutral and earth wires

- The **live wire** carries a high voltage around the house.

- The **neutral wire** completes the circuit, providing a return path for the current.

- The **earth wire** is connected to the case of an appliance to prevent it becoming live.

- A **fuse** contains wire which melts, breaking the circuit, if the current becomes too large.

- No current can flow, preventing overheating and further damage to the appliance.

- Earth wires and fuses stop a person receiving an electric **shock** if they touch a faulty appliance. As soon as the case becomes 'live', a large current flows in the earth and live wires and fuse 'blows'.

- A re-settable fuse (**circuit-breaker**) doesn't need to be replaced to restore **power**; it can be re-set.

Remember!
Double insulated appliances do not need an earth wire as the outer case is not a conductor.

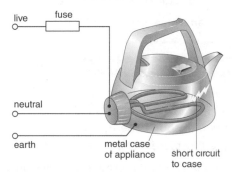

live   fuse
neutral
earth   metal case of appliance   short circuit to case

The arrangement of wires in metal-cased appliance

Electrical power

- The rate at which an appliance transfers energy is its power rating:

power = voltage × current

- The formula for electrical power can be used to calculate the correct fuse to use in an electrical device, e.g.

power	= voltage × current
current	= $\dfrac{\text{power}}{\text{voltage}}$
mains voltage	= 230 V
power of kettle	= 2500 W
current	= $\dfrac{2500}{230}$ = 10.9 A

Therefore a 13 A fuse is required.

 Improve your grade

Electrical safety
Explain how the fuse and earth wire operate to protect a user. *AO1* [2 marks]

Ultrasound

Longitudinal waves

- **Ultrasound** is sound above 20 000 **Hz** which is a higher **frequency** than humans can hear.
  - It travels as a **pressure wave** containing **compressions** and **rarefactions**.
- Compressions are regions of higher pressure and rarefactions are regions of lower pressure.

Remember!
All sound, including ultrasound, is produced by vibrating particles.

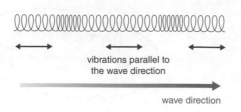

vibrations parallel to the wave direction

wave direction

A longitudinal wave

vibrations from side to side

right angles to direction of wave

wave direction

A transverse wave

- The features of **longitudinal** sound waves are:
  - They can't travel through a **vacuum**. The denser the medium, the faster a sound **wave** travels.
  - The higher the frequency or **pitch**, the smaller the **wavelength**.
  - The louder the sound, or the more powerful the ultrasound, the more energy is carried by the wave and the larger its **amplitude**.

- In a longitudinal wave the vibrations of the particles are parallel to the direction as the wave.
- In a **transverse** wave the vibrations of the particles are at right angles to the direction of the wave.

Uses of ultrasound

- When ultrasound is used to break down kidney stones:
  - a high-powered ultrasound beam is directed at the kidney stones
  - the ultrasound energy breaks the stones down into smaller pieces
  - the tiny pieces are then excreted from the body in the normal way.

- When ultrasound is used in a body scan, a pulse of ultrasound is sent into the body.
  - At each boundary between different tissues some ultrasound is **reflected** and the rest is transmitted.
  - The returning **echoes** are recorded and used to build up an image of the internal structure.

- Ultrasound can be used for body scans because:
  - when ultrasound is reflected from different interfaces in the body, the depth of each structure is calculated by using the formula distance = speed × time, knowing the speed of ultrasound for different tissue types and the time for the echo to return
  - the proportion of ultrasound reflected at each interface depends on the densities of each of the adjoining tissues and the speed of sound in the adjoining tissues
  - if the tissues are very different (e.g. blood and bone) most of the ultrasound is reflected, leaving very little to penetrate further into the body
  - the information gained is used to produce an image of the part of the body scanned.

- Ultrasound is preferred to **x-rays** because:
  - it is able to produce images of soft tissue
  - it doesn't damage living cells.

Improve your grade

Ultrasound and imaging

(a) State one similarity and one difference between sound and ultrasound waves.

(b) Give two reasons why doctors may decide to use ultrasound instead of x-rays to get images of inside the body. *AO1* [4 marks]

What is radioactivity?

Radioactive decay

- **Radioactive** substances decay naturally, giving out **alpha**, **beta** and **gamma** radiation.
- Nuclear radiation causes **ionisation** by removing **electrons** from **atoms** or causing them to gain electrons.
- Radioactive decay is a **random** process; it isn't possible to predict exactly when a nucleus will decay.
- There are so many atoms in even the smallest amount of **radioisotope** that the average **count rate** will always be about the same. Radioisotopes have unstable nuclei. Their nuclear particles aren't held together strongly enough.
- The **half-life** of a radioisotope is the average time for half the nuclei present to decay. The half-life cannot be changed.

The nucleus

- A **nucleon** is a particle found in the **nucleus**. So, **protons** and **neutrons** are nucleons.
- The nucleus of an atom can be represented as:

  $^A_Z X$ where
  - A = atomic mass (or nucleon number)
  - Z = **atomic number** (or proton number)
  - X = chemical symbol for the element

  Z = the number of protons in the nucleus, so the number of neutrons = (A − Z).

  The carbon **isotope** $^{14}_6C$ has 6 protons and 8 neutrons in its nucleus.

Remember!
Nucleons cannot be lost. Charge is always conserved.

What are alpha and beta particles?

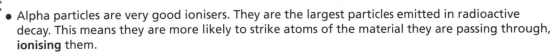

- When an **alpha** or a **beta particle** is emitted from the nucleus of an atom, the remaining nucleus is a different element.

- Alpha particles are very good ionisers. They are the largest particles emitted in radioactive decay. This means they are more likely to strike atoms of the material they are passing through, **ionising** them.

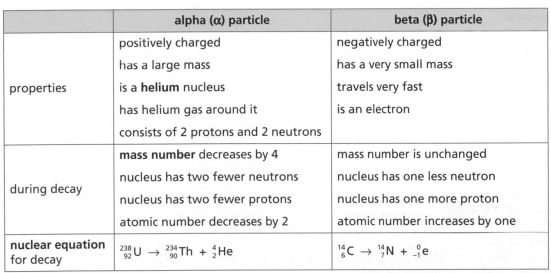

	alpha (α) particle	beta (β) particle
properties	positively charged has a large mass is a **helium** nucleus has helium gas around it consists of 2 protons and 2 neutrons	negatively charged has a very small mass travels very fast is an electron
during decay	**mass number** decreases by 4 nucleus has two fewer neutrons nucleus has two fewer protons atomic number decreases by 2	mass number is unchanged nucleus has one less neutron nucleus has one more proton atomic number increases by one
nuclear equation for decay	$^{238}_{92}U \rightarrow {}^{234}_{90}Th + {}^4_2He$	$^{14}_6C \rightarrow {}^{14}_7N + {}^{0}_{-1}e$

EXAM TIP
You must know the difference between alpha, beta and gamma radiation.

Improve your grade

Nuclear radiation

When carbon-14 undergoes beta decay what happens to the atom and what new element is formed? *AO1* [2 marks]

Uses of radioisotopes

Background radiation

D–C

- Background **radiation** is due to:
  - **radioactive** substances present in rocks (especially **granite**) and soil
  - **cosmic rays** from Space
  - man-made sources including **radioactive waste** from industry and hospitals.

B–A*

- Most **background radiation** is from natural sources but some comes from human activity. This is shown in the pie chart.

Remember!
Always subtract background radiation from any measurement of activity of a radioactive source.

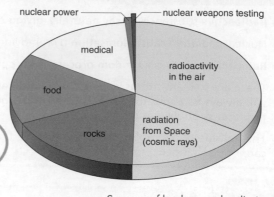

Sources of background radiation

Tracers

D–C

- When using a **tracer** to locate a leak in an underground pipe:
  - a very small amount of a **gamma** emitter is put into the pipe
  - a detector is passed along the ground above the path of the pipe
  - an increase in activity is detected in the region of the leak and little or no activity is detected after this point.

B–A*

- Gamma sources are used as tracers because the radiation is able to penetrate to the surface.

Smoke detectors

D–C

- A **smoke detector** contains an **isotope** which emits **alpha particles**.
  - Without smoke, the alpha particles **ionise** the air which creates a tiny **current** that can be detected by the circuit in the smoke alarm.
  - With smoke, the alpha particles are partially blocked so there is less ionisation of the air. The resulting change in current is detected and the alarm sounds.

Dating rocks

D–C

- Some rock types such as granite contain traces of **uranium**, a radioactive material.
  - The uranium isotopes present in the rocks go through a series of decays, eventually forming a **stable** isotope of **lead**.
  - By comparing the amounts of uranium and lead present in a rock sample, its approximate age can be found.

B–A*

- Uranium-238 decays, with a very long **half-life** of 4500 million years.
  - $^{238}_{92}U \rightarrow {}^{234}_{90}Th \ (+ {}^{4}_{2}He) \rightarrow {}^{234}_{91}Pa \ (+ {}^{0}_{-1}e) \rightarrow$ .... $\rightarrow {}^{206}_{82}Pb$ (stable)
- The proportion of lead increases as time increases. If there are equal quantities of $^{238}_{92}U$ and $^{206}_{82}Pb$, the rock is 4500 million years (one half-life) old.

Radiocarbon dating

D–C

- **Carbon-14** is a radioactive isotope of carbon that is present in all living things. By measuring the amount of carbon-14 present in an archaeological find, its approximate age can be found.

B–A*

- Carbon dating can only be used on objects that were once alive.
  - When an object dies, no more carbon-14 is produced.
  - As the carbon-14 decays, the activity of the sample decreases.
  - The ratio of current activity from living matter to the sample activity provides a reasonably accurate date.

Improve your grade

Radioisotope dating

Carbon-14 has a half-life of 5700 years.

(a) What is meant by the half-life of a radioactive sample?

(b) A sample of bone was found to have a 25% of the amount of carbon-14 found in a living organism. How old was this bone?
AO1 [1 mark], *AO2* [2 marks]

Treatment

Using radiation

- Radiation emitted from the **nucleus** of an unstable **atom** can be **alpha** (α), **beta** (β) or **gamma** (γ).
  - Alpha radiation is absorbed by the skin so is of no use for diagnosis or **therapy**.
  - Beta radiation passes through skin, but not bone. Its medical applications are limited but it is used, for example, to treat the eyes.
  - Gamma radiation is very penetrating and is used in medicine. Cobalt-60 is a gamma-emitting **radioactive** material that is widely used to treat **cancers**.
- When nuclear radiation passes through a material it causes **ionisation**. Ionising radiation damages living cells, increasing the risk of cancer.
- Cancer cells within the body can be destroyed by exposing the affected area to large amounts of radiation. This is called **radiotherapy**.
- Materials can be made radioactive when their **nuclei** absorb extra **neutrons** in a nuclear reactor.

D–C

Comparing x-rays and gamma rays

- When **x-rays** pass through the body the tissues absorb some of this ionising radiation. The amount absorbed depends on the thickness and the **density** of the absorbing material.
- Gamma rays and x-rays have similar **wavelengths** but are produced in different ways.
- X-rays are made by firing high-speed **electrons** at metal targets.
- An x-ray machine allows the rate of production and **energy** of the x-rays to be controlled, but you can't change the gamma radiation emitted from a particular radioactive source.
- When the nucleus of an atom of a radioactive substance decays, it emits an alpha or a beta particle and loses any surplus energy by emitting gamma rays.

D–C

B–A*

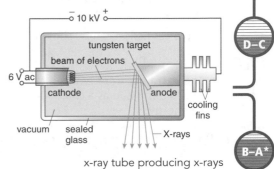

x-ray tube producing x-rays

Tracers

- A radioactive **tracer** is used to investigate inside a patient's body without surgery.
  - Technetium-99m is a commonly used tracer. It emits only gamma radiation.
  - **Iodine**-123 emits gamma radiation. It is used as a tracer to investigate the **thyroid gland**.
  - The radioactive tracer being used is mixed with food or drink or injected into the body.
  - Its progress through the body is monitored using a detector such as a gamma camera connected to a computer.

D–C

Treating cancer

- A **radioisotope** is used to destroy a **tumour** in the body.
- Three sources of radiation, each providing one-third of the required dose, are arranged around the patient with the tumour at the centre.
  - The healthy tissue only receives one-third of the dose, which limits damage to it.
- Or each radiation source is slowly rotated around the patient. The tumour receives constant radiation but healthy tissue receives only intermittent doses.

B–A*

source of radiation

source of radiation

source of radiation

tumour

Treating a brain tumour with gamma radiation

Improve your grade

Medical tracers

Explain what a medical tracer is and how suitable materials are chosen to be used as tracers. *AO1* [3 marks]

Fission and fusion

Nuclear power stations

- Natural **uranium** consists of two isotopes, uranium-235 and uranium-238.
  - The 'enriched uranium' used as fuel in a **nuclear power station** contains a greater proportion of the uranium-235 **isotope** than occurs naturally.
- **Fission** occurs when a large **unstable nucleus** is split up and **energy** is released as heat.
  - The heat is used to boil water to produce steam.
  - The pressure of the steam acting on the **turbine** blades makes it turn.
  - The rotating turbine turns the **generator**, producing electricity.
- When uranium fissions, a **chain reaction** starts. A nuclear bomb is an example of a chain reaction that is not controlled.

- In a nuclear power station, **atoms** of uranium-235 are bombarded with **neutrons**. This causes the nucleus to split, releasing energy.
- A typical fission can be shown as:

$$^{235}_{92}U + {}^{1}_{0}n \rightarrow {}^{90}_{36}Kr + {}^{143}_{56}Ba + 3({}^{1}_{0}n) + \gamma\text{-rays}$$

- The extra neutrons emitted cause further uranium nuclei to split. This is described as a chain reaction and produces a large amount of energy.

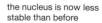

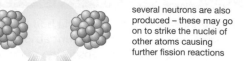

a neutron is absorbed by the nucleus of a uranium-235 atom

the nucleus is now less stable than before

it splits into two parts and releases energy

several neutrons are also produced – these may go on to strike the nuclei of other atoms causing further fission reactions

this is called a chain reaction

Uranium-235 undergoing a chain reaction

Controlling nuclear fission

- The output of a nuclear reactor can be controlled.
  - A **graphite moderator** between the **fuel rods** slows down the fast-moving neutrons emitted during fission. Slow-moving neutrons are more likely to be captured by other uranium nuclei.
  - **Boron control rods** can be raised or lowered. Boron absorbs neutrons, so fewer neutrons are available to split more uranium nuclei. This controls the rate of fission.

Boron control rods can be lowered into the reactor to absorb neutrons

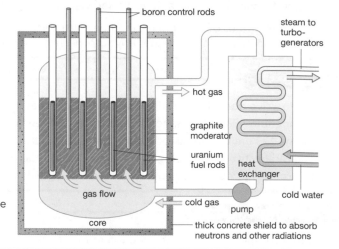

Fusion

Nuclear fusion happens when two light nuclei fuse (join) together releasing large amounts of heat energy.

- Fusion requires extremely high temperatures which have proved difficult to achieve and to manage safely on Earth and so its use for large-scale power generation remains a dream.
- Research in this area is very expensive so it is carried out as an international joint venture to share costs, expertise and benefits.
- In stars, fusion takes place under extremely high temperatures and pressures.
- **Fusion bombs** are started with a fission reaction which creates exceptionally high temperatures.

- So far attempts to replicate these conditions safely on Earth have been unsuccessful.
  - Scientists are still trying to solve the safety and practical challenges presented.
  - **Cold fusion** is still not accepted as realistic since any results are impossible to verify so far.

Improve your grade

Controlling nuclear fission

Explain how the output of a nuclear reactor is controlled. *AO1* [3 marks]

P4 Summary

Electrostatics

The chance of getting an electric shock can be reduced by:
- correct earthing
- standing on insulating mats
- wearing shoes with insulating soles
- bonding fuel tanker to aircraft.

Electrostatic effects are caused by the transfer of electrons.
A positively charged object lacks electrons.
A negatively charged object has extra electrons.

There are two kinds of electric charge, positive and negative:
- like charges repel
- unlike charges attract.

Uses of electrostatics include:
Defibrillators, paint and crop sprayers, dust precipitators and photocopiers.

Using electricity safely

In a three pin plug:
- live wire is at a high voltage
- neutral wire completes the circuit
- earth wire is connected to the case to prevent it becoming live.

Double insulated appliances do not need an earth wire.

power = voltage × current
The fuse (or circuit breaker) is in the live wire. If the current is greater than the rating, the fuse will melt (or circuit breaker switch) breaking the circuit.

$$resistance = \frac{voltage}{current}$$

Longer wires have more resistance.
Thinner wires have more resistance.

Ultrasound

Medical uses include:
- scans to see inside the body without surgery
- measuring the rate of blood flow
- breaking up kidney stones.

Ultrasound is sound above 20 000 Hz which is a higher frequency than humans can hear.

Longitudinal waves – e.g. sound and ultrasound.
Transverse waves – e.g. light.
Wavelength (λ) is the distance occupied by one complete wave.
Frequency (f), measured in hertz (Hz), is the number of complete waves in 1 second.

Nuclear radiation

Nuclear equations are used to represent α decay:
$$^{238}_{92}U \rightarrow \,^{234}_{90}Th + \,^{4}_{2}He$$
or β decay:
$$^{14}_{6}C \rightarrow \,^{14}_{7}N + \,^{0}_{-1}e$$

Gamma rays are given out from the nucleus of certain radioactive materials.
x-rays are made by firing high speed electrons at metal targets.

Nuclear radiation is emitted from the nuclei of radioisotopes:
- α particle is a helium nucleus
- β particle is a fast-moving electron
- γ radiation is a short wavelength electromagnetic wave.

Medical uses of radioactivity include:
- diagnosis, as a tracer
- sterilising equipment
- treating cancers.

Only β and γ can pass through skin. γ radiation is most widely used as it is the most penetrating.

Other uses include:
- smoke detectors
- industrial tracers
- dating rocks and archaeological finds.

Fission is the splitting up of a large nucleus.
Fusion is the joining together of smaller nuclei.
Both release a lot of energy.
Fission is currently used to generate electricity.
Fission leads to a chain reaction which must be carefully controlled.

The half-life is the average time for half of the nuclei present to decay.

B1 Improve your grade Understanding organisms

Page 4 Heart disease

Excess saturated fat or excess salt in the diet can increase the risk of heart disease. Explain why. *AO1/2* [4 marks]

If there is too much saturated fat in the diet, cholesterol builds up in the wall of the arteries, so the heart gets less blood. This means that the cells die. Too much salt will increase the blood pressure, making heart disease more likely.

Answer grade: D–C. There is a correct link between saturated fats in the diet and cholesterol build-up in arteries. There is also correct reference to salt raising the blood pressure. However, the answer refers to the heart and not heart muscle. For full marks, explain that high blood pressure is likely to damage the walls of the blood vessels. This produces blood clots or a thrombosis, which blocks the blood vessels.

Page 5 EAR for protein

Explain the importance of knowing your EAR for protein. *AO1/2* [4 marks]

The EAR is the estimated average requirement in your diet each day.
The protein is needed for growth.
Too much or too little will cause a problem.

Answer grade: D–C. The answer states what is meant by EAR and knows protein is important for growth. For full marks, explain that too little protein can cause kwashiorkor and that the body cannot store excess protein. For full marks, answer would also explain that the EAR is only an estimated amount for the average person and that a fast-growing teenager would need a greater amount of protein.

Page 6 Immunisation

Immunisation carries a small risk. Despite this risk, why is immunisation important? *AO1* [4 marks]

Immunisation will protect the individual against a particular disease. It will also prevent many other people getting that disease.

Answer grade: D–C. The answer correctly states that being immunised will protect that individual. However, it does not mention that the small risk of side effects from immunisation is offset by the avoidance of a potentially fatal disease. For full marks, also explain that when a high percentage of the population is immunised there is a very low risk of infection from that disease.

Page 7 Vision

In old age, muscles lose their ability to contract and relax quickly. Ligaments become less flexible. The lens becomes less elastic. Explain the effects of these changes on vision. *AO2* [4 marks]

The eye muscles and ligaments cannot work properly. The eye now cannot accommodate. This will affect vision, especially in older people.

The answer states correctly that the changes would affect accommodation, but does not explain it, and does not explain how vision would be affected. For full marks, explain that if the ciliary muscles cannot contract and if the suspensory ligaments are not flexible then the lens would not change shape and accommodate (change focus) for near and distant objects. A less elastic lens would also result in the inability of the lens to change shape and focus.

Page 8 Affects of alcohol

Drinking alcohol increases the risk of accidents. Explain why. *AO2* [4 marks]

Alcohol is a depressant drug. It affects reaction time to danger. A slow reaction time will increase the risk of an accident. Alcohol affects nerve transmission across synapses.

The answer correctly links up alcohol being a depressant drug with slow reaction times and accident risk. It also states that alcohol has an effect on the nerve transmission across synapses but did not say what effect. Details of the blocking of receptor molecules should be included for full marks.

Page 9 Types of diabetes

Joe has Type 1 diabetes and Charlie has Type 2 diabetes. Explain why they require different treatments. *AO1* [4 marks]

Joe needs regular insulin doses, whereas Charlie can regulate his diet. Both treatments regulate blood sugar levels.

There is no explanation of the causes of the two types of diabetes. For full marks, explain that Joe's pancreas is unable to make any insulin and so needs regular insulin doses. Charlie's body is producing some insulin so if he lowers his sugar/carbohydrate intake he will not need so much insulin. He will also need to match his sugar/carbohydrate intake with his amount of exercise.

Page 10 Phototropism

Plant shoots grow straight upwards in the dark but will grow towards a light source. Explain how and why they do this. *AO1* [4 marks]

Plant shoots will grow upwards and towards the light so the plant's leaves will photosynthesise faster. Plant shoots are called positively phototropic because they grow towards the light. The curvature is controlled by auxins.

Answer grade: D–C. There is a correct link between light and photosynthesis and stating that the curvature towards light is caused by auxins. For full marks, state that a higher amount of auxin in the shaded part of the shoot results in longer cells in that area, and that in the dark there would be an equal distribution of auxin in the shoot, so the shoot would grow straight and not curved.

Page 11 Inherited disorders

Rabeena finds out that her mother has cystic fibrosis, an inherited disorder caused by a faulty allele (c). Her father does not have this faulty allele. Explain why Rabeena hopes her future husband will not be heterozygous for cystic fibrosis. *AO2/3* [4 marks]

Rabeena will not have cystic fibrosis but she will be heterozygous (Cc). If her future husband is also heterozygous for cystic fibrosis, some of their children could have cystic fibrosis.

Answer grade: D–C. The answer correctly identifies Rabeena as being heterozygous and the risk of their children having cystic fibrosis. For full marks, include details of possible children. If her husband was heterozygous, there would be a 1 in 4 chance of a child having cystic fibrosis, an equal chance of a child carrying the faulty allele and a 1 in 4 chance of not carrying any faulty alleles for cystic fibrosis. This could be shown in a Punnett square.

B2 Improve your grade Understanding our environment

Page 13 Classifying newly discovered organisms

Two similar types of animals have been discovered living close together in a jungle. Describe how scientists could find out how closely related the two animals are. *AO2* [3 marks]

The scientists could breed the animals together to see if they are the same species.

Answer grade: D–C. The answer gives a method but does not explain the possible outcomes of the cross. For full marks, the student would need to state that they would be the same species if offspring were fertile. The base sequence in their DNA could also be compared.

Page 14 Pyramids of biomass

Explain one advantage and one disadvantage of using pyramids of biomass to show feeding relationships. *AO2* [3 marks]

Pyramids of biomass are useful because they are always pyramids, unlike pyramids of numbers. However, it is hard to get the information to draw them.

Answer grade: D–C. The answer includes an advantage and a disadvantage but neither is explained properly. The explanation should say that pyramids of biomass are always pyramids because the energy decreases at each trophic level. They therefore give an idea of the energy loss. However, they involve measuring dry mass, which is a destructive process.

Page 15 The nitrogen cycle

Explain how nitrogen in a protein molecule in a dead leaf can become available again to a plant. *AO1* [4 marks]

The protein is broken down by decomposers such as bacteria and fungi in the soil. Nitrates are produced, which are taken up by the roots of plants.

Answer grade: D–C. The answer states that decomposers break down proteins and that plants take up nitrates. The intermediate steps are missing, however. For full marks, state that: bacteria and fungi in the soil decompose the dead leaves, they convert protein into ammonia, which is then converted into nitrates by nitrifying bacteria and the nitrates can be taken up by the roots of plants.

Page 16 Niches

The scientist Gauss put forward a theory that said organisms of two different species cannot share an identical ecological niche.
(a) Explain what is meant by the term 'ecological niche'. *AO1* [2 marks]
(b) Suggest why Gauss said that two species cannot share the same niche. *AO2* [2 marks]

(a) The term ecological 'niche' describes what an organism eats.
(b) The two organisms would be eating the same food.

Answer grade: D–C. What an organism eats is only part of the definition of ecological niche. For full marks, state that the term describes where an organism lives and its role in the ecosystem and that if two organisms had the same niche, then one would out-compete the other.

Page 17 Living in hot, dry conditions

An elephant has a large body, large ears, skin with few hairs and the ability to produce concentrated urine. Explain which of these features are advantages or disadvantages when living in hot, dry areas. *AO2* [4 marks]

The large ears will help the elephant lose heat. The lack of hair will stop the animal over-heating as it will not prevent heat loss. Producing a concentrated urine will help conserve water.

Answer grade: D–C. The answer gives some advantages with some explanation but has not given any disadvantages. For full marks, discuss the large size of the elephant, which gives it a small surface area to volume ratio, making it harder to lose heat. However, the large ears help to increase the surface area. Thin hair will allow heat loss but does not give much insulation from the sun's rays.

Page 18 Explanations for evolution

Human ancestors had more hair than modern humans. This can be explained by saying that scratching the skin due to parasites has gradually over many generations made some of the hair fall out.
(a) Explain why this explanation uses Lamarck's ideas. *AO2* [2 marks]
(b) How might Darwin's ideas be used to explain why modern humans have less hair? *AO2* [2 marks]

(a) This suggestion uses the idea that characteristics can change.
(b) Humans with less hair are less likely to get parasites and so this is an advantage.

Answer grade: D–C. The answer to (a) does not say that the characteristics are acquired during the organism's life and then passed on. In part (b) the advantage is a reasonable idea, but the ideas of variation and being able to survive to pass on the genes for shorter hair must also be included.

Page 19 Population and pollution

Explain the reasons for the increase in carbon dioxide levels in the atmosphere and explain why people are concerned about this. *AO1* [4 marks]

More carbon dioxide is released from the increased burning of fossil fuels such as coal and oil. People are worried that this might cause global warming.

Answer grade: D–C. The answer gives a reason for the increase of carbon dioxide and a possible consequence but neither is fully explained. For full marks the student should also say why fossil fuels are being burned and explain some of the possible consequences of global warming.

Page 20 Saving endangered species

The Hawaiian goose is only found on the islands of Hawaii. In the mid-1900s only about 30 were left alive. The space on the islands is restricted and humans have introduced a number of animals to the islands. Write about the problems facing scientists who are trying to save the goose from extinction in Hawaii. *AO2* [4 marks]

There are not many geese left to breed with and there is also little room for them to live on the islands. The animals on the island may harm them.

Answer grade: D–C. The answer highlights the small number of geese that are left and two of the problems on the island. For full marks the student should also refer to the lack of genetic variation left in the geese, and to the lack of suitable habitats, possible predation and competition from other animals.

Page 22 Fractional distillation

Why can crude oil be separated using fractional distillation?
AO1 [3 marks]

Because fractions have different boiling points, the fractions with the lowest boiling points 'exit' the tower at the top. Fractions with the highest boiling points exit at the bottom. There is a temperature gradient up the tower from high temperature at the bottom to lower temperature at the top.

Answer grade: D–C. The statements are correct but do not explain the molecular forces involved. For full marks the student should add ideas about larger molecules (in bitumen, for example) having stronger forces of attraction between them so more energy is needed to break the intermolecular forces. This information should be used to explain that larger molecules have higher boiling points than smaller molecules.

Page 23 Choosing a fuel

Use the table at the top of the page to decide which fuel should be used in a car engine. Justify your answer from the evidence. *AO3* [2 marks]

Because, although both have good availability, we have a system of garages around the country that sell petrol which is easier to obtain than coal.

No marks would be awarded for this answer. For full marks, evidence, not prior knowledge, is needed. The evidence needed to justify the selection is in the table:
- petrol flows easily around engines (ease of use)
- petrol is volatile and so ignites well (storage).

Page 24 Evolution of the atmosphere

Describe a theory, put forward by scientists, for the formation of the atmosphere we have today. *AO1* [5 marks]

Gases of the atmosphere were released by volcanoes. There was no oxygen at first. Plants used the carbon dioxide for photosynthesis and produced oxygen.

Answer grade: D–C. For full marks the answer should explain how water vapour from early volcanoes condensed to form oceans and carbon dioxide dissolved in the water; photosynthetic organisms developed which produced increasing levels of oxygen by photosynthesis; levels of unreactive nitrogen increased because it did not react with other gases.

Page 25 Interpreting displayed formulae

Explain why this molecule is unsaturated and whether it can be made into a polymer. *AO2* [3 marks]

```
      H   H   H   H
      |   |   |   |
  H — C — C — C = C — H
      |   |
      H   H
```

The molecule has a double covalent bond between the carbon atoms which means it must be an alkene. It is unsaturated because it has a double carbon bond. Only alkane molecules can be made into polymers.

Answer grade: D–C. Sentences 1 and 2 are correct. Sentence 3 is incorrect. For full marks the answer must state that unsaturated molecules (alkenes) are the molecules that can be made into polymers.

Page 26 Properties and uses of polymers

Explain why each of the polymers used for drain pipes, electrical cable covers and socks is suitable for its purpose and explain how the structure of the polymer causes it to have these properties. Use ideas about intermolecular forces. *AO2* [6 marks]

Drain pipes need to be waterproof and rigid. Electrical cable covers need to be electrical insulators and flexible. Material for socks needs to be flexible and able to be stretched.

Answer grade: D–C. For full marks the answer should include ideas about the molecular structure of the materials. The polymer used for drain pipes has strong cross-links between the polymer molecules, making it rigid. The polymer for electrical cable covers has relatively weak intermolecular forces between the polymer molecules so that they can slide over each other easily, making the plastic flexible. The polymer for sock fibre has relatively weak intermolecular forces between the polymer molecules so that they can slide over each other easily, making the material flexible, suitable for creating fibres and consequently, able to be stretched.

Page 27 Using baking powder

Construct the balanced symbol equation for the decomposition of baking powder. This is sodium hydrogencarbonate, $NaHCO_3$, which decomposes to give sodium carbonate, Na_2CO_3, carbon dioxide and water. *AO1* [2 marks]

$$NaHCO_3 \rightarrow Na_2CO_3 + CO_2 + H_2O$$

Answer grade: D–C. The answer gives the reactants and products in the correct place and the correct formula for carbon dioxide and water. For full marks, the equation must be balanced by using 2 molecules of $NaHCO_3$.

It is important to be able to construct this balanced symbol equation without being given any of the formulae.

$$2NaHCO_3 \rightarrow Na_2CO_3 + CO_2 + H_2O$$

Page 28 Esters and perfumes

A perfume must evaporate easily. Explain *how* it evaporates easily, using ideas on kinetic theory and why a perfume needs other specific properties. *AO1* [3 marks]

It must evaporate easily so that the perfume particles can reach the nose.

Answer grade: D–C. For full marks, the answer needs to explain that in order to evaporate, particles of a liquid need sufficient kinetic energy to overcome the forces of attraction to other molecules in the liquid. This can happen in perfumes as only weak attractions exist between particles of the liquid perfume. It is easy to overcome these weak forces as the molecules have sufficient kinetic energy.

Page 29 Drying paint

Explain how paints dry in different ways. *AO1* [3 marks]

Emulsion paints are water-based paints and they dry when the solvent evaporates. Oil paints dry when the oil evaporates.

Answer grade: D–C. For full marks the answer must explain that the oil left is oxidised by atmospheric oxygen to make a protective skin.

C2 Improve your grade Chemical resources

Page 31 Plate tectonic theory
Wegener's continental drift theory of plate tectonics was not accepted in 1914 but it is now widely accepted. Explain why. *AO1* [4 marks]

Because evidence has been found to back up the idea. Continents looked like they were joined. Fossils on both sides of the Atlantic ridge were the same.

Answer grade: D–C. Sentence 1 states, but does not describe, the evidence. Sentences 2 and 3 give reasons and are both worth marks. For full marks ideas should be included about why the theory was not accepted due to a lack of a testable mechanism, and why it is now accepted using new repeatable evidence of sea floor spreading.

Page 32 Rock hardness
Explain why granite, marble and limestone have different degrees of hardness. *AO1* [4 marks]

Granite is the hardest, and limestone weakest. Limestone is made of sediments which are only squashed together. Both granite and marble contain crystals holding them together better.

Answer grade: D–C. Sentences 1 and 2 explain why limestone is weak. Sentence 3 is a correct fact, but not an explanation.

For full marks, the student should add that granite has interlocking crystals, making it harder than marble, and marble is harder than limestone as it has been baked, forming a metamorphic rock.

Page 33 Recycling copper
Suggest advantages and disadvantages of recycling copper. *AO1* [4 marks]

It melts at a low temperature but it's hard to sort it. Copper is a metal so it's worth recycling.

Answer grade: D–C. Sentence 1 gives one advantage and disadvantage, but lacks detail. The second sentence adds little. For full marks, two bullet points for each would be useful along with a reason for each.

Page 34 Recycling materials from cars
Give two advantages and two disadvantages of recycling materials from cars. *AO1* [4 marks]

Advantages:
- *saves mining new metals*
- *less landfill.*

Disadvantages:
- *fewer mining jobs*
- *fewer mines built.*

Answer grade: D–C. All the answers are valid, but three are focused on mining so are too similar. For full marks include different ideas – bullet points are a good way to organise these.

Page 35 The Haber process
Explain how a high yield can be obtained using the Haber process. *AO1* [5 marks]

Use a low temperature and a high pressure. A catalyst would also speed it up. Recycle the unreacted reactants.

Answer grade: D–C. Four correct ideas have been given as statements rather than explanations. For full marks details should be given about why each factor improves the yield.

Page 36 Acids and bases
(a) Predict the name of the salt formed when nitric acid reacts with potassium hydroxide. *AO1* [1 mark]
(b) Write a word equation and a balanced symbol equation for the reaction between sulfuric acid and copper oxide. *AO1 and AO2* [4 marks]

(a) Potassium nitric
(b) Sulfuric acid + copper oxide → copper sulfate + water
$$2H_2SO_4 + Cu_2O \rightarrow 2CuSO_4 + 2H_2O$$

Answer grade: D–C. Part (a) is incorrect; it should be potassium nitrate. In part (b) the word equation is correct, and some of the symbol equation. For full marks the correct equation should be:
$$H_2SO_4 + CuO \rightarrow CuSO_4 + H_2O$$

Page 37 Fertilisers
Describe how to prepare a sample of ammonia phosphate. *AO1* [3 marks]

Use phosphoric acid and ammonia solution. Put the ammonia solution and indicator into a flask. Add acid until the indicator changes colour.

Answer grade: D–C. Correct reactants have been used, but only a partial method is given.

For full marks, the student should mention the need to repeat to obtain a precise reading, and then making a sample without an indicator and crystallising it.

Page 38 Electrolysis of sodium chloride solution
Describe and explain using equations what is seen during the electrolysis of sodium chloride solution and explain what is happening. *AO1* [4 marks]

When a direct current is applied, gases form at both sides. One is hydrogen and the other chlorine.

Chlorine is a pale green gas which bleaches damp indicator paper. Hydrogen burns with a pop.

Answer grade: D–C. This answer contains lots of correct facts, but doesn't fully answer the question. Tests are not asked for. For full marks, the student should link hydrogen formation to the cathode, and chlorine to the anode. At B-A* grade equations could be added.

Page 40 Specific heat capacity

Ed uses a stainless steel saucepan to heat his soup from 17 °C to 94 °C. The saucepan has a mass of 1.1 kg and a specific heat capacity of 510 J/kg °C. Energy is required to heat the soup.

(a) Calculate the extra energy required to raise the temperature of the saucepan. *AO2* [2 marks]

(b) Ed reads that a 1.1 kg copper saucepan will be more energy efficient. Explain why. *AO2* [2 marks]

(a) energy = 1.1 × 510 × 94 = 52 734 J

(b) Copper is a better conductor of heat than stainless steel so less energy will be lost in heating the saucepan.

Answer grade: D–C. In (a) the final temperature has been used instead of the temperature change. The student should show their workings as this might gain some marks. In (b) the statement is correct but does not answer the question. For full marks a description of the difference in the specific heat capacities is needed.

Page 41 Energy loss in a cavity wall

The Johnson's house has cavity walls. They decide to have foam injected into the cavity to reduce energy loss.

Explain how energy is transferred to the roof space from the cavity. *AO1* [3 marks]

The air in the cavity moves into the roof space by convection. The foam traps the air so it cannot move.

Answer grade: D–C. Sentence 1 states, but does not explain, the process. Sentence 2 explains the reason why foam is used but does not answer the question. For full marks, the student should include a description of convection as being due to the changing density of air as it is warmed.

Page 42 Diffraction effects

Light is diffracted as it passes through a narrow slit. Describe how the amount of diffraction depends on the wavelength of the light and the width of the slit. *AO1* [2 marks]

The wider the slit, the less the diffraction.

Answer grade: C–B. The sentence is correct but it is not a complete answer. For full marks, the student should include the fact that maximum diffraction happens when the slit width and wavelength are a similar size.

Page 43 Sending signals

Adam is standing on top of a hill in line of sight and 10 km away from Becky who is on top of another hill. They can communicate either by using light, radio or electrical signals. Suggest one advantage and one disadvantage of using each type of signal. *AO1* [3 marks]

● *Light signals need a code*
● *Electrical signals need wires*
● *Radio signals can be intercepted.*

Answer grade: D–C. Bullet points can sometimes help to organise ideas. Disadvantages are given but no advantage. For full marks, the student should include a unique advantage and disadvantage.

Page 44 Microwave transmitters

The Telecom Tower in London is one of the tallest buildings in the city. There are many microwave aerials surrounding the top of the tower.

Explain why they are sited so high up. *AO1* [2 marks]

Microwaves do not diffract very much around buildings.

Answer grade: D–C. The sentence is correct but not a complete answer. For full marks, the student should include the fact that aerials need to be in line of sight.

Page 45 Advantages of using digital signals and optical fibres

Explain the advantages of using digital signals and optical fibres compared with analogue signals and electrical cables for data transmission. *AO1* [4 marks]

Analogue signals can show a lot of interference. Digital signals do not exhibit any interference. You can only send one signal down copper wires but an optical fibre allows multiplexing.

Answer grade: D–C. Sentence 1 states a reason for not using analogue signals. Sentence 2 does not give enough information. For full marks, the student should include reference to digital signals not showing interference because they only have 2 values. Sentence 3 is not totally true. It is possible to multiplex along copper wires, but less so than through optical fibres. For full marks, the student should explain the meaning of multiplexing.

Page 46 Radio communication

The picture shows a transmitter and receiver on the Earth's surface, out of line of sight.

(a) Explain how long wave radio signals travel from the transmitter to the receiver. *AO1* [3 marks]

(b) Explain how microwave signals travel from the transmitter to the receiver. *AO1* [2 marks]

(a) Long wave signals are reflected from the ionosphere, then from the sea, then the ionosphere again and so on until they reach the receiver.

(b) Microwave signals have too high a frequency to be reflected back by the atmosphere. They are received and amplified by a satellite before being reflected back to Earth.

Answer grade: C–B. In (a), reflection from the sea is mentioned as well as from the ionosphere. For full marks, the student should refer to refraction and total internal reflection from the atmosphere. In (b), sentence 1 is irrelevant to the question. Sentence 2 describes microwaves being received but they are not reflected. For full marks, the student should refer to the satellite retransmitting them.

Page 47 Earthquake waves

An Earthquake occurs with its epicentre at **E**. It is detected at two monitoring stations **A** and **B**.

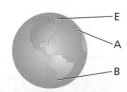

Describe and explain the appearance of the seismograph traces at **A** and **B**. *AO1/AO2* [4 marks]

The trace at A will show a P wave and an S wave. The one at B will only show a P wave. S waves will not travel through the Earth's core so are not received at station B.

Answer grade: D–C. Sentences 1 and 2 describe the traces that will be seen. For full marks, the student should include the fact that the P wave is received before the S wave at A and that the P wave is received at B later than at A. Sentence 3 states why the wave is not received but does not explain why. For full marks, the student should explain that the core contains liquid.

P2 Improve your grade Living for the future

Page 49 Photocells

A photocell contains two pieces of silicon joined together to make a p-n junction.

Explain how light falling on a photocell produces an electric current. *AO1* [3 marks]

n-type silicon has extra electrons which move to produce current.

Answer grade: D–C. The explanation of the p-n junction is incomplete. The answer can be improved by adding the description of p-type silicon as having an impurity added which leads to an absence of electrons and then explaining that energy from the Sun's photons allows the electrons to move across from n-type to p-type silicon.

Page 50 Generators

Describe three ways to increase the output of an electrical dynamo. *AO1* [3 marks]

Longer wire, bigger magnets and faster movement.

Answer grade: D–C. The phrase 'bigger magnets' is too vague and longer wire is incorrect. To improve this answer, add the use of a stronger magnetic field and more turns on the coil.

Page 51 The greenhouse effect

Describe the greenhouse effect and explain how it contributes to global warming. *AO1* [4 marks]

The electromagnetic radiation from the Sun is absorbed by and warms the Earth. The Earth then re-radiates the energy which warms the atmosphere.

Answer grade: D–C. The description of what is happening is correct but the explanation of why this takes place is missing. For full marks, the student should include the idea that the Sun's radiation is at a shorter wavelength than that re-radiated by the Earth and also mention that the greenhouse gases absorb this longer wavelength radiation and this warms the atmosphere.

Page 52 Cost of electricity

Tracey's flat has electric storage heaters which heat up at night and release the heat slowly during the day. Why is this cheaper for Tracey? *AO1* [2 marks]

She pays less for electricity during the night.

Answer grade: D–C. This would gain one mark for a correct statement but the question is worth 2 marks, so a fuller explanation is required. For full marks, the student should include the fact that electrical power stations will not be switched off but will always be producing power and when the demand is lower at night the electricity will be sold more cheaply.

Page 53 Ionising radiation

Give two advantages and two disadvantages of nuclear power. *AO1* [4 marks]

Nuclear power does not pollute the environment and it is renewable
Nuclear power is expensive and risky.

Answer grade: D–C. Answer lacks detail and is vague. For full marks, the student should include clear advantages – nuclear power does not produce greenhouse gases; and disadvantages – costs of maintenance are high and there is a risk of nuclear accidents. Nuclear power is a non-renewable energy source.

Page 54 Unmanned space travel

Why do we send unmanned space probes to explore our Solar System? *AO1* [3 marks]

We explore our Solar System to find out more about the other planets and to search for alien life.

Answer grade: D–C. Whilst this answer may be true it is not addressing the question specifically focusing on unmanned space exploration. A better answer would have explained the advantages of not having humans on board, e.g. no risk to life, no food water or oxygen needed, lower costs, longer journeys possible, no need to return.

Page 55 Asteroids

What evidence have scientists gathered to show how the Moon was formed? *AO1* [3 marks]

The Moon was formed when there was a collision between the Earth and another planet. During the collision, iron became concentrated in the Earth's core.

Answer grade: D–C. This answer correctly describes how the Moon was formed. However, the question is asking for scientific evidence. For full marks, the student should include the fact that there is no iron on the Moon as it all merged with the Earth's core and the fact that the Moon has the same oxygen composition as the Earth.

Page 56 The Big Bang

What is red shift and how does it provide evidence for the Big Bang? *AO1* [4 marks]

Red shift is when the light from stars is shifted to the red end of the spectrum and it shows galaxies are expanding.

Answer grade: D–C. This gives an idea about red shift but the answer is not clearly explained. For full marks, the student should explain that red shift is when light from distant galaxies contains a spectrum of lines that are shifted towards a longer wavelength. This happens when the object emitting the light is moving away from the observer. Since light from most distant galaxies exhibits red shift and the further away they are the greater the red shift, this indicates that all galaxies are moving apart which is evidence for the Big Bang.

B3 Improve your grade Living and growing

Page 58 Protein synthesis
Describe where and how proteins are coded for and made. *AO1* [6 marks]

Proteins are coded for by DNA. They are made in the cytoplasm on ribosomes.

> **Answer grade: D–C.** Sentence 1 correctly states that DNA codes for proteins but does not say how. Sentence 2 correctly gives the site of production but not how the information reaches there. For full marks, include details of how the order of bases codes for the order of amino acids and discuss the function of mRNA.

Page 59 Mutations and enzymes
A mutation can occur in a gene that codes for an enzyme. Explain how a mutation could lead to an enzyme failing to work properly. *AO1* [6 marks]

The mutation could change the bases in the DNA of the gene. This could mean that the protein is made with a different shape.

> **Answer grade: D–C.** Sentence 1 correctly states that the bases might change but does not explain how this changes the protein. Sentence 2 suggests that the shape changes but does not explain how this changes the function. For full marks, a change in the order of amino acids needs to be stated and some reference to a change in shape preventing the substrate fitting into the active site.

Page 60 Respiration and enzymes
An experiment was set up to measure the oxygen uptake of insects. The data shows that the maximum uptake was at about 35 °C. Above and below that temperature less oxygen was used.

Explain what the oxygen is used for and why it is used fastest at 35 °C. *AO1* [2 marks], *AO2* [2 marks]

The oxygen is used for respiration. 35 °C is closest to body temperature and so this is the optimum temperature for respiration.

> **Answer grade: D–C.** Sentence 1 links oxygen to respiration but does not say which type. Sentence 2 uses the term 'optimum' but does not explain why there is an optimum. For full marks, the word 'aerobic' should be used to describe respiration and the role of enzymes should be discussed.

Page 61 Meiosis and mitosis
The two processes mitosis and meiosis occur in the human body. Compare each process, writing about where they occur and any differences in the process. *AO1* [6 marks]

Mitosis happens all over the body but meiosis makes gametes. They both make new cells but in mitosis they have the same number of chromosomes and in meiosis they have half.

> **Answer grade: D–C.** Sentence 1 correctly states where mitosis occurs but not where meiosis happens. Sentence 2 compares the outcome but not the process. For full marks the answer should say that meiosis occurs in the sex organs. Also, in mitosis there is one cell division, during which the copies of the chromosomes separate. In meiosis, there are two divisions. In the first, the chromosomes pair up and separate. In the second, copies move to the opposite poles of the cell.

Page 62 Valves
Valves are found in veins, at the start of the arteries leaving the heart and in the heart. Write about the importance of these different valves. *AO1/2* [2 marks]

The job of valves is to stop blood flowing backwards. In the veins it keeps it moving back to the heart. In the heart it makes sure that it does not flow back into the atria.

> **Answer grade: D–C.** The answer explains the job of valves in general but does not say why they are needed in veins. Also, there is not enough detail about the valves in the heart and arteries. For full marks, include an explanation involving the low pressure of the blood in veins and the role of the semilunar valves in stopping the blood flowing back into the heart when the heart muscle relaxes.

Page 63 A plant growth curve
Katie wants to plot a growth curve for a broad bean plant using dry mass. Given 100 broad bean seeds, explain how you would collect the data to plot the graph and explain why she wants to plot dry mass. *AO1* [5 marks]

I would plant the seeds and every week dry out one of the seedlings and weigh it. Dry mass gives the best measure of growth.

> **Answer grade: D–C.** Sentence 1 states that the 100 seeds are sampled every month but does not give details of how they are dried. Sentence 2 does not give a reason for why dry mass is preferable. For full marks, explain that the seedlings have to be dried in an oven at about 80 °C, which prevents burning. Also, dry mass measures permanent growth in all directions, not just changes in water content.

Page 64 Spider silk
Spider silk is very strong and could be very useful in industry. Goats have now been produced that make spider silk in their milk.

Describe how this could be done and suggest reasons why this method of production might be more useful. *AO2* [4 marks]

The spider silk DNA is put into goats. This is useful because the silk is easier to collect from the milk.

> **Answer grade: D–C.** The answer briefly describes the process and gives one possible advantage. For full marks, it would need to say that the spider gene for silk is isolated and transferred. Also, more than one reason is needed, such as a comparison of the quantity made.

Page 65 Cloning plants
A garden centre wants to sell an attractive coloured geranium plant. They decide to produce many clones of the plant using tissue culture.
(a) What are the possible disadvantages of this method of reproduction? *AO1* [2 marks]
(b) Why is this method not possible for producing goldfish for garden ponds? *AO2* [2 marks]

(a) The plants might all die at once.
(b) Because animals cannot be cloned in this way.

> **Answer grade: D–C.** (a) is incomplete because it does not explain why. For full marks, it should state that the lack of genetic variation (and so the susceptibility to disease) is important. (b) is also incomplete because no reason is given. It should state the lack of ability of animal cells to differentiate.

Page 67 Distribution of animals

Rick wants to find out about the zonation of different types of limpets down a sea shore. Explain:
(a) how he should do this
(b) what can cause this zonation. *AO1/2* [5 marks]

(a) He puts a long line down the sea shore and counts the limpets at different stages.
(b) The limpets at the top of the shore will be more exposed at the top of the shore and may not survive.

Answer grade: D–C. For full marks, (a) should describe using a quadrat at measured intervals and state that different species should be investigated. (b) should also say that the abiotic factors would have a different effect on different limpet species.

page 68 Exchange of gases in plants

When gas exchange in plants is analysed, they seem to respire only at night. Explain why. *AO2* [4 marks]

A plant will carry out respiration during both day and night. Plants respire by taking in oxygen and releasing carbon dioxide. The oxygen is used to release energy from glucose.

Answer grade: D–C. For full marks, include details of photosynthesis taking place in the light, using up carbon dioxide and producing oxygen. Also, explain that the rate of gas exchange in photosynthesis is more than that in respiration, so respiration in plants during daylight is difficult to isolate and measure.

Page 69 Absorption of light

Look at the graph.
(a) Which parts of the light spectrum are
i) reflected and
ii) used by leaves?
AO2 [2 marks]
(b) Explain why a leaf has many different pigments.
AO1 [2 marks]

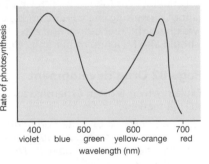

(a) i) A wide range of colours are used by leaves. The middle part of the spectrum is reflected.
ii) The middle part is not used by leaves.
(b) The leaf has many different pigments so it can be a variety of colours. It can also use a lot of different colours in the spectrum.

Answer grade: D–C. (a) i) gives a vague answer ('middle part') instead of quoting a colour (green) or actual numbers. In ii) needs to state what *is* used and the colours (violet/blue and yellow/orange) or numbers in nm. For full marks in (b), quote names of pigments as well as using the word 'photosynthesis'.

Page 70 Osmosis in plant and animal cells

Describe and explain the effects of water entry into plant and animal cells. *AO1/2* [5 marks]

When water enters plant and animal cells the turgor pressure increases, helping to maintain the cell shape. Because animal cells do not have a cell wall, they can swell up if too much water enters.

Answer grade: D–C. For full marks, Sentence 1 should also state that the process involves osmosis, and describe it and how it works. The answer states that an animal cell would swell up, but should also explain that in the absence of a cell wall, the cell could burst and die (lysis).

Page 71 Transport in plants

Marram grass grows in exposed sand dunes. It has narrow leaves, stomata sunk in pits, many hairs on its leaves and leaves that can curl up into a tube. Explain how these adaptations help it to survive. *AO2* [4 marks]

As Marram grass grows in exposed places it must reduce its water loss, or its roots will not be able to take up enough water to maintain turgor and carry out photosynthesis. All these adaptations will reduce its water loss and help it to survive.

Answer grade: D–C. The answer is good, as it explains why Marram grass will have problems in retaining enough water. It also makes a link between reducing water loss and survival. For full marks, explain how each of the adaptations would decrease water loss, e.g. narrow leaves have a smaller surface area and fewer stomata than broad leaves.

Page 72 Mineral uptake

Look at the graph.

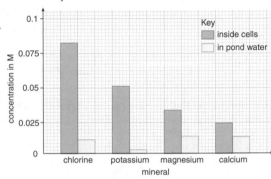

Mineral uptake by algae growing in water

What conclusions can be made from this data? *AO3* [4 marks]

The graph shows that chlorine, potassium, magnesium and calcium are taken up by the algae. Chlorine is taken up the most and calcium the least.

Answer grade: D–C. For full marks, the answer should also say that the data shows that minerals are taken up against a concentration gradient and in different amounts. This indicates a system of active transport, a system of carriers using energy to transport minerals into the cell's cytoplasm.

Page 73 Growing mushrooms

Lynn wants to grow mushrooms (a fungus). Explain what conditions she should provide for optimum growth. *AO1/2* [3 marks]

Lynn should grow them in soil with plenty of water, minerals, a reasonable temperature and enough oxygen.

Answer grade: D–C. The question asked for an explanation. For full marks, state that a temperature of 25 °C is required for optimum respiratory rate; water is required by a saprophyte such as mushrooms for extracellular digestion; plenty of oxygen is required for aerobic respiration for optimum growth and reproduction.

Page 74 Hydroponics

Look at the diagram of a hydroponics system. Explain how this system is useful for growing lettuce in glasshouses. *AO1/2* [4 marks]

Lettuce are small plants and do not need support in a hydroponics system. The system supplies the water, minerals and air the lettuce needs to grow. The system does not use soil.

Answer grade: D–C. For full marks, include details about being able to: grow a large number of lettuces in a small space (can regulate and recycle the amount of minerals supplied for optimum growth); use the system in glasshouses and avoid the crop being eaten by animals; control climate conditions, so grow a succession of crops all year round; and to grow lettuce in areas of poor soil.

Page 76 Rate of reaction

When limestone reacts with excess hydrochloric acid, carbon dioxide gas is made. Explain why the rate of reaction changes during the reaction. Use ideas about reacting particles. *AO2 [4 marks]*

The rate is fast at the start, and gradually slows down. When all the limestone has reacted, no gas is made.

Answer grade: D–C. Both sentences are correct and worth a mark each but do not explain why. For full marks, the answer needs to explain that the limestone is the limiting reactant, that the amount of carbon dioxide made is proportional to the amount of limiting reactant used and that at the start of the reaction there are a large number of particles that collide and react whereas towards the end of the reaction there are fewer particles that can collide and react.

Page 77 Temperature and rate of reaction

Explain, using the reacting particle model, why raising the temperature increases the rate of a chemical reaction. *AO1 [3 marks]*

Heating raises the temperature. This means that particles move faster and so collide more often. More collisions mean that the reaction is faster.

Answer grade: D–C. Sentence 1 does not contain enough detail for a mark. Sentences 2 and 3 start to explain the process, but lack detail. For full marks, ideas should be included about how more kinetic energy, leads to harder and more frequent collisions, increasing the rate of successful collisions – the collision frequency.

Page 78 Surface area and rate of reaction

Draw a sketch graph as on page 76. On this graph sketch further lines to show the effect of changing the surface area of reactants on:
(a) the amount of product formed in a reaction
(b) the rate of reaction. *AO2 [3 marks]*

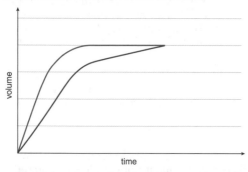

Answer grade: B. The sketch graph correctly shows the two lines required with different rates but the same final volume. The axis labels are also correct.

For full marks, the graph needs a key or labels. These labels would indicate that the blue line represents a reaction involving reactants with a large surface area (more spread out/smaller pieces), and that the red line represents a reaction involving reactants with larger pieces.

Page 79 Conservation of mass

Show that mass is conserved in this reaction.
$$ZnCO_3 \rightarrow ZnO + CO_2$$
A_r (Zn = 65, C = 12, O = 16)
AO2 [3 marks]

Zinc carbonate = 65 + 12 + 48 = 125
Zinc oxide = 65 + 16 = 81
Carbon dioxide = 12 + 32 = 44

Answer grade: C. All three relative formula masses are correct, but the answer does not show that mass is conserved. For full marks, the two products should be added together to show that they are equal to the reactant.

Page 80 Calculating atom economy

Zinc oxide can be made by this reaction:
$$ZnCO_3 \rightarrow ZnO + CO_2$$
CO_2 is a waste product. ZnO is the desired product. Calculate the atom economy for this reaction.
A_r (Zn = 65, C = 12, O = 16)
AO2 [2 marks]

64.8%

Answer grade: C. The answer is correct but it is good practice to show how to reach the answer. The method of calculating M_r should be shown, the formula should be written out and the steps taken clearly laid out. If a mistake is made in the final calculation, marks may be awarded for correct method.

Page 81 Measuring energy transfer

Describe how you could compare the energy released when different liquid fuels burn. *AO1 [4 marks]*

Burn the fuel in a spirit burner.
Heat water, and measure the temperature rise. Work out the energy using $E = mc\Delta T$.
Repeat for the other fuel.

Answer grade: C. The content of the answer is correct, but it lacks detail. For full marks, information should be included about measuring the volume (mass) of the water used, weighing the fuel before and after burning. Reference should also be made to fair and reliable testing.

Page 82 Drug development

Explain why it is often expensive to develop a new drug. *AO1 [4 marks]*

The drug needs to be tested on people for a long time to show it is safe. People can volunteer for this and get paid. It could be expensive to make it. The raw materials might be expensive.

Answer grade: D. Sentence 1 would gain one mark for the ideas about testing; this would be the same marking point as sentence 2. Sentences 3 and 4 would need to have more detail to gain marks. For full marks, ideas should include: how new compounds are found, extracted and purified; the need to prove the effectiveness of a new drug; the lack of serious side effects when the drug is tested.

Page 83 Structure of allotropes

Diamond and graphite are both giant carbon structures. Explain why graphite layers slide over each other, while diamond is hard in every direction. *AO1 [4 marks]*

Atoms in diamond are joined to four others, but in graphite only to three others. Graphite forms in layers that can slide, and that's why it is used in pencils.

Answer grade: D–C. Sentence 1 is correct. Further explanation in sentence 2 would improve the grade. For full marks, information should be added about the bonding in graphite forming strong layers, with only weak forces holding the different layers together and the 3-D tetrahedral bonding in diamond giving strength in all directions.

C4 Improve your grade The periodic table

Page 85 Arrangement of electrons

How many electrons does a potassium atom have and why is it in the fourth row of the periodic table? (Use page 248). *AO2* [2 marks]

Because potassium has an atomic number of 19 this means that an atom has 19 electrons.

Answer grade: D–C. One mark would be awarded for this answer. For full marks it should also say that the electronic structure of potassium is 2.8.8.1 so electrons occupy four shells and therefore potassium (K) is in the fourth row.

Page 86 Structure and bonding

Explain why the melting point of magnesium oxide is so high. *AO1* [2 marks]

Because magnesium forms a positive ion by donating two electrons to an oxygen atom which forms a negative ion. They attract each other strongly. A lot of energy is needed to break them apart.

Answer grade: D–C. The three sentences would gain one mark. For full marks, the answer needs to explain that magnesium ions are very small and can get in close to the oxygen ions.

Page 87 Electrical conductors

Explain why carbon dioxide does not conduct electricity. *AO1* [2 marks]

Because a carbon atom has 6 electrons, it has four electrons in the outer shell. It must share these electrons as they cannot be transferred. They are shared with the electrons of two oxygen atoms. So CO_2 cannot conduct electricity

Answer grade: D–C. Although this answer is correct, for full marks, the answer needs to explain that there are no spare electrons or ions.

Page 88 Reactivity in alkali metals

Explain why the reaction of rubidium with water is more vigorous than the reaction of sodium with water. *AO1* [2 marks]

The reaction of sodium with water produces sodium hydroxide and hydrogen. Because rubidium is lower in the group than sodium the reaction is more vigorous.

Answer grade: D–C. Only the second sentence of the answer scores a mark. For full marks the answer must explain that: the easier it is for the group 1 metal to become an ion, the more vigorous the reaction will be; to become an ion, the electron in the outer shell is lost; the further away this electron is from the nucleus the more reactive the element; the outer electron of rubidium is further away than that of sodium, so the reaction would be more vigorous.

Page 89 Displacement reactions of halogens

Explain why a red–brown colour is seen if chlorine is bubbled into a solution of potassium iodide. *AO1* [2 marks]

Chlorine is more reactive than iodine and so will displace iodine from a solution of its ions. Iodine is red-brown.

Answer grade: D–C. The information given is correct. For full marks, the answer must explain that chlorine is more reactive than iodine because its outer shell of electrons is nearer the nucleus, so it is easier for chlorine to gain electrons.

Page 90 Precipitation reactions

Explain how you can show a solution contains copper ions. *AO1* [2 marks]

Add sodium hydroxide to the solution.

Answer grade: D–C. This sentence is correct and for full marks the answer should also explain that if copper ions are present, then you would get a blue precipitate.

Page 91 Electrical conductivity of metals

Explain how metals conduct electricity. *AO1* [2 marks]

Metals conduct electricity because electrons are free to move through the metal ions.

Answer grade: D–C. This sentence is correct, but for full marks it should also explain that it is the delocalised electrons that move.

Page 92 Testing water

Explain, with the aid of a balanced equation, how you would test whether magnesium chloride is present in a sample of water. *AO1* [3 marks]

Add silver nitrate to a sample of the water. A white precipitate will form if there are chloride ions in it.

Answer grade: D–C. Both sentences are correct. For full marks, the answer must include a balanced symbol equation of the reaction.

Page 94 Calculating time

How many hours will it take to travel 560 km at an average speed of 25 m/s? *AO1* [1 mark] *AO2* [2 marks]

$$Time = \frac{distance}{speed} = \frac{560}{25 \times 60 \times 60} = 6.2 \times 10^{-3}\, h$$

Answer grade: D–C. The method for the calculation is correct and the conversion from seconds to hours completed. For full marks it is important to check that all units are consistent. Metres should be converted to kilometres.

Page 95 Calculating acceleration

A car is travelling at 10 m/s. It accelerates at 4 m/s² for 8 s. How fast is it then going? *AO1* [1 mark] *AO2* [2 marks]

change in speed = acceleration × time = 4 × 8 = 32 m/s

Answer grade: D–C. The working for the change in speed is correct but this needs to be added to the original speed of 10 m/s. For full marks, it is important to read the question carefully and double check the calculations.

Page 96 Stopping distance

Explain why brakes and tyres are checked when a car has its annual MOT Test. *AO2* [2 marks]

If they are worn, the braking distance increases.

Answer grade: D–C. The sentence is correct but does not provide a full explanation. For full marks, the answer should explain that worn brakes and worn tyres reduce friction and hence the braking force.

Page 97 Calculating driving force

A Ford Focus has a power rating of 104 kW.
(a) Calculate the resultant force acting when the car is travelling at 90 km/h.
(b) Explain how this force compares with the driving force of the engine. *AO1* [3 marks] *AO2* [1 mark]

(a) force = $\frac{power}{speed} = \frac{104 \times 1000 \times 3600}{90 \times 1000} = 4160\,N$

(b) It would be larger.

Answer grade: D–C. The calculation is correct and the working is clearly shown. The answer to part (b) is ambiguous and is only a description, not an explanation. For full marks, the answer should avoid the use of the word 'it' and explain that the driving force is larger because there is drag and other frictional forces to overcome.

Page 98 Carbon dioxide emissions

Some scientists suggest that carbon dioxide emissions from burning bio-fuels may be at least 20% lower than those from fossil fuels. Some scientists argue that overall the emissions may be higher than from fossil fuels. Suggest why emissions may be higher. *AO3* [3 marks]

Deforestation and processing the bio-fuel.

Answer grade: D–C. The sentence gives two correct reasons but not full explanations. For full marks, the answer should explain that deforestation leads to an increase in carbon dioxide levels and electricity is needed in the processing of the bio-fuel, which is generated in a power station and releases carbon dioxide.

Page 99 Seat belts

Some people think that wearing a seat belt should be up to the individual, not the law. Explain how the wearing of seat belts can help to avoid injury but may not always do so. *AO1* [3 marks]

Seat belts stretch and absorb energy in a crash. They increase the time it takes for a person to come to rest therefore reducing the force since

$$force = \frac{change\ in\ momentum}{time}$$

Answer grade: C–B. Both sentences are correct and are a good answer to the first part of the question. For full marks, the answer should explain the circumstances where a seat belt may be a disadvantage, e.g. trapping the person in a burning or submerged car.

Page 100 Gravitational field strength

(a) Explain why a 1 kg ball, dropped from a height of 10 m above the ground at the North Pole, takes less time to reach the ground than an identical 1 kg ball dropped from 10 m above the ground at the equator.
(b) How would the time taken be different if the ball was taken to the top of a mountain and dropped from a height of 10 m? *AO1* [2 marks], *AO2* [1 mark]

(a) The acceleration due to gravity is greater at the poles than it is at the equator.
(b) The acceleration due to gravity is less so it takes less time to fall.

Answer grade: C–B. The answer to part (a) is correct but does not give the reason why the acceleration is greater. The first sentence in part (b) is correct but the second contradicts the first. For full marks, it should explain that the acceleration at the poles is greater because the Earth is not a perfect sphere and the poles are nearer the centre. Answers should always be read carefully to make sure there are no contradictions. If acceleration is less, the time will be greater.

Page 101 Energy changes during free fall

Mel is a free-fall parachutist. During her time in free fall she reaches terminal speed. Explain how her gravitational potential energy and kinetic energy change during her descent. *AO1* [3 marks]

As she falls, her gravitational potential energy decreases as it is transferred into kinetic energy. When she reaches terminal speed, her kinetic energy remains constant.

Answer grade: C–B. The first sentence implies that KE increases as she falls and is worthy of credit. The second sentence does not explain what happens to the GPE if it is not transferred into KE. For full marks, the answer should explain that as she leaves the plane, her KE is zero, GPE is a maximum and that this reduces as the KE increases. It should also explain that the GPE is used to do work against friction, resulting in the heating of the air.

Page 103 Static charge

Connor is in the library walking on a nylon carpet. He touches a metal bookshelf and receives an electric shock. Explain how he became charged and why he received a shock. *AO2* [3 marks]

He becomes charged and when he touches the shelf he discharges.

Answer grade: D–C. The basic ideas are correct but there is insufficient detail in the answer. The answer should explain that it is the friction of walking on the nylon carpet that leads to the transfer of electrons charging him up. Then the charges flow to earth through the metal shelf causing a shock.

Page 104 Spray painting

Static electricity is useful in spray-painting cars.

Explain how by writing about:
- electrostatic charge
- electrostatic force
- why it is used. *AO1* [3 marks]

The paint experiences an electrostatic force because it is electrostatically charged. Paint will stick to the car better.

Answer grade: D–C. The answer is very brief and lacks detail. Charge on the plates is induced by the dust particles which are then attracted because opposite charges attract.

The answer simply repeats the words in the question without showing understanding. For full marks, it should explain that the spray gun is charged so that this in turn charges the paint and that the car is charged oppositely to the paint so that the paint is attracted and sticks to the car, and less is wasted.

Page 105 Electrical safety

Explain how the fuse and earth wire operate to protect a user. *AO1* [2 marks]

If a fault occurs the fuse blows stopping the current.

Answer grade: D–C. The answer is correct but does not include an explanation of the role of the earth wire. For full marks, the answer should explain that a fault would cause a large current to flow through the earth wire causing the fuse to overheat and melt.

Page 106 Ultrasound and imaging

(a) State one similarity and one difference between sound and ultrasound waves.
(b) Give two reasons why doctors may decide to use ultrasound instead of x-rays to get images of inside the body.

AO1 [4 marks]

(a) Both are waves but ultrasound is at a much higher frequency.
(b) Ultrasound is not harmful and it can make images.

Answer grade: D–C. (a) the fact that these are waves is given in the question. They are both longitudinal waves would be a better answer. It is also a good idea to add the fact that ultrasound is above 20 000 Hz.

X rays are an ionising radiation and this is why they could be harmful. Both can produce images but the ultrasound shows soft tissue better.

Page 107 Nuclear radiation

When carbon-14 undergoes beta decay what happens to the atom and what new element is formed? *AO1* [2 marks]

In the nucleus of the atom a neutron changes into a proton and an electron is emitted at high speed. The carbon atom is now carbon-13.

Answer grade: C–B. The first part of the answer is correct and quite detailed. In the second part a mistake has been made – the atomic number, not the mass number, should be changed so a new element is formed nitrogen-14.

Page 108 Radioisotope dating

Carbon-14 has a half-life of 5700 years.
(a) What is meant by the half-life of a radioactive sample?
(b) A sample of bone was found to have a 25% of the amount of carbon-14 found in a living organism. How old was this bone?

AO1 [1 mark], *AO2* [2 marks]

(a) The half-life of a radioactive sample is the time taken for half of the nuclei to decay.
(b) The sample was 25% of 5700 = 1425 years old.

Answer grade: D–C. The answer to part (a) is correct but should say the average time taken. In (b) it has not been understood that 25% is one quarter, which means the amount of carbon-14 nuclei would have halved and then halved again. So 2 half-lives must have passed or 11 400 years.

Page 109 Medical tracers

Explain what a medical tracer is and how suitable materials are chosen to be used as tracers. *AO1* [3 marks]

A tracer is a substance put into the body to find out if there is a problem. It must be able to be detected outside of the body and must not harm the patient.

Answer grade: D–C. The answer gives two correct statements but not full explanations. For full marks, the answer should explain that a tracer is a radioisotope that will emit radiation that can be detected from the outside of the body. It should also explain that since it will be emitting potentially harmful radiation it must have a short half-life so that this is limited, and finally that a gamma emitter is used as it is the most penetrating and the least ionising radiation.

Page 110 Controlling nuclear fission

Explain how the output of a nuclear reactor is controlled. *AO1* [3 marks]

The graphite moderator slows down the neutrons to make them more likely to be captured by uranium. Boron rods absorb neutrons to slow down the rate of fission.

Answer grade: C–B. This is a good answer giving the two ways the chain reaction is controlled. For full marks, the answer should explain that the boron control rods can be raised or lowered to absorb different numbers of neutrons when required.

Understanding the scientific process

As part of your Science assessment, you will need to show that you have an understanding of the scientific process – How Science Works.

This involves examining how scientific data is collected and analysed. You will need to evaluate the data by providing evidence to test ideas and develop theories. Some explanations are developed using scientific theories, models and ideas. You should be aware that there are some questions science that cannot answer and some that science cannot address.

Collecting and evaluating data

You should be able to devise a plan that will answer a scientific question or solve a scientific problem. In doing so, you will need to collect data from both primary and secondary sources. Primary data will come from your own findings – often from an experimental procedure or investigation. While working with primary data, you will need to show that you can work safely and accurately, not only on your own but also with others.

Secondary data is found by research, often using ICT – but do not forget books, journals, magazines and newspapers are also sources. The data you collect will need to be evaluated for its validity and reliability as evidence.

Presenting information

You should be able to present your information in an appropriate, scientific manner. This may involve the use of mathematical language as well as using the correct scientific terminology and conventions. You should be able to develop an argument and come to a conclusion based on recall and analysis of scientific information. It is important to use both quantitative and qualitative arguments.

Changing ideas and explanations

Many of today's scientific and technological developments have both benefits and risks. The decisions that scientists make will almost certainly raise ethical, environmental, social or economic questions. Scientific ideas and explanations change as time passes and the standards and values of society change. It is the job of scientists to validate these changing ideas.

Glossary

A

ABS braking system, known as advance braking system, which helps to control a skidding car 99

acceleration a measurement of how quickly the speed of a moving object changes (when speed is in m/s the acceleration is in m/s^2) 95, 96, 98, 99, 100

accommodation the eye's ability to change focus 7

acid solution with a pH of less than 7 28, 36, 37

acid rain rain water which is made more acidic by pollutant gases 19, 24, 34

acrosome part of the sperm that contains enzymes 61

active immunity you have immunity if your immune system recognises a pathogen and fights it 6

active site the place on an enzyme where the substrate molecule binds 59

active transport in active transport, cells use energy to transport substances through cell membranes against a concentration gradient 72

adaptations features that organisms have to help them survive in their environment 17, 18

aerobic respiration respiration that involves oxygen 60, 73

air bags cushions which inflate with gas to protect people in a vehicle accident 99

air resistance the force exerted by air to any object passing through it 100, 101

alcohol substance made by the fermentation of yeast 4, 8, 28

algal bloom a thick mat of algae near the surface of water which stops sunlight getting through 37

alkali metals very reactive metals in group 1 of the periodic table, e.g. sodium 88, 89

alkalis substances which produce OH– ions in water 36, 37, 88

alkanes a family of hydrocarbons found in crude oil with single covalent bonds, e.g. methane 22, 25

alkenes a family of hydrocarbons with one double covalent bond between carbon atoms 25

allele inherited characteristics are carried as pairs of alleles on pairs of chromosomes. Different forms of a gene are different alleles 11, 64

allotropes different forms of the same element 83

alloy mixture of two or more metals – used to make coins, for example 33, 34

alpha particles radioactive particles which are helium nuclei – helium atoms without the electrons (they have a positive charge) 107, 108, 109

alternating current or voltage an electric current that is not a one-way flow 50

amalgam an alloy which contains mercury 33

amino acids small molecules from which proteins are built 5, 58, 59, 72

amplitude the amplitude of a wave is the maximum displacement of a wave from its rest position 42, 106

anaerobic respiration respiration without using oxygen 60

analogue signal a signal that shows a complete range of frequencies; sound is analogue 45

anode electrode with a positive charge 33, 38

antibiotic therapeutic drug acting to kill bacteria which is taken into the body 6, 18

antibody protein normally present in the body or produced in response to an antigen which it neutralises, thus producing an immune response 6, 62

antigen any substance that stimulates the production of antibodies – antigens on the surface of red blood cells determine blood group 6

antiviral drug therapeutic drug acting to kill viruses 6

arteries blood vessels that carry blood away from the heart 4, 62

asexual reproduction reproduction involving only one parent 13

asteroid composed of rock or metallic material orbiting the Sun in a region between Mars and Jupiter 55

atmosphere mixture of gases above the lithosphere, mainly nitrogen and oxygen 46, 49, 51, 53, 54

atom the basic 'building block' of an element which cannot be chemically broken down 25, 26, 33, 53, 79, 81, 83, 85, 86, 87, 88, 89, 91, 103, 107, 109, 110

atom economy a way of measuring the amount of atoms that are wasted or lost when a chemical is made 80

atomic number the number of protons found in the nucleus of an atom 85, 107

ATP molecule used to store energy in the body 60

attract move towards, for example, unlike charges attract 56, 83, 86, 91, 103, 104

auxin a type of plant hormone 10

average speed total distance travelled divided by the total time taken for a journey 94

axon part of neurone that carries nerve impulse 7

B

background radiation ionising radiation from space and rocks, especially granite, that is around us all the time but is at a very low level 53, 108

bacteria single-celled micro-organisms which can either be free-living organisms or parasites (they sometimes invade the body and cause disease) 4, 6, 13, 15, 18, 37, 53, 64, 73, 92

balanced symbol equation a symbolic representation showing the kind and amount of the starting materials and products of a reaction 23, 27, 35, 68, 79, 80, 88, 89, 90, 92

barium chloride testing chemical for sulfates in water 92

basalt a rock forms when iron rich magma cools 31

base a substance that will react with acids 36

batch process a process used to make small fixed amounts of substances, like medicines, with a clear start and finish 82

battery two or more electrical cells joined together 98

beta particles particles given off by some radioactive materials (they have a negative charge) 107, 109

Big Bang the event believed by many scientists to have been the start of the universe 56

binocular vision the ability to maintain visual focus on an object with both eyes, creating a single visual image 7

binomial system the scientific way of naming an organism 13

biodegradable a biodegradable material can be broken down by micro-organisms 26

biodiversity range of different living organisms in a habitat 67

biofuels fuels made from plants – these can be burned in power stations 98

biological catalyst molecules in the body that speed up chemical reactions 59

biological control a natural predator is released to reduce the number of pests infesting a crop 74

biomass waste wood and other natural materials which are burned in power stations 14

bitumen thick tar-like substance that does not boil in a fractionating column 22

black hole a region of space from which nothing, not even light, can escape 54, 56

blind trial a drugs trial where volunteers do not know which treatment they are receiving 6

blood pressure force with which blood presses against the walls of the blood vessels 4

blood sugar level amount of glucose in the blood 9, 59

body mass index (BMI) measure of someone's weight in relation to their height 5

boiling point temperature at which the bulk of a liquid changes into a gas 22, 82, 83, 88, 89, 91

boron control rods rods that are raised or lowered in a nuclear reactor to control the rate of fission 110

braking distance distance travelled while a car is braking 96, 98, 99

brass an alloy which contains copper and zinc 33

bromine an orange, corrosive halogen, used to test alkenes 25, 89

buckminsterfullerene a very stable sphere of 60 carbon atoms joined by covalent bonds. An allotrope of carbon 83

Glossary

C

cancer life-threatening condition where body cells divide uncontrollably 6, 109

capillaries small blood vessels that join arteries to veins 62

captive breeding breeding a species in zoos to maintain the wild population 20

carbon a very important element, carbon is present in all living things and forms a huge range of compounds with other elements 15, 22, 83, 87

carbon cycle a natural cycle through which carbon moves by respiration, photosynthesis and combustion in the form of carbon dioxide 24

carbon dioxide (CO_2) gas present in the atmosphere at a low percentage but important in respiration, photosynthesis and combustion; a greenhouse gas which is emitted into the atmosphere as a by-product of combustion 9, 15, 19, 23, 24, 27, 51, 60, 62, 68, 69, 70, 78, 87, 90, 97, 98

carbon footprint the total amount of greenhouse gases given off by a person in a given time 19

carbon monoxide poisonous gas made when fuels burn in a shortage of oxygen 4, 23, 24

carbon-14 radioactive isotope of carbon 85, 108

carotene plant pigment involved in photosynthesis 69

catalyst a chemical that speeds up a reaction but is not changed or used up by the reaction 24, 25, 35, 78, 83, 90

catalytic converters boxes fitted to vehicle exhausts which reduce the level of nitrogen oxides and unburnt hydrocarbons in the exhaust fumes 24

cathode electrode in a battery with a negative charge 33, 38

cell differentiation when cells become specialised 61

cell membrane layer around a cell which helps to control substances entering and leaving the cell 70, 72

cement the substance made when limestone and clay are heated together 32, 51

central nervous system (CNS) collectively the brain and spinal cord 7

centripetal force force acting on a body, travelling in a circle, which acts toward the centre of the circle and keeps the body moving in a circle 54

CFCs gases which used to be used in refrigerators and which harm the ozone layer 19, 47

chain reaction a reaction where the products cause the reaction to go further or faster, e.g. in nuclear fission 110

charge(s) a property of matter charge exists in two forms, positive and negative, which attract each other 85, 86, 88, 103, 104

chemical properties the characteristic reactions of substances 87, 91

chlorination addition of chlorine to water supplies to kill micro-organisms 92

chlorophyll pigment found in plants which is used in photosynthesis (gives green plants their colour) 68, 69, 72

cholesterol fatty substance which can block blood vessels 4

chromatography a method for splitting up a substance to identify compounds and check for purity 82

chromosomes thread-like structures in the cell nucleus that carry genetic information 11, 58, 61, 63

circuit-breakers resettable fuses 105

clone genetically identical copy 65

close-packed metal ions structure of a metal 91

cold fusion attempts to produce fusion at normal room temperature that have not been validated since other scientists could not reproduce the results 110

collagen protein used for support in animal cells 59

collision frequency the number of successful collisions between reacting particles that happen in one second 77, 78

colloid a liquid with small particles dispersed throughout it, forming neither solution nor sediment 29

combustion process where fuels react with oxygen to produce heat 15, 23, 24

comets lumps of rock and ice found in space – some orbit the Sun 54, 55

community all the plants and animals living in an ecosystem, e.g. a garden 67

complete combustion when fuels burn in excess of oxygen to produce carbon dioxide and water only 23

composite material a material which consists of identifiably different substances 32

compost dead and decaying plant material 74

compound two or more elements which are chemically joined together, e.g. H_2O 25, 79, 82, 83, 88

compressions particles pushed together, increasing pressure 106

computer modelling using a computer to 'model' situations to see how they are likely to work out if you do different things 99

concentration the amount of chemical dissolved in a certain volume of solution 73, 77

concrete a form of artificial stone 32

conservation a way of protecting a species or environment 20

conservation of mass the total mass of reactants equals the total mass of products formed 78, 79

core the centre part of the a planet, made of iron 31, 47, 55, 56

corrode to lose strength due to chemical attack 34

cosmic rays radiation from space that contributes to background radiation 108

count rate average number of nuclei that decay every second 107

covalent bond bond between atoms where an electron pair is shared 25, 26, 83, 87

cracking the process of making small hydrocarbon molecules from larger hydrocarbon molecules using a catalyst 22

crenation when red blood cells shrink in concentrated solutions, they look partly deflated 70

critical angle angle at which a light ray incident on the inner surface of a transparent glass block just escapes from the glass 43

crop rotation system of growing crops in sequence 74

cross-sectional area the area displaced by a moving object 100

crude oil black material mined from the Earth from which petrol and many other products are made 22, 26, 34

cruise control system that automatically controls the speed of a vehicle 99

crumple zones areas of a car that absorb the energy of a crash to protect the centre part of the vehicle 99

crust surface layer of the Earth made of tectonic plates 31, 32, 55

current flow of electrons in an electric circuit 49, 50, 52, 53, 105, 108

curved line line of changing gradient 76, 94

D

decay to rot 15, 51, 73

deceleration a measurement of how quickly the speed of a moving object decreases 98

decolourise turn from a coloured solution to a colourless solution 25

decomposer organisms that break down dead animals and plants 14, 15

decomposed chemically broken down 27

defibrillator machine which gives the heart an electric shock to start it beating regularly 104

deforestation removal of large area of trees 24, 51, 98

degassing gases coming out of volcanoes 24

dehydration result of body losing too much water 9

delocalised electrons electrons which are free to move away through a collection of ions – as in a metal 91

denitrifying bacteria bacteria that convert nitrates into nitrogen gas 15

density the density of a substance is found by dividing its mass by its volume 34, 41, 55, 56, 91, 109

depressant a drug that slows down the working of the brain 8

Glossary

detritivore an organism that eats dead material, e.g. an earthworm 73

detritus the dead and semi-decayed remains of living things 73

di-bromo compound colourless compound resulting from an alkene and bromine solution 25

diastolic pressure the lowest point that your blood pressure reaches as the heart relaxes between beats 4

diet what a person eats 4, 5, 6, 9

diffraction a change in the directions and intensities of a group of waves after passing by an obstacle or through an opening whose size is approximately the same as the wavelength of the waves 42, 44, 46

diffuse when particles diffuse they spread out 7, 69, 70, 71, 72

digital signal a signal that has only two values – on and off 43, 45, 46

diploid cells that have two copies of each chromosome 61

displacement reaction chemical reaction where one element displaces or 'pushes out' another element from a compound 89

displayed formula when the formula of a chemical is written showing all the atoms and all the bonds 25

disposal getting rid of unwanted substances such as plastics 20, 26

distance–time graph a plot of the distance moved against the time taken for a journey 94

distillation the process of evaporation followed by condensation 92

DNA molecule found in all body cells in the nucleus – its sequence determines how our bodies are made (e.g. do we have straight or curly hair), and gives each one of us a unique genetic code 13, 53, 58, 59, 61, 63, 72

DNA bases four chemicals that are found in DNA, they make up the base sequence and are given the letters A, T, G and C 58, 61

dominant allele/characteristic an allele that will produce the characteristic if present 11

dot and cross model a drawn model representing the number of electrons in the outside shell of bonding atoms or ions 86, 87

double circulatory system where the blood is pumped to the lungs then returned to the heart before being pumped round the body 62

double covalent bond covalent bond where each atom shares two electrons with the other atom 25, 87

drag energy losses caused by the continual pushing of an object against the air or a liquid 100

dummies used in crash testing to learn what would happen to the occupants of a car in a crash 99

dynamo a device that converts energy in movement into energy in electricity 50

E

EAR for protein estimated average daily requirement of protein in diet 5

earth wire the third wire in a mains cable which connects the case of an appliance to the ground so that the case cannot become charged and cause an electric shock 103, 105

earthed (electrically) connected to the ground (at 0V) 104

echoes reflection of sound (or ultrasound) 106

ecological niche the role of an organism within an ecosystem 16

efficiency ratio of useful energy output to total energy input; can be expressed as a percentage 4, 14, 41, 50, 74

egestion the way an animal gets rid of undigested food waste called faeces 14

electrical conductivity the measurement of the ability to conduct electricity 88

electrical conductors materials which let electricity pass through them 83

electrolysis when an electric current is passed through a solution which conducts electricity 33, 38

electrolyte the liquid in which electrolysis takes place 33

electromagnet a magnet which is magnetic only when a current is switched on 91

electromagnetic spectrum electromagnetic waves ordered according to wavelength and frequency – ranging from radio waves to gamma rays 42, 44

electron shells the orbit around the nucleus likely to contain the electron 85

electronic structure the number of electrons in sequence that occupy the shells, e.g. the 11 electrons of sodium are in sequence 2.8.1 85, 88, 89

electrons small particles within an atom that orbit the nucleus (they have a negative charge) 83, 85, 86, 87, 88, 89, 91, 103, 104, 107, 109

electrostatic attraction attraction between opposite charges, e.g. between Na^+ and Cl^- 86

elements substances made out of only one type of atom 79, 83, 85, 87, 89

elliptical orbit a path that follows an ellipse – which looks a bit like a flattened circle 54

endangered a species where the numbers are so low they could soon become extinct 20

endoscope device using optical fibres which allows doctors to look inside the human body 43

endothermic reaction chemical reaction in which heat is taken in 81

energy the ability to 'do work', for example the human body needs energy to function 20, 29, 40, 41, 52, 81, 98, 99, 101

enriched uranium uranium containing more of the U-235 isotope than occurs naturally 110

enzymes biological catalysts that increase the speed of a chemical reaction 9, 17, 58, 59, 60, 61, 68, 72, 73

escape lane rough-surfaced uphill path adjacent to a steep downhill road enabling vehicles with braking problems to stop safely 99

essential elements the three elements, nitrogen, phosphorus and potassium that are essential for the growth of plants 37

eutrophication when waterways become too rich with nutrients (from fertilisers) which allows algae to grow wildly and use up all the oxygen 37

evaporation when a liquid changes to a gas, it evaporates 9, 28, 71

evolution the gradual change in organisms over millions of years caused by random mutations and natural selection 13, 18

excretion the process of getting rid of waste from the body 14

exhaust gases gases discharged into the atmosphere as a result of combustion of fuels 97

exothermic reaction chemical reaction in which heat is given out 81

explosion a very fast reaction making large volumes of gas 31, 56, 78, 103

exponential growth the ever-increasing growth of the human population 19

extinct when all members of a species have died out 20

extrapolation making an estimate beyond the range of results 76, 77

F

fertilisation when a sperm fuses (joins with) an egg 11

fertiliser chemical put on soil to increase soil fertility and allow better growth of crop plants 19, 35, 37, 74, 80, 90, 92

filtration the process of filtering river or ground water to purify it for drinking water 92

finite resource resources such as oil that will eventually run out 22, 34

first-class protein proteins from meat and fish which contain all essential amino acids 5

fission splitting apart, especially of large radioactive nuclei such as uranium 110

flaccid floppy 70

flame test test where a chemical burns in a Bunsen flame with a characteristic colour – tests for metal ions 88

force a push or pull which is able to change the velocity or shape of a body 83, 91, 95, 96, 97, 99, 100

fossil fuels fuels such as coal, oil and gas 15, 19, 20, 22, 23, 51

free-fall a body falling through the atmosphere without an open parachute 100

Glossary

frequency the number of waves passing a set point per second 42, 45, 46, 50, 106

friction energy losses caused by two or more objects rubbing against each other 96, 98, 101

fuel consumption the distance travelled by a given amount of fuel, e.g. in km/100 litres 97, 98

fuel rods rods of enriched uranium produced to provide fuel for nuclear power stations 110

fullerenes cage-like carbon molecules containing many carbon atoms, e.g. buckyballs 83

fungicide chemical used to kill fungi 74

fuse(s) a special component in an electric circuit containing a thin wire which is designed to melt if too much current flows through it, breaking the circuit 105

fusion the joining together of small nuclei, such as hydrogen isotopes, at very high temperatures with the release of energy 56, 110

fusion bombs hydrogen bombs or H-bombs based on fusion reactions 110

G

gametes the male and female sex cells (sperm and eggs) 11, 61, 64

gamma rays ionising electromagnetic waves that are radioactive and dangerous to human health – but useful in killing cancer cells 107, 108, 109

gene section of DNA that codes for a particular characteristic 11, 18, 58, 59, 61, 64

gene pool the different genes available within a species 64

gene therapy medical procedure where a virus is used to 'carry' a gene into the nucleus of a cell (this is a new treatment for genetic diseases) 64

generator device that converts rotational kinetic energy to electrical energy 50, 64

genetic engineering transfer of genes from one organism to another 64

genotype the genetic makeup of an organism 11

geotropism a plant's growth response to gravity 10

giant ionic lattice sodium chloride forms a lattice, also called a giant ionic structure 86

global warming the increase in the Earth's temperature due to increases in carbon dioxide levels 19, 49, 51, 92

gradient rate of change of two quantities on a graph; change in y/change in x 76, 94

granite an igneous rock containing low levels of uranium 32, 108

graphite a type of carbon used as a moderator in a nuclear power station 83, 110

gravitational field strength the force of attraction between two masses 97, 100, 101

gravitational potential energy the energy a body has because of its position in a gravitational field, e.g. an object 101

gravity an attractive force between objects (dependent on their mass) 10, 56, 97, 100

greenhouse gas any of the gases whose absorption of infrared radiation from the Earth's surface is responsible for the greenhouse effect, e.g. carbon dioxide, methane, water vapour 19, 51, 53, 97

group 1 metals metals in the group 1 of the periodic table, e.g. lithium, sodium and potassium 88

group 7 elements non-metals in group 7 of the periodic table, e.g. fluorine, bromine and iodine 89

H

Haber process industrial process for making ammonia 35, 90

habitat where an organism lives, e.g. the worm's habitat is the soil 13, 16, 17, 20, 38, 67, 74

haemoglobin chemical found in red blood cells which carries oxygen 4, 59, 62

half-life average time taken for half the nuclei in a radioactive sample to decay 107, 108

hallucinogen a drug, like LSD, that gives the user hallucinations 8

halogens reactive non-metals in group 7 of the periodic table , e.g. chlorine 89

haploid cells that have only one copy of each chromosome 61

heat stroke result of body being too hot; skin is cold, pulse is weak 9

helium second element in periodic table; an alpha particle is a helium nucleus 56, 107

herbicide chemical used to kill weeds 64, 74

hertz (Hz) units for measuring wave frequency 42, 50, 106

heterozygous a person who has two different alleles for an inherited characteristic, e.g. someone with blond hair may also carry an allele for red hair 11

homozygous a person who has two alleles that are the same for an inherited feature, e.g. a blue-eyed person will have two blue alleles for eye colour 11

hormones chemicals that act on target organs in the body (hormones are made by the body in special glands) 9, 10, 59, 62

hybrid the infertile offspring produced when two animals of different species breed 13

hydrated iron(III) oxide the chemical name for rust 34

hydroponics growing plants in mineral solutions without the need for soil 74

hypothalamus small gland in brain, detects temperature of blood 9

hypothermia a condition caused by the body getting too cold, which can lead to death if untreated 9

I

igneous rock rock which has formed when liquid rock has solidified 31, 32

inbreeding breeding closely related animals 64

indicator species organisms used to measure the level of pollution in water or the air 19

infrared waves non-ionising waves that produce heat – used in toasters and electric fires 44, 45, 51

insecticide a chemical that can kill an insect 6, 74

instantaneous speed the speed of a moving object at one particular moment 94

insulation a substance that reduces the movement of energy; heat insulation in the loft of a house slows down the movement of warmth to the cooler outside 17, 41

insulin hormone made by the pancreas which controls the level of glucose in the blood 9, 59, 64

intensive farming farming that uses a lot of artificial fertilisers and energy to produce a high yield per farm worker 74

interference waves interfere with each other when two waves of different frequencies occupy the same space; interference occurs in light and sound and can produce changes in intensity of the waves 44, 45, 46

intermolecular force force between molecules 22, 26, 87

interpolation making an estimate within the range of results 76

iodine radioactive isotopes of iodine are used in diagnosing and treating thyroid cancer 109

ionic bond a chemical bond between two ions of opposite charges 86

ionic equation an equation representing the formation of ions by the transfer of electrons 88, 89

ionisation the formation of ions (charged particles) 53, 107, 108, 109

ionises adds or removes electrons from an atom, leaving it charged 53

ionosphere a region of the Earth's atmosphere where ionisation caused by incoming solar radiation affects the transmission of radio waves; it extends from 70 kilometres (43 miles) to 400 kilometres (250 miles) above the surface 46

ions charged particles (can be positive or negative) 33, 36, 38, 53, 68, 86, 88, 89, 90

isotopes atoms with the same number of protons but different numbers of neutrons 68, 85, 107, 108, 110

Glossary

J

joule unit of work done and energy 40, 101

K

kilowatt 1000 watts 52

kilowatt-hour unit of electrical energy equal to 3 600 000 J 52

kinetic energy the energy that moving objects have 40, 41, 44, 77, 98, 99, 101

kite diagram method of displaying results from a transect line 67

kwashiorkor an illness caused by protein deficiency due to lack of food. Sufferers often have swollen bellies caused by retention of fluid in the abdomen 5

L

laser a special kind of light beam that can carry a lot of energy and can be focused very accurately; lasers are often used to judge the speed of moving objects or the distance to them 43

latent heat the energy needed to change the state of a substance 40

lead heaviest element having a stable isotope; all isotopes of the elements above it in the periodic table are unstable 92, 108

legume a plant that has bacteria inside root nodules that provide the plant with nitrates 16

light-emitting diode (LED) a very small light in electric circuits that uses very little energy 45

light-year a unit of distance equal to the distance light travels through space in one year 54

limestone a sedimentary rock, made of calcium carbonate 15, 24, 32

limiting reactant chemical used up in a reaction that limits the amount of product formed 76

linear a line of constant gradient on a graph 96

lithosphere the cold rigid outer part of the Earth which includes the crust and the upper part of the mantle 31

live wire carries a high voltage into and around the house 105

longitudinal (wave) wave in which the vibrations are in the same direction as the direction in which the wave travels 47, 106

lysis to split apart 70

M

magma molten rock found below the Earth's surface 31, 32

malleable bendable 34, 91

mammal animals that have fur and produce milk for their young 11, 13, 61

mantle semi-liquid layer of the Earth beneath the crust 31

marble a metamorphic rock, made of calcium carbonate 32

mass describes the amount of something; it is measured in kilograms (kg) 76, 78, 79, 81

mass number number of protons and neutrons in a nucleus 85, 107

meiosis cell division that results in haploid cells 61

melanin the group of naturally occurring dark pigments, especially the pigment found in skin, hair, fur, and feathers 47

melting point temperature at which a solid changes into a liquid 26, 33, 82, 83, 86, 87, 88, 91

meristem tips of roots and shoots where cell division and elongation takes place 63

messenger RNA copy of a section of DNA used to carry the gene code to the ribosomes 58

metabolic rate the speed at which the amount of energy the body needs is released 60

metal halide a compound of a halogen and a metal, e.g. potassium bromide 89

metallic bonding the bonding between close-packed metal ions due to delocalised electrons 90, 91

metallic properties the physical properties specific to a metal, such as lustre and electrical conductivity 87, 90, 91

metals solid substances that are usually lustrous, conduct electricity and form ions by losing electrons 86, 88, 90, 91

metamorphic rock rock which has been changed after it has formed 32

meteors bright flashes in the sky caused by rocks burning in Earth's atmosphere 54

microbes tiny microscopic organisms 53, 65, 92

microorganism very small organism (living thing) which can only be viewed through a microscope – also known as a microbe 15

microwaves non-ionising waves used in satellite and mobile phone networks – also in microwave ovens 42, 43, 44, 46

minerals natural solid materials with a fixed chemical composition and structure, rocks are made of collections of minerals; mineral nutrients in our diet are things like calcium and iron, they are simple chemicals needed for health 68, 71, 72, 74

mitochondria structures in a cell where respiration takes place 58, 61, 63

mitosis cell division that results in genetically identical diploid cells 61

moderator material used to slow down neutrons in a nuclear power station 110

molecular formula the formula of a chemical using symbols in the periodic table, e.g. methane has a molecular formula of CH_4 23

molecule two or more atoms which have been chemically combined 58, 59, 60, 61, 68, 70, 83, 87, 89

molten liquid a solid that has just melted, usually referring to rock, ores, metals or salts with very high melting points 86

momentum the product of mass and velocity 99

monohybrid cross a cross between two organisms that differ by a single characteristic. Used to follow the inheritance of a single pair of genes 11

Morse code a code consisting of dots and dashes that code for each letter of the alphabet 43

motor neurone nerve cell carrying information from the central nervous system to muscles 7

multicellular organism organisms made up of many specialised cells 61

multiplexing combination of multiple signals into one signal transmitted over a shared medium 45

mutation where the DNA within cells have been altered (this happens in cancer) 11, 59

mutualism a relationship in which both organisms benefit 16

N

nanometre units used to measure very small things (one billionth of a metre) 83

nanotube carbon atoms formed into a very tiny tube 83

National Grid network that carries electricity from power stations across the country (it uses cables, transformers and pylons) 52

natural selection process by which 'good' characteristics that can be passed on in genes become more common in a population over many generations ('good' characteristics mean that the organism has an advantage which makes it more likely to survive) 18

near-Earth object asteroid, comet or large meteoroid whose orbit crosses Earth's orbit 55

negative ion an ion made by an atom gaining electrons 86, 89

neutral a neutral substance has a pH of 7 37, 85

neutral wire provides a return path for the current in a mains supply to a local electricity substation 105

neutralisation reaction between H^+ ions and OH^- ions (acid and base react to make a salt and water) 36

neutrons small particle which does not have a charge found in the nucleus of an atom 85, 107, 109, 110

newtons unit of force (abbreviated to N) 96

nitinol a smart alloy which contains nickel and titanium 33

nitrogen-fixing bacteria bacteria that convert nitrogen into ammonia or nitrates 15, 16

non-metals substances that are dull solids, liquids or gases that do not conduct electricity and form ions by gaining electrons 86, 87

non-renewable energy energy which is used up at a faster rate than it can be replaced e.g. fossil fuels 22

Glossary

nuclear equation equation showing changes to the nuclei in a nuclear reaction 107

nuclear power stations power stations using the energy produced by nuclear fission to generate heat 110

nuclear transfer type of cloning that involves taking a nucleus from a body cell and placing it into an egg cell 65

nucleons protons and neutrons (both found in the nucleus) 107

nucleus central part of an atom that contains protons and neutrons 53, 58, 62, 63, 65, 85, 86, 88, 89, 103, 107, 109, 110

o

obesity a medical condition where the amount of body fat is so great that it harms health 5

ohms units used to measure resistance to the flow of electricity 105

optical fibre a flexible optically transparent fibre, usually made of glass or plastic, through which light passes by successive internal reflections 43, 45

optimum conditions the conditions under which a reaction works most effectively 35

optimum temperature the temperature range that produces the best reaction rate 9, 35

oscilloscope a device that displays a line on a screen showing regular changes (oscillations) in something. an oscilloscope is often used to look at sound waves collected by a microphone 50

osmosis a type of diffusion which depends on the presence of a partially-permeable membrane that allows the passage of water molecules but large molecules such as glucose 68, 70, 71, 72

oxidation a chemical reaction in which a substance gains oxygen and/or loses electrons 33, 34, 38, 88

oxygen debt the debt for oxygen that builds up in the body when demand for oxygen is greater than supply 60

ozone layer layer of the Earth's atmosphere that protects us from ultraviolet rays 19, 47

P

paddle shift controls controls attached to the steering wheel of a car so that the driver can use them without taking their eyes off the road 99

paddles charged plates in a defibrillator that are placed on the patient's chest 104

painkiller a drug that stops nerve impulses so pain is not felt 8

palisade cells tightly packed together cells found on the upper side of a leaf 69

parasite organism which lives on (or inside) the body of another organism 6, 14, 16

partially-permeable membrane a membrane that allows some small molecules to pass through but not larger molecules 70

pathogen harmful organism which invades the body and causes disease 6

payback time the time it takes for the original cost outlay to be recovered in savings 41

percentage yield the percentage of actual product made in a chemical reaction compared to the amount which ideally could be made 35, 80

performance enhancer a drug used to improve performance in a sporting event 8

period a row in the periodic table 85, 87

periodic table a table of all the chemical elements based on their atomic number 79, 85

pesticide residue unwanted residues sometimes found in water contaminated by local pesticide use 73, 92

petrol volatile mixture of hydrocarbons used as a fuel 22, 23, 98

pH scale scale in which acids have a pH below 7, alkalis a pH of above 7 and a neutral substance a pH of 7 15, 36, 59, 60

pharmaceuticals medical drugs 82

phase fraction of a complete wave that one wave disturbance is different to another 43

phenotype the characteristic that is shown/expressed 11

phloem specialised transporting cells which form tubules in plants to carry sugars from leaves to other parts of the plant 71

photocell a device which converts light into electricity 49

photon a photon is a unit or particle of electromagnetic energy; photons travel at the speed of light but have no mass 49

photosynthesis process carried out by green plants where sunlight, carbon dioxide and water are used to produce glucose and oxygen 15, 16, 24, 59, 67, 68, 69, 71, 72, 73

phototropism a plant's growth response to light 10

physical property property that can be measured without changing the chemical composition of a substance, e.g. hardness 86, 91

pitch whether a sound is high or low on a musical scale 106

placebo a dummy pill 6

plant hormones hormones that control various plant processes such as growth and germination 10

plaque build up of cholesterol in a blood vessel (which may block it) 4

plasma yellow liquid found in blood 62

plasmolysis the shrinking of a plant cell due to loss of water, the cell membrane pulls away 70

plutonium a radioactive metal often formed as a bi-product from a nuclear power station – sometimes used as a nuclear fuel 53

p–n junction the boundary between two special types of silicon in a photocell and other electronic components 49

pollination the process of transferring pollen from one plant to another 16

pollutants unwanted residues found that can sometimes cause damage 19, 24, 92

pollute contaminate or destroy the environment 37, 49, 97, 98, 104

pollution contaminating or destroying the environment as a result of human activities 19, 23, 24, 33, 34, 49, 98

polymer a number of short-chained molecules joined together to form a long-chained molecule 25

population group of organisms of the same species in a habitat 19, 20, 23, 24, 37, 64, 67, 74

positive ion an ion made by an atom losing electrons 86, 88

potential difference another word for voltage (a measure of the energy carried by the electric current) 105

power the rate that a system transfers energy, power is usually measured in watts (W); electric power = voltage × current 97, 105

power station facility that generates electricity on a large scale 50, 98, 110

power transmission transmission of electricity 91

precipitate solid formed in a solution during a chemical reaction 90, 92

precipitation reaction chemical test in which a solid precipitate is formed – tests for metal ions 90, 92

predator animal which preys on (and eats) another animal 16, 74

pressure wave vibrating particles in a longitudinal wave creating pressure variations 47, 106

prey animals which are eaten by a predator 16

primary safety features help to prevent a crash, e.g. ABS brakes, traction control 99

probe unmanned space vehicle designed to travel beyond Earth's orbit 54

producers organisms in a food chain that make food using sunlight 14

product molecules produced at the end of a chemical reaction 76, 77, 79, 80, 82, 90

protons small positive particles found in the nucleus of an atom 53, 85, 86, 107

P wave longitudinal seismic wave capable of travelling through solid and liquid parts of the Earth 47

R

radiation thermal energy transfer which occurs when something hotter than its surroundings radiates heat from its surface 41, 49, 51, 53, 59, 107, 108, 109

Glossary

radio waves non-ionising waves used to broadcast radio and TV programmes 46

radioactive material which gives out radiation 107, 108, 109

radioactive waste waste produced by radioactive materials used at nuclear power stations, research centres and some hospitals 53, 108

radiotherapy using ionising radiation to kill cancer cells in the body 109

random having no regular pattern 11, 70, 107

rarefactions particles further apart than usual, decreasing pressure 106

rate of reaction the speed with which a chemical reaction takes place 59, 76, 77, 78

reactants chemicals which are reacting together in a chemical reaction 76, 77, 78, 79, 80, 81, 89

reaction time the time it takes for a driver to step on the brake after seeing an obstacle 8, 96

receiver device which receives waves, e.g. a mobile phone 42, 44

recessive allele/characteristic two recessive alleles needed to produce the characteristic 11

recharging battery being charged with a flow of electric current 98

recycle to reuse materials 15, 33, 35

red blood cells blood cells which are adapted to carry oxygen 6, 62

red shift when lines in a spectrum are redder than expected – if an object has a red shift it is moving away from us 56

reduction a chemical reaction in which a substance loses oxygen and/or gains electrons 33, 38, 39

reflected radiation rebounding off a surface 106

reflex a muscular action that we take without thinking about 7

refraction when a light ray travelling through air enters a glass block and changes direction 7, 42, 43, 46, 47

reinforced concrete concrete with steel rods or mesh running through it 32

relative atomic mass the mass of an atom compared to $\frac{1}{12}$ of a carbon atom 79

relative velocity vector difference between the velocities of two objects 95

renewable energy energy that can be replenished at the same rate that it's used up e.g. biofuels 49

repel move away, for example, like charges repel 91, 103, 104

reptile cold blooded vertebrate having an external covering of scales or horny plates 13

reservoir a water resource where large volumes of water are held 92

resistance measurement of how hard it is for an electric current to flow through a material 91, 105

respiration process occurring in living things where oxygen is used to release the energy in foods 14, 15, 51, 58, 59, 60, 67, 68, 72, 73

respiratory quotient (RQ) the result of dividing carbon dioxide produced by oxygen used, can be used to determine the types of molecule used in respiration 60

resultant force the combined effect of forces acting on an object 96

rheostat a variable resistor 105

rhyolite a rock which forms when silica rich magma cools 31

ribosome structures in a cell where protein synthesis takes place 58

rust the substance made when iron corrodes, hydrated iron(III) oxide 34

s

salt the substance formed when any acid reacts with a base 34, 36, 37, 38

saprophyte an organism that breaks down dead organic matter, usually used to refer to fungi 73

satellite a body orbiting around a larger body; communications satellites orbit the Earth to relay television and telephone signals 44, 46, 47

sea water water containing high levels of dissolved salts making it undrinkable 92

seat belts harness worn by occupants of motor vehicles to prevent them from being thrown about in a collision 99

second-class protein proteins from plants which only contain some essential amino acids 5

sedimentary rock rock which has formed when fragments of older rock or living things have stuck together, or by precipitation 32

sedimentation a process during water purification where small solid particles are allowed to settle 92

seismic wave vibration transmitted through the Earth 31

selective breeding process of breeding organisms with the desired characteristics 64

sensor device that detects a change in the environment 99

sensory neurone nerve cell carrying information from receptors to central nervous system 7

sex chromosomes a pair of chromosomes that determine gender, XX in female, XY in male 11

shock occurs when a person comes into contact with an electrical energy source so that electrical energy flows through a portion of the body 103, 104, 105

silver nitrate a chemical used for testing halide ions in water 92

single covalent bond bond between atoms where each the atoms share an electron pair 25, 87

smart alloy an alloy which will return to a previous shape 33

smoke detector device to detect smoke, some forms of which contain a source of alpha radiation 108

Solar System the collection of planets and other objects orbiting around the Sun 54, 55

solar power energy provided by the Sun 98

solder an alloy which contains lead and tin 33

soluble a soluble substance can dissolve in a liquid, e.g. sugar is soluble in water 36, 73, 92

solution when a solute dissolves in a solvent, a solution forms 86, 89, 90, 92

sparks type of electrostatic discharge briefly producing light and sound 78, 103

species basic category of biological classification, composed of individuals that resemble one another, can breed among themselves, but cannot breed with members of another species 11, 13, 16, 18, 20, 64, 67, 74

specific heat capacity the amount of energy needed to raise the temperature of 1 kg of a substance by 1 degree Celsius 40, 81

specific latent heat the amount of energy needed to change the state of a substance without changing its temperature; for example the energy needed to change ice at 0 °C to water at the same temperature 40

speed how fast an object travels: speed = distance ÷ time 4, 42, 43, 49, 55, 59, 72, 82, 90, 94, 95, 96, 97, 98, 99, 100, 101, 106, 109, 111

speed–time graph a plot of how the speed of an object varies with time 95

spongy mesophyll cells found in the middle of a leaf with an irregular shape and large air spaces between them 69, 70, 71

stable electronic structure a structure where the outer electron shell of an atom is full 86, 88, 89, 108

star bright object in the sky which is lit by energy from nuclear reactions 54, 56

stem cells unspecialised body cells (found in bone marrow) that can develop into other, specialised, cells that the body needs, e.g. blood cells 63, 65

stimulant a drug that speeds up the working of the brain 8

stomata (_singular_ **stoma)** small holes in the surface of leaves which allow gases in and out of leaves 69, 70, 71

stopping distance sum of the thinking and braking distances 96

straight line line of constant gradient 76, 94, 95

stratosphere a layer in the atmosphere starting at 15 km above sea level and extending to 50 km above sea level; the ozone layer is found in the stratosphere 47

Glossary

stroke sudden change in blood flow to the brain – can be fatal 4

subduction where one plate sinks below another 31

subsidence settling of the ground caused, for example, by mining 38

substrate the substance upon which an enzyme acts 59

superconductors materials that conduct electricity with little or no resistance 91

sustainable development managing a resource so that it does not run out 20

S wave transverse seismic wave capable of travelling through solid but not liquid parts of the Earth 47

synapse gap between two neurons 7, 8

systolic pressure the highest point that your blood pressure reaches as the heart beats to pump blood through your body 4

T

tectonic plate a large section of the lithosphere which can move across the surface of the Earth 31

temperature a measure of the degree of hotness of a body on an arbitrary scale 9, 17, 22, 24, 29, 31, 33, 35, 40, 41, 51, 59, 60, 68, 71, 73, 77, 81, 83, 84, 89, 91, 110

temperature coefficient (Q_{10}) equation used to calculate the effect of temperature on the rate of an enzyme-controlled reaction 59

terminal speed or velocity the top speed reached when drag matches the driving force 100, 101

therapy treatment of a medical problem 109

thermal decomposition the breaking down of a compound into two or more products on heating 32, 90

thermogram a picture showing differences in surface temperature of a body 40

thinking distance distance travelled while the driver reacts before braking 96

thrombosis blood clot in a blood vessel causing it to be blocked 4

thyroid gland gland at the base of the neck which makes the hormone thyroxin 109

tissue culture process that uses small sections of tissue to clone plants 65

total internal reflection the reflection of light inside an optically denser material at its boundary with an optically less dense material (usually air) 43, 46

toxic a toxic substance is one which is poisonous, e.g. toxic waste 8, 23, 26, 34

toxin a poisonous substance 6

tracer a radioactive, or radiation-emitting, substance used in a nuclear medicine scan or other research where movement of a particular chemical is to be followed 108, 109

transect line across an area to sample organisms 67

transformer device by which alternating current of one voltage is changed to another voltage 52

transition element an element in the middle section of the periodic table, between the group 1 and 2 block and the group 3 to group 0 block 90

transmitter a device which gives out some form of energy or signal, usually used to mean a radio transmitter which broadcasts radio signals 44, 46

transverse (wave) wave in which the vibrations are at right angles to the direction in which the wave travels 42, 47, 106

tread pattern on part of tyre that comes in contact with road surface to provide traction 96

trials tests to find if something works and is safe 6, 82

trophic level the stages in a food chain 14

tsunami huge waves caused by earthquakes – can be very destructive 55

tumour abnormal mass of tissue that is often cancerous 6, 109

turbine device for generating electricity – the turbine moves through a magnetic field and electricity is generated 49, 50, 110

turgid plant cells which are full of water with their walls bowed out and pushing against neighbouring cells 70

turgor pressure the pressure exerted on the cell membrane by the cell wall when the cell is fully inflated 70, 71

U

ultrasound high-pitched sounds which are too high for detection by human ears 106

ultraviolet radiation electromagnetic waves given out by the Sun which damage human skin 47

unbalanced (forces) forces acting in opposite directions that are unequal in size 95, 96

units of alcohol measurement of alcoholic content of a drink 8

universe the whole of space 56

unstable nucleus liable to decay 110

uranium radioactive element with a very long half-life used in nuclear power stations 108, 110

V

vacuum space containing hardly any particles 41, 73, 106

variable resistor a resistor whose resistance can change 105

variation the differences between individuals (because we all have slight variations in our genes) 11, 18, 20, 61, 64, 65

vascular bundle group of xylem and phloem cells 69, 71

vasoconstriction in cold conditions the diameter of small blood vessels near the surface of the body decreases – this reduces the flow of blood 9

vasodilation in hot conditions the diameter of small blood vessels near the surface of the body increases – this increases the flow of blood 9

vector an animal that carries a pathogen without suffering from it 6

vegan a type of diet; a person who does not eat animals or animal products 5

vegetarian a type of diet; a person who does not eat meat or fish 5

veins blood vessels that carry blood back to the heart 62, 69

velocity how fast an object is travelling in a certain direction: velocity = displacement ÷ time 95, 98

voltage a measure of the energy carried by an electric current (also called the potential difference) 49, 50, 52, 103, 104, 105

voltmeter instrument used to measure voltage or potential difference 104, 205

volts (V) units used to measure voltage 52, 105

W

watt (W) a unit of power, 1 watt equals 1 joule of energy being transferred per second 51, 52

wave oscillatory motion 42, 44, 45, 46, 47, 106

wavelength (λ) distance between two wave peaks 42, 49, 51, 56, 106, 109

weight the force of gravity acting on a body 97, 100

white blood cells blood cells which defend against disease 6

work done the product of the force and distance moved in the direction of the force 97, 98, 99, 101

X

x-rays ionising electromagnetic waves used in x-ray photography (where x-rays are used to generate pictures of bones) 58, 106, 109

xanthophylls plant pigments involved in photosynthesis 69

xylem cells specialised for transporting water through a plant; xylem cells have thick walls, no cytoplasm and are dead, their end walls break down and they form a continuous tube 71

Exam tips

The key to successful revision is finding the method that suits you best. There is no right or wrong way to do it.

Before you begin, it is important to plan your revision carefully. If you have allocated enough time in advance, you can walk into the exam with confidence, knowing that you are fully prepared.

Start well before the date of the exam, not the day before!

It is worth preparing a revision timetable and trying to stick to it. Use it during the lead up to the exams and between each exam. Make sure you plan some time off too.

Different people revise in different ways and you will soon discover what works best for you.

Remember

There is a difference between *learning* and *revising*.

When you revise, you are looking again at something you have already learned. Revising is a process that helps you to remember this information more clearly.

Learning is about finding out and understanding new information.

Using the Workbook

This Workbook allows you to work at your own pace and check your answers using the detachable Answer section on pages 249–270. In addition to the exam practice questions, the Workbook also contains questions that require longer answers (Extended response questions). You will find one question that is similar to these in each section of your written exam papers. The model answers supplied for these questions give guidance about the content that should be included, but do not necessarily provide a complete response for the questions concerned.

Some general points to think about when revising

- Find a quiet and comfortable space at home where you won't be disturbed. You will find you achieve more if the room is ventilated and has plenty of light.

- Take regular breaks. Some evidence suggests that revision is most effective when tackled in 30 to 40 minute slots. If you get bogged down at any point, take a break and go back to it later when you are feeling fresh. Try not to revise when you're feeling tired. If you do feel tired, take a break.

- Use your school notes, textbook and this Revision guide.

- Spend some time working through past papers to familiarise yourself with the exam format.

- Produce your own **summaries** of each module and then look at the summaries in this Revision guide at the end of each module.

- Draw mind maps covering the key information on each topic or module.

- Review the **Grade booster checklists** on pages 242–247.

- Set up revision cards containing condensed versions of your notes.

- Prioritise your revision of topics. You may want to leave more time to revise the topics you find most difficult.

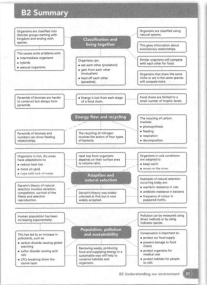

1 Joe's mum has her blood pressure checked by a nurse. The nurse tells her the results and she fills in a questionnaire for the nurse.

D–C

a Suggest **two** changes Joe's mum should make to lower her blood pressure.

....

....

....

....

....

.... **[2 marks]**

Blood pressure questionnaire			
Questions	Notes	Answers Yes	No
1 Do you take regular exercise?	Strong heart muscles will lower blood pressure		✓
2 Do you eat a healthy balanced diet?	Reducing salt intake will lower blood pressure		✓
3 Are you overweight?	Being overweight by 5 kg raises blood pressure by 5 units	✓	
4 Do you regularly drink alcohol?	A high alcohol intake will damage liver and kidneys	✓	
5 Are you under stress?	Relaxation will lower blood pressure	✓	

B–A*

b Explain why Joe's mum should lower her blood pressure.

....

.... **[2 marks]**

2 Look at the data about the risk factors linked to heart disease (USA data, 1961–2004).

— Overweight
- - - Hypertension
— — Smoking
----- High cholestrol

D–C

a Describe the general trends shown by this data.

....

.... **[2 marks]**

b What trend would you have expected to find in data about the **actual** heart disease in the same time period?

....

.... **[2 marks]**

D–C

3 Joe is an athlete and he is very fit. However, he still catches a cold. Explain why being fit may not keep you healthy.

.... **[2 marks]**

B–A*

4 Gas boilers in houses must be regularly checked to make sure they are not releasing carbon monoxide. Explain the effects of carbon monoxide on the body.

....

....

....

.... **[4 marks]**

Human health and diet

1 Children in developing countries sometimes have a swollen abdomen, a condition caused by a low-protein diet.

a Write down the name of this condition.

.... [1 mark]

b Describe two possible reasons for the lack of protein in developing countries.

....

.... [2 marks]

2 a Simon has a mass of 40 000 g. Calculate his estimated average daily requirement (EAR) for protein in grams. Use this formula: EAR in g = 0.6 × body mass in kg.

Show your working.

EAR = ... g   [2 marks]

b Simon has a sister called Karen. He is worried that she is not eating enough food.

Suggest why Karen may have chosen not to eat enough food to meet her daily requirements.

....

....

.... [2 marks]

c Simon is a vegetarian. Although he eats enough protein every day to meet his EAR, he needs to think carefully about his protein intake.

Explain why.

....

....

....

.... [2 marks]

3 Glycogen is a large molecule made up of hundreds of glucose molecules joined together.

glucose units   glycogen is a complex carbohydrate

a What group of food chemicals does glycogen belong to?

.... [1 mark]

b Proteins are also made of hundreds of small molecules that are joined together in a chain.

Write down the name of these molecules.

............Amino acids... [1 mark]

c Sometimes a person takes in more glycogen and protein than they need.

Compare how the body deals with the extra glycogen and protein that is taken in.

The glycogen is stored in the adipose tissue
and the protein cannot be stored   [2 marks]

Staying healthy

1 Complete the sentences about infectious diseases. Use words from this list.

antibodies   antibiotics   antigens   hormones   pathogens   toxins   vectors

a The symptoms of an infectious disease are caused by ..... *Pathogens* ....
They produce chemical waste, which contains ..... *toxins* ....

b Each disease-causing microorganism has its own .... *antigens* .... so
the body needs specific .... *antibodies* .... to combat them.   **[2 marks]**

2 Look at the diagram. It shows how mosquitoes spread malaria.

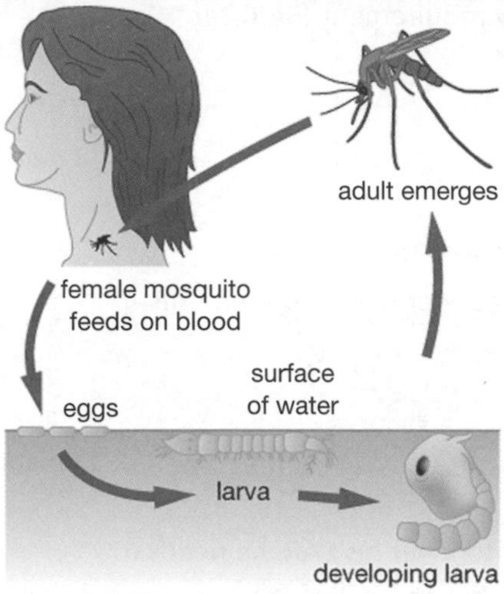

adult emerges

female mosquito
feeds on blood

surface
of water

eggs

larva

developing larva

a Using the diagram, name a parasite and its host.

....

....   **[1 mark]**

b Use the diagram to explain **two** ways in which the
spread of malaria could be controlled.

..... *By covering the surface of the* .....
..... *stagnant water with oil, and* .....
..... *draining stagnant bodies of* .....
..... *water.* .....

....   **[4 marks]**

3 The graph shows the levels of immunity to
a disease using passive and active
immunity methods.

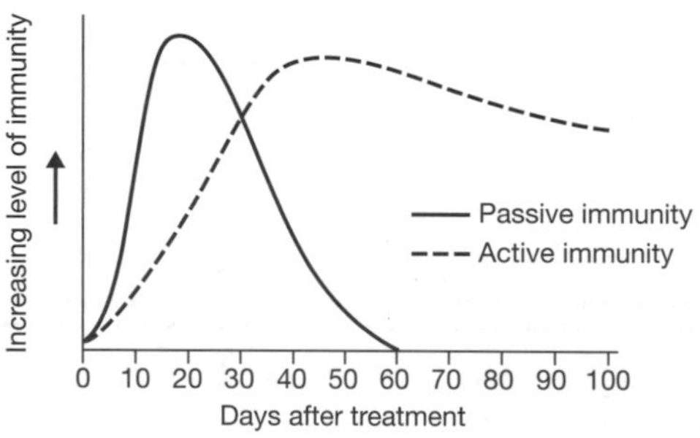

Increasing level of immunity

—— Passive immunity
--- Active immunity

0 10 20 30 40 50 60 70 80 90 100
Days after treatment

a Describe the difference in immunity:

i at 20 days ....   **[1 mark]**

ii after 60 days ....   **[1 mark]**

b Explain why there is a difference in immunity between passive and active immunity.

....

....

....   **[4 marks]**

c Which method of immunity should be used in an outbreak of infectious disease following a
natural disaster such as a tsunami? Explain your choice.

....   **[1 mark]**

d Using MRSA as an example, explain why doctors are concerned about the overuse of
antibiotics.

....

....

....   **[3 marks]**

The nervous system

1 a Rashid does his homework on how the eye works. This is what he writes. There are **three** mistakes.

> *The light rays are reflected by the cornea and lens. An image is formed on the optic nerve. Nerve impulses are then sent to the spinal cord. The amount of light entering the eye is controlled by the iris.*

Write down the incorrect word(s) and their correct replacements.

.... should be .... ;

.... should be .... ;

.... should be .... [3 marks]

D–C

b Write down **two** ways in which vision will be affected if only one eye is used to look at an object.

.... [2 marks]

c Use the information in the diagram to explain how the eye accommodates.

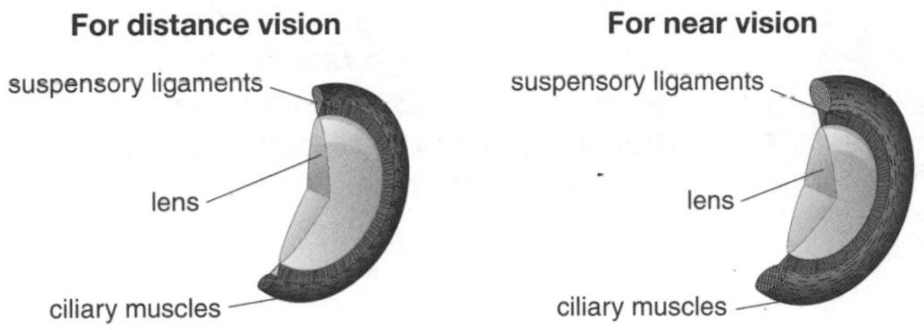

For distance vision

suspensory ligaments

lens

ciliary muscles

For near vision

suspensory ligaments

lens

ciliary muscles

B–A*

For distance vision the ....

....

....

.... [4 marks]

2 a Look at the diagram of a motor neurone.

Write down the names of the parts A, B, and C.

A Receptor ....

B ....

C ....

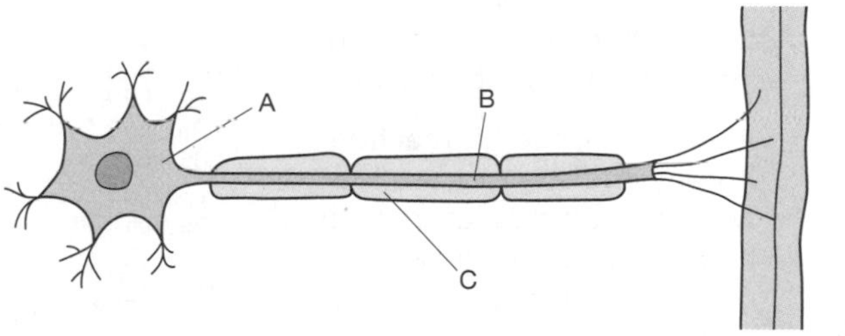

A   B   C

[3 marks]

b Which part A, B or C, carries the nerve impulse? ...B................... [1 mark]

D–C

c Look at the sequences A, B, C and D in a reflex arc. Only one sequence is correct.

A effector⇒sensory neurone⇒motor neurone⇒central nervous system⇒response

B stimulus⇒receptor⇒sensory neurone⇒motor neurone⇒central nervous system⇒response

C stimulus⇒receptor⇒sensory neurone⇒central nervous system⇒motor neurone⇒response

D response⇒receptor⇒sensory neurone⇒central nervous system⇒motor neurone⇒stimulus

Which sequence is correct? B .... [1 mark]

d Explain how a nerve impulse is transmitted from one neurone to the next.

....

....

.... [5 marks]

B–A*

Drugs and you

1 Draw a straight line from each type of drug to a correct example.

Types of drug	Example
depressant	LSD
painkiller	alcohol
stimulant	paracetamol
hallucinogen	caffeine

D–C

[3 marks]

2 a

standardised death rates from lung cancer for 100,000 men per year

non-smokers: 10
cigars only:
pipes only: 39
cigarettes pipes cigars: 67
cigarettes only: 135

D–C

Look at the graph. It shows death rates in men from lung cancer in the 1950s, in relation to the type of tobacco they smoked.

i Work out the death rate per 100 000 men in the UK for all types of smoking.

.... [1 mark]

ii The present death rate from lung cancer in the UK is about 60 deaths per 100 000. Explain the reasons for the change in death rate.

....
....

B–A*

[2 marks]

b Nicotine in cigarette smoke is a stimulant. Describe the effect of stimulants on synapses.

....
....

[3 marks]

3 A driver's reaction time is the time between seeing a danger and reacting to it by braking.

A volunteer had his reaction distances (distance travelled while reacting) measured when he was trying to read a message on his phone, when texting and when his reaction was impaired by alcohol. The volunteer's normal reaction distances were 4ft at 35mph and 8ft at 70 mph. The tests were done on a test circuit, not on public roads.

Reading 35 mph: 188
Texting 35 mph: 90
Impaired 35 mph: 7
Reading 70 mph: 189
Texting 70 mph: 319
Impaired 70 mph: 17

Reaction distances (ft)

D–C

a Comment on his results, identifying trends and conclusions.

....
....
....

[4 marks]

b Comment on the reliability of these results.

B–A*

....

[2 marks]

Staying in balance

1 a What is meant by homeostasis?

To maintain a constant thermal enviroment [1 mark]

b The body can increase or decrease heat transfer to the outside environment. What is the importance of the body temperature being 37 °C?

37 °C is the optimal temparature for enzyme activity [1 mark]

D–C

c If the body gets too hot you can suffer from dehydration. Explain why.

Because the ....

.... [2 marks]

d The temperature of an incubator for premature babies is controlled by a negative feedback mechanism. If the temperature gets too low, the heater is switched on.

i Explain what is meant by negative feedback.

To act to cancel out a change to restore the normal environ [2 marks]

B–A*

II Explain how negative feedback mechanisms are used in homeostasis.

....

.... [3 marks]

2 a Type 1 diabetes is treated by hormone injections.

Explain why Type 2 diabetes can be controlled by diet.

....

....

.... [3 marks]

D–C

b Look at the graph showing the blood sugar levels of two people after drinking a glucose solution.

i Why did the blood glucose levels rise in the first hour?

....

....

.... [1 mark]

ii Explain why the blood glucose levels for the 'normal' person returned to its original level after two hours.

....

....

....

....

....

....

.... [3 marks]

B–A*

iii Explain why the blood glucose levels remained high in the diabetic person.

....

.... [2 marks]

Controlling plant growth

1 Charlotte grows fruit trees.

a She takes cuttings of the best trees and dips the ends in rooting powder.

Describe and explain the effect the rooting powder has on the cuttings.

....

.... **[2 marks]**

b Charlotte has many weeds in her lawn. She uses a selective weedkiller.

i Explain what is meant by a selective weedkiller.

....

.... **[2 marks]**

ii How does a selective weedkiller work?

....

.... **[2 marks]**

2 a Complete the sentences about tropisms. Use words from this list.

auxins   antigens   geotropic   gravity   heat   light   negatively   positively

Since shoots grow towards the light they are called ....
phototropic.

Since roots grow with the pull of .... they are called

positively ....

These reactions involve plant hormones called .... **[4 marks]**

b Look at the diagram. It shows the result of an experiment on a shoot tip.

i What will the agar block contain after the shoot tip has been placed on it?

.... **[1 mark]**

ii Using this information, suggest which part of a plant is sensitive to light.

.... **[1 mark]**

iii Describe and explain what happens to the shoot when the agar block is placed on it.

....

....

....

....

....

.... **[6 marks]**

Variation and inheritance

1 Mutations can cause genetic variation in organisms.

   a What is a mutation?

     .... **[1 mark]**

   b Explain **two** other causes of genetic variation.

     1 ....

     2 .... **[4 marks]**

2 Complete the sentences about chromosomes.

   Use words from this list.

     12   23   46   the same   a different   a random

   Most human body cells have ..... *the same* ..... number of chromosomes.

   Most human body cells have ..... *46* ..... pairs of chromosomes.

   Other species have ..... *a different* ..... number of chromosomes. **[4 marks]**

3 a Look at the diagram showing how gender is inherited.

Complete the diagram by writing a letter at the end of each of the eight arrows at the bottom, to show the sex chromosomes and gender of the possible genetic combinations. **[3 marks]**

   b Use the diagram to explain why there is an equal chance of a baby being male or female.

     ..... *Because it is random* ....

     ....

     ....

     .... **[3 marks]**

   c Cystic fibrosis is an inherited condition. Darren has cystic fibrosis but his parents have not.

i Complete the genetic diagram to show how Darren inherited cystic fibrosis. **[2 marks]**

ii Put a ring around Darren's genotype in the diagram. **[1 mark]**

Blood pressure

Four adult friends measure their blood pressure. They then look at a blood pressure chart.

	Blood pressure in mmHg	
	Systolic	Diastolic
Alan	160	95
Rick	110	70
Shabeena	130	95
Toni	100	85

In the blood pressure chart, only one of the numbers needs to be higher/lower than it should be to count as either high or low blood pressure.

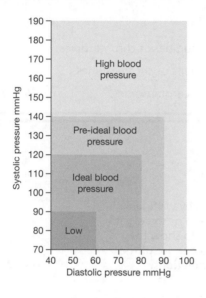

Suggest reasons for the friends' blood pressure readings and explain the possible consequences.

❶ The quality of written communication will be assessed in your answer to this question.

....

....

....

....

....

....

....

....

....

....

....

....

....

....

....

....

.... **[6 marks]**

Classification

1 a This table shows the classification of lions. Complete it by inserting the missing groups.

kingdom	Animal
	Chordate
	Mammal
	Carnivore
family	Felidae
	Panthera
	leo

[5 marks]

b There are seven different types of lion alive today. Describe one advance in science that could be used to help classify them that would not have been available 60 years ago.

....
.... [2 marks]

2 a What is meant by the term species?

.... [2 marks]

b Lions and tigers belong to the same family of cats. Look at the table. It shows the Latin names of some different cats.

Common name	Latin name
bobcat	*Felix rufus*
cheetah	*Acinonyx jubatus*
lion	*Panthera leo*
ocelot	*Felix pardalis*

i Two of the cats are more closely related. Write down the common names of these two cats.

.... [1 mark]

ii Explain your answer to part (bi).

.... [1 mark]

3 Scientists discovered a fossil of an extinct animal called *Archaeopteryx*. It had some bird features and some reptile features. Explain why there are always going to be some organisms like this that are difficult to classify.

....
.... [2 marks]

4 A zorse is a cross between two different species: a zebra and a horse.

a What name is used to describe a cross between two different species?

.... [1 mark]

b The zorse is difficult to classify. Explain why.

.... [1 mark]

5 Camels live in deserts in Asia. Llamas live in dry areas in South America. An animal similar to the llama and camel lived in North America about 11 million years ago. The appearance of a camel and a llama has a number of similarities. Explain possible reasons why.

....
.... [3 marks]

Energy flow

1 Look at the food chain.

```
                    ┌──────────┐          ┌────────┐          ┌────────┐
              ┌────▶│ beetles  │─────────▶│  mice  │─────────▶│  owls  │
┌──────────┐  │     └──────────┘          └────────┘          └────────┘
│ rose bush│──┤           │                                        ▲
└──────────┘  │           ▼                                        │
              │     ┌───────────┐                                  │
              └────▶│ hedgehogs │──────────────────────────────────┘
                    └───────────┘
```

a It is possible to construct a pyramid of biomass for this food chain.

i What does a pyramid of biomass show?

....

.... [2 marks]

ii Explain why the pyramid of biomass for this food chain would be a different shape than a pyramid of numbers.

....

.... [2 marks]

b Suggest why is it easier to construct a pyramid of numbers for this food chain.

....

.... [2 marks]

2 Look at the diagram. It shows the energy transfer from crops to a cow.

1022 kJ in heat loss
Sun
1909 kJ in waste
3056 kJ energy in 1 m²

a The cow loses 1022 kJ in heat loss.

Write down the process occurring in the cow that generates this heat.

.... [1 mark]

b Farmers often try to reduce the energy lost from cows.

i Suggest one way that they can do this.

.... [1 mark]

ii Suggest why they want to do this.

.... [1 mark]

c i Calculate the amount of energy used for growth by the cow.

Energy used for growth = ... kJ [2 marks]

ii Calculate the efficiency of energy transfer to the cow.

Energy efficiency = ... % [2 marks]

iii Some people think that it is better to grow crops for food instead of producing meat to eat.

Use your answer to cii to explain why.

....

.... [2 marks]

B2 Understanding our environment

Recycling

1 The diagram shows the carbon cycle.

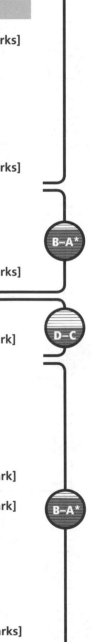

carbon dioxide in atmosphere

higher-level consumers

plants

fossil fuels

primary consumers

death and decay

death and decay

death and decay

detrivores   detritus

a Finish labelling the cycle by writing the correct processes in the three blank labels.   **[3 marks]**

b Decomposers return carbon dioxide to the air.

They will release carbon dioxide more slowly from waterlogged soils.

Explain why.

....

....   **[2 marks]**

c Animals called corals live in many oceans. The carbon in coral is recycled over millions of years.

Explain how this occurs.

....

....   **[3 marks]**

2 a Why do plants need nitrates?

....   **[1 mark]**

b The diagram shows part of the nitrogen cycle.

nitrogen in organic compounds in plant leaves	decomposers →	compound X	bacteria Y →	nitrates

Write down the name of:

i compound X ....   **[1 mark]**

ii bacteria Y ....   **[1 mark]**

c Nitrogen gas in the air is unreactive.

Describe how it can be converted into a form that can be used by plants.

....

....

....   **[4 marks]**

Interdependence

1 Read the information about the red and grey squirrels.

> There are two main species of squirrel living in Britain, the red squirrel and the grey squirrel. The grey was introduced from America. The number of reds has gone down in Britain over the last 60 years. They are now rare. There have been a number of studies to try to find out why the number of reds is declining.
>
> In coniferous woodland the reds are lighter animals and so spend more time up in the trees, feeding on the seeds from pine cones.
>
> It is in deciduous woodland that the reds are disappearing. Both types try to feed on acorns on the floor of the forest but the greys can digest the acorns more easily.

a What do red squirrels compete with each other for that they do not compete with grey squirrels for?

.... [1 mark]

b i The red squirrel has a particular ecological niche. What is meant by 'ecological niche'?

....

.... [2 marks]

ii Use the term ecological niche to explain why red and grey squirrels compete more in deciduous forests.

.... [2 marks]

2 The graph shows the predator–prey relationship of lemmings and snowy owls.

a Write about why the number of lemmings change, as shown in the graph.

....

.... [2 marks]

b Why do the owl numbers peak a short time after the lemming numbers?

.... [2 marks]

3 Oxpeckers are birds that eat insects on buffaloes. Buffalo often suffer from insect parasites. The oxpecker and the buffalo both benefit from the relationship.

Write down the name given to this type of relationship.

.... [1 mark]

4 Pea plants contain bacteria in special nodules on their roots. Pea plants are pollinated by bees. These bees often have small animals called mites that feed on fluid from the bee's body.

a Write down an organism that is acting as a parasite in these feeding relationships.

.... [1 mark]

b The pea plant and the bee both gain from their relationship. Explain how.

.... [2 marks]

c How do the bacteria and the pea plant both gain from their relationship?

....

.... [2 marks]

Adaptations

1 a Polar bears are adapted to live in the cold.

Suggest ways that the polar bear may be adapted to live in the cold.

....

....

.... [3 marks]

D–C

b The black bear and the polar bear both live in Canada. Both types of bears spend the winter months buried in holes or dens.

Explain why they do this.

....

....

.... [3 marks]

c The black bear lives further south than the polar bear. A female black bear has a mass of 40–80 kg, whereas a male polar bear is 150–250 kg. Black bears also have much larger ears than polar bears.

Explain how these differences are related to the different habitats of the bears.

....

....

.... [3 marks]

B–A*

2 a Cacti live in deserts.

The cactus has spines instead of leaves. Explain why.

....

.... [2 marks]

D–C

b Lizards called gila monsters live in the same habitat as cacti. They hunt small mammals.

In the morning they lie out in the Sun for some time before they hunt. Explain why.

....

.... [2 marks]

3 Raccoons are mammals about the size of small dogs that live throughout most of North America.

They are omnivores, eating a very wide range of foods.

Explain how this allows them to survive changing conditions and live over such a wide area.

B–A*

....

.... [2 marks]

Natural selection

1 Question 1 on page 46 includes a passage about competition between the red and grey squirrel. Re-read the passage before attempting these questions.

D–C

a What adaptation allows the grey squirrels to be better adapted to living in deciduous woodland?

.... [1 mark]

b Many scientists hope that a new strain of the red squirrel will evolve that can digest acorns. Use Darwin's theory of natural selection to explain how this might come about.

....

.... [2 marks]

B–A*

c Red squirrels now only live in isolated populations in different parts of the country. How might this lead to new species of squirrels developing?

....

....

....

.... [4 marks]

2 The following article gives information on the superbug MSRA. Read it carefully.

D–C

> ### Where did it come from?
> MRSA evolved because of natural selection. There are lots of different strains of the bacteria. Each strain has slightly different DNA. The DNA is also constantly mutating as the bacteria reproduce. Some of these mutations will be more resistant to antibiotics than others. When people take antibiotics, the less resistant strains die first. The more resistant strains are harder to destroy. If people stop taking the antibiotics too soon, the resistant strains survive.

Use Darwin's theory of natural selection to explain how MRSA has evolved.

....

.... [2 marks]

3 On his voyage on the *Beagle*, Charles Darwin visited many small islands. He made a number of observations that helped him to develop his theory of natural selection. One observation was that on small islands, animals often evolve to produce smaller animals than on the mainland.

D–C

a When Darwin returned from his voyage he was rather worried about publishing his ideas about natural selection. Suggest why that was.

....

.... [2 marks]

B–A*

b Lamarck's theory of evolution could also have explained why animals on an island are usually smaller. How would Lamarck's theory have explained this observation?

....

.... [2 mark]

c Why would this explanation be considered to be incorrect by most scientists now?

.... [1 mark]

Population and pollution

1 a The rise in human population is causing an increased level of carbon dioxide in the air. Suggest **two** effects this increase may have on the environment.

.... [2 marks]

b The ozone layer in the Earth's atmosphere protects us from harmful ultraviolet rays. Chemicals are destroying the ozone layer.

i Write down the name of the chemicals.

.... [1 mark]

ii The overuse of these chemicals has caused an increase in skin cancer. Suggest a reason why.

.... [1 mark]

D–C

2 Look at the graph. It shows the past, present and predicted future world human population.

a Human population is in the rapid growth stage. Write down the name given to this stage of growth.

....

....

year 2000
6.1 billion

underdeveloped countries

developed countries

[1 mark]

D–C

b In developed countries such as America the population is constant.

In underdeveloped countries the population may be higher than in developed countries, yet they cause less pollution. Suggest two reasons why.

....

.... [2 marks]

B–A*

3 Scientists can look for the variety of animal species living in a stream when they want to measure how polluted the water is.

a Write down the name given to species that are used to measure levels of water pollution.

.... [1 mark]

b Look at the table. It shows the sensitivity of different animals to pollution.

A river sample contained mussels, damsel fly larvae and bloodworms, but no mayfly or stonefly larvae. Use this information to explain how you can tell that the river is polluted.

....

....

....

Animal	Sensitivity to pollution
stonefly larva	sensitive
water snipe fly	sensitive
alderfly	sensitive
mayfly larva	semi-sensitive
freshwater mussel	semi-sensitive
damselfly larva	semi-sensitive
bloodworm	tolerates pollution
rat-tailed maggot	tolerates pollution
sludgeworm	tolerates pollution

[2 marks]

D–C

c If a river is polluted, the water usually contains less oxygen.
Scientists could use oxygen probes to give an indication of the level of pollution.

Give **one** advantage and **one** disadvantage of this method rather than looking at the variety of animals in the river.

....

.... [2 marks]

B–A*

Sustainability

1 a Pandas live in a remote part of China. Their habitat is being destroyed. Some people want to save the panda from extinction.

Suggest **two** reasons why saving the panda might help the people who live in the same habitat.

....

.... [2 marks]

b Pandas are being bred in zoos in China and in other countries. A careful record is kept about the family tree of each panda and this is consulted before they are allowed to mate.

Explain why this is.

....

.... [2 marks]

2 Some countries want to hunt whales for food.

a Suggest **one** argument for and **one** argument against hunting whales.

....

.... [2 marks]

b It is very difficult to stop people hunting whales.

Suggest **one** reason why.

.... [1 mark]

3 The fishing industry is trying to follow sustainable development.

a What is meant by sustainable development?

....

.... [2 marks]

b To try and achieve this, the government has set fish quotas.
Fishermen can only catch a set amount of fish at any one time.
Also, the size of the individual fish that they catch has to be above a certain level.
These regulations should help maintain the population of fish in the sea.

Explain why.

....

.... [2 marks]

c The population of Brazil was estimated at 150 million in 1990 and 195 million in 2010.

Explain the problems of achieving sustainable development when the population is rising this rapidly.

....

....

.... [3 marks]

B2 Extended response question

The diagram shows the flow of energy through a food chain.

```
                        9000 units          800 units
                   ┌──→               ┌──→
                   │                  │
sunlight ──→ ┌──────────┐ ──→ ┌──────────┐ ──→ ┌──────────┐
             │  wheat   │     │  cows    │     │  humans  │
             │10000 units│     │1000 units│     │200 units │
             └──────────┘     └──────────┘     └──────────┘
```

Use your knowledge of how energy is lost from food chains to explain the figures shown and discuss how the figures could be used to argue for vegetarianism.

❗ The quality of written communication will be assessed in your answer to this question.

....

....

....

....

....

....

....

....

....

....

....

....

....

....

....

....

....

....

....

....

....

....

....

....

....

....

....

....

....

....

....   [6 marks]

....

Making crude oil useful

1 a All the oils of crude oil are hydrocarbons. What is a hydrocarbon?

....

.... [2 marks]

b The hydrocarbons are separated by fractional distillation.

i Label the diagram A where the crude oil is heated. [1 mark]

ii Label the diagram B where the fraction bitumen 'exits' from. [1 mark]

iii Label the diagram C at the coldest part. [1 mark]

iv Which fraction 'exits' from the coldest part?

....

....

.... [1 mark]

c Explain why crude oil can be separated by fractional distillation. Use ideas about molecular size, forces of attraction and boiling points in your answer.

....

....

....

....

.... [3 marks]

2 Explain how damage is caused to wildlife if oil tanker ships are damaged.

....

.... [2 marks]

3 a Cracking breaks down long-chain molecules called alkanes. When a large alkane is cracked it becomes a smaller alkane and an alkene. Explain why an alkene is a different type of hydrocarbon to an alkane.

.... [1 mark]

b What are alkenes useful for making?

.... *Polymers* .... [1 mark]

c A country produces 25% more than its demand of heavy oil from crude oil distillation. However, its supply of petrol from the distillation is only 68% of its need. What percentage shortfall is there in petrol needed?

.... [1 mark]

d Suggest how they could solve this problem other than by importing petrol from elsewhere.

....

.... [3 marks]

C1 Carbon chemistry

Using carbon fuels

1 a Look at the table.

Characteristic	Coal	Petrol
energy value	high	high
availability	good	good
storage	bulky and dirty	volatile
toxicity	produces acid fumes	produces less acid fumes
pollution caused	acid rain, carbon dioxide, soot	carbon dioxide, nitrous oxides
ease of use	easier to store for power stations	flows easily around engines

i Use the information to compare how coal and petrol are different in terms of storage and toxicity.

Petrol is less toxic, but is more dangerous to store than coal **[2 marks]**

ii From the information on the table, give two reasons why coal and petrol are equally suitable for use as a heating fuel.

....

.... **[2 marks]**

b Explain why the use of petrol and diesel is increasing and how this is contributing to global problems.

....

.... **[2 marks]**

2 a An experiment can show that two products are made in the complete combustion of a hydrocarbon fuel.

i Write down a **word equation** for the complete combustion of a hydrocarbon fuel.

.... **[1 mark]**

ii Describe an experiment to show what the products of combustion are when a candle burns in a plentiful supply of air.

.... **[2 marks]**

b Complete combustion is better than incomplete combustion. Explain why.

....

....

.... **[3 marks]**

c Write a balanced symbol equation for the complete combustion of pentane, C_5H_{12}.

.... **[2 marks]**

d Write a balanced symbol equation for the incomplete combustion of propane, C_3H_8, producing carbon.

.... **[2 marks]**

Clean air

1 a Label the diagram with the percentages of nitrogen, oxygen and carbon dioxide found in clean air. (Ignore noble gases).

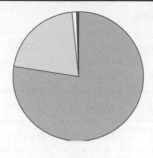

[3 marks]

b Mark in the percentages of the gases.

[3 marks]

c These percentages do not change very much because there is a balance between three of the processes that use up or make carbon dioxide and use or make oxygen. Explain how the balance is maintained.

....

....

.... [6 marks]

2 a Over the last few centuries the percentage of carbon dioxide in air has increased slightly. This is due to a number of factors. Explain the impact of the two factors in the table.

	Impact on levels of carbon dioxide
increasing population	
deforestation	

[2 marks]

b Explain one theory of how the Earth's atmosphere has evolved. Include in your answer: how the original gases got into the atmosphere, why the levels of nitrogen increased, how the levels of oxygen evolved to their current levels.

....

.... [3 marks]

3 a It is important to control levels of atmospheric pollution. Use the table to explain why pollution should be controlled from petrol-powered cars.

Pollutant	Environmental effects
Carbon monoxide	Poisonous gas formed by incomplete combustion of petrol- or diesel-powered motor vehicles
Oxides of nitrogen	Photochemical smog and acid rain formed by reaction of nitrogen and oxygen in an internal combustion engine
Sulfur dioxide	Acid rain formed from sulfur impurities when fossil fuels burn

....

.... [3 marks]

b Which gas is controlled by a catalytic converter where it is turned into carbon dioxide?

.... [1 mark]

c Explain why nitrogen is converted to oxides of nitrogen inside an internal combustion engine.

.... [1 mark]

d A reaction between nitric oxide (NO) and carbon monoxide takes place on the surface of the catalyst in a catalytic converter. The reaction forms nitrogen and carbon dioxide. Why is it important that these two gases are made?

.... [1 mark]

Making polymers

1 a Butanol C_4H_5OH is **not** a hydrocarbon.
Explain why.

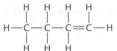

.... [1 mark]

b Butene is an alkene.
Explain how you know.

.... [1 mark]

c Bromine water is used to test for an alkene. What do you see if an alkene is tested?

....
.... [2 marks]

D–C

2 a What is an unsaturated compound?

.... [1 mark]

b Bromine water is used to test for unsaturation. Explain how it does this and what is formed.

....
....
.... [3 marks]

B–A*

3 a Which molecule is a polymer? Put a ring around A, B, C or D.

[1 mark]

D–C

b Write down **two** conditions needed for polymerisation.

....
.... [2 marks]

c What does a monomer need in its structure to undergo addition polymerisation?

.... [1 mark]

d Construct the **displayed formula** of the addition
polymer from this monomer.

[2 marks]

e Draw the monomer that makes this addition polymer.

[2 marks]

B–A*

f Explain what happens in an addition polymerisation reaction.

....
.... [3 marks]

Designer polymers

1 a Polymers are better than other materials for some uses. Give one example of the use of a polymer and explain why that polymer was suitable for the use.

....

.... **[2 marks]**

b i What properties of nylon make it suitable for use in outdoor clothing?

....

.... **[2 marks]**

ii Nylon has a disadvantage when used for outdoor clothing. What is it?

.... **[1 mark]**

c i GORE-TEX® type materials are made to make clothing waterproof and breathable. The inner layer of the clothing is made from expanded PTFE (polytetrafluoroethene). Explain why this makes the material waterproof and breathable.

....

.... **[2 marks]**

ii The inner layer needs to be laminated onto another material. Explain why it needs to be laminated.

.... **[1 mark]**

2 a Disposing of non-biodegradable polymers causes problems. Explain the problems for each of the disposal methods listed below.

i Landfill sites .... **[1 mark]**

ii Burning waste plastics .... **[1 mark]**

iii Recycling .... **[1 mark]**

b Scientists are developing new types of polymers that are more easily disposed of after use. Explain what is different about the ways that these plastics can be disposed of.

....

.... **[2 marks]**

3 Look at the diagram of polymer molecules.

a i Label the strong covalent bonds. **[1 mark]**

ii Label the intermolecular forces of attraction. **[1 mark]**

b Some plastics have low melting points and can be stretched easily whereas other plastics have high melting points and cannot be stretched easily; they are rigid. Explain why, using ideas about forces between molecules in your answer. You may use a diagram to help you.

....

....

....

.... **[4 marks]**

Cooking and food additives

1 a What happens to the protein molecules in eggs and meat when eggs and meat are cooked and what is this process called?

....

....

.... **[2 marks]**

b Potatoes are easier to digest when they are cooked. Explain why.

....

....

.... **[2 marks]**

D–C

c When sodium hydrogencarbonate (baking powder) is heated it decomposes.

Write down the word equation for this reaction.

....

.... **[1 mark]**

d Write down the balanced symbol equation for the decomposition of sodium hydrogencarbonate to make sodium carbonate, carbon dioxide and water. You may look up the formulae for sodium hydrogencarbonate and sodium carbonate at Grades D–C.

.... **[2 marks]**

2 a Look at the diagram. Explain the meaning of the labels and identify what kind of molecule this is.

hydrophilic head

hydrophobic tail

....

....

....

.... **[3 marks]**

D–C

b Explain how these types of molecules keep oil and water mixed. You can use labelled diagrams to help you.

....

....

.... **[3 marks]**

B–A*

Smells

1 To make a perfume, a type of chemical is made by mixing alcohol with an acid.

   a Write down a word equation for the reaction to make this type of chemical.

   .... **[1 mark]**

   b Describe how you would make a sample of this chemical in a school laboratory.
       Draw a diagram of the apparatus you would use.

   ....

   ....

   ....

[3 marks]

2 A good perfume needs to have several properties. These are listed in the boxes. Draw a straight line to match the best reason to the property needed.

evaporates easily	it can be put directly on the skin
non-toxic	it does not react with perspiration
insoluble in water	it does not poison people
does not irritate the skin	it cannot be washed off easily
does not react with water	its particles can reach the nose

[4 marks]

3 A solute and a solvent that do not separate is a ....   **[1 mark]**

4 For a perfume to be effective, its particles must be able to reach the nose easily – the substance must be volatile. Explain why a substance is volatile using ideas from kinetic theory. You may use a diagram to help you.

....

....

....

[4 marks]

5 Give two uses of esters.

....

[1 mark]

6

water molecules

nail varnish molecule

Water does not dissolve nail varnish. Use the diagram to help explain why, including ideas about the forces of attraction between molecules in your answer.

....

....

....

....

[4 marks]

Paints and pigments

1 Paint is a **colloid.** Explain the meaning of the term 'colloid'.

.... [1 mark] D–C

2 Why do the components of a colloid not separate?

.... [1 mark] B–A*

3 Emulsion paint is a water-based paint. It is made of tiny droplets of a liquid in water, which is called an emulsion. When emulsion paint has been painted on to a surface as a thin layer, it is left to dry. What happens as it dries?

.... [1 mark] D–C

4 Explain how oil-based paints dry.

pigment particles

painted surface

....
....
....
....
....
.... [2 marks] B–A*

5 A thermochromic pigment changes colour at 45 °C. Suggest two objects that the pigment could be used in, explaining your reasons.

....
.... [2 marks] D–C

6 a Most thermochromic pigments change from having a colour to being colourless when they are heated. Thermochromic paints come in a limited range of colours. How do manufacturers ensure that a larger range of colours is available?

.... [1 mark]

b If a green paint becomes yellow when heated explain what the mixture contains and how it changes colour.

....
....
.... [3 marks] B–A*

7 Why do phosphorescent pigments glow in the dark?

....
.... [2 marks] D–C

8 What have phosphorescent pigments on luminous clock faces replaced and why?

....
.... [2 marks] B–A*

C1 Extended response question

Perfumes should have a pleasant smell and evaporate easily. Synthetic perfumes are made from esters.

Explain how a perfume evaporates easily, using ideas about kinetic theory. Describe three other properties it must have, including reasons why it must have these properties.

❗ The quality of written communication will be assessed in your answer to this question.

....
....
....
....
....
....
....
....
....
....
....
....
....
....
....
....
....
....
....
....
....
....
....
....
....
....
....
....
....
....
....

.... **[6 marks]**

The structure of the Earth

1 a The outer part of the Earth is called the **lithosphere**. What makes up the lithosphere?

....

.... [2 marks]

b Describe why it is difficult to study the structure of the Earth.

....

.... [2 marks]

c What process moves tectonic plates?

.... [1 mark]

d The diagram shows a **subduction zone**. Describe how rock is recycled to form mountains.

....

....

....

.... [5 marks]

2 a Suggest why most scientists now accept plate tectonic theory.

....

.... [2 marks]

b Suggest why plate tectonic theory was not accepted when Wegener first proposed it.

....

.... [2 marks]

3 a Suggest two reasons why geologists study volcanoes.

....

.... [2 marks]

b Why can different types of igneous rock form?

.... [1 mark]

c Explain why some volcanic eruptions are fairly safe, but others are destructively explosive.

....

.... [3 marks]

Construction materials

1 a Put these materials into order of hardness.

        granite   limestone   marble

Least hard ...., ...,

.... Hardest   **[1 mark]**

b Aluminium, brick and glass are all manufactured. Finish the table to show the raw materials used to make each one.

Building material	aluminium	brick	glass
Raw material			

                                                  [3 marks]

c Cement is used as a building material. Describe how cement is made.

....

....

.... **[1 mark]**

d Granite and marble are different types of rock. Write sentences to describe each rock by using the words in this list. Some words apply to both types of rock.

crystals   igneous   interlocking   limestone   heat   pressure   metamorphic   solidifies

Granite is ....

....

.... **[4 marks]**

Marble is ....

....

.... **[4 marks]**

2 a Calcium carbonate decomposes at high temperatures. Write a word equation for this reaction.

....

.... **[1 mark]**

b Describe how concrete is made, and how it can be strengthened.

....

....

.... **[2 marks]**

c Explain why reinforced concrete is a better construction material than just using concrete. You can include a diagram to help explain your points. Include ideas about forces in your answer.

....

....

....

....

.... **[3 marks]**

Metals and alloys

1 Copper can be purified by electrolysis.

d.c. power supply

impure copper

electrolyte

pure copper

impurities

a Label the cathode on the diagram. [1 mark]

b Describe how the electrolysis process purifies copper.

....

....

.... [4 marks]

c Give **two** advantages and **two** disadvantages of recycling copper.

Advantages: ....

.... [2 marks]

Disadvantages: ....

.... [2 marks]

d Write half-equations to show what happens at the two electrodes when copper is purified by electrolysis.

i anode

.... [1 mark]

ii cathode

.... [1 mark]

e Use the equations in (d) to explain why electrolysis involves both oxidation and reduction.

....

....

.... [3 marks]

2 **a** Most metals form alloys. Draw a straight line to match each alloy to the metals it is made from.

amalgam		contains copper and zinc
brass		contains mercury
solder		contains lead and tin

[1 mark]

b Suggest why a smart alloy is useful for making spectacle frames.

....

....

.... [2 marks]

Making cars

1 a In winter, icy roads are treated with rock salt. Why is this a problem for car bodies made from steel?

.... [1 mark]

b Aluminium does not corrode in moist air. Explain why.

.... [1 mark]

c Write a word equation for rusting.

.... [2 marks]

2 a Steel is an alloy of iron.

i What is an alloy? .... [1 mark]

ii Give **two** advantages of steel over iron.

.... [2 marks]

b Steel and aluminium can be used to make car bodies. Write down **two** disadvantages of using aluminium.

.... [2 marks]

c What properties would be needed in a material used to make a car windscreen?

....

.... [2 marks]

d Explain *two* advantages of using aluminium alloys for making car bodies.

....

.... [2 marks]

3 Look at the table and answer the questions below.

Condition / Metal	At start	Dry air	Moist clean air	Moist acidic air
Aluminium	shiny silver	shiny silver	shiny silver	dull silver
Copper	shiny salmon-pink	shiny salmon-pink	small patches of green on surface	green layer on surface
Iron	shiny silver	shiny silver	small patches of brown on surface	lots of brown flakes on surface

a Which conditions are likely to cause the most corrosion on a car body?

.... [1 mark]

b Which would be the best material for a car body. Give a your reason for your answer.

.... [2 marks]

4 Why are old cars recycled?

....

.... [2 marks]

C2 Chemical resources

Manufacturing chemicals – making ammonia

1 a Suggest **three** conditions that increase yield in the Haber process.

....

.... [3 marks]

D–C

b Ammonia (NH_3) is made by reacting nitrogen (N_2) and hydrogen (H_2). Construct a balanced symbol equation for this reaction.

.... [2 marks]

2 Look at this graph.

Percentage of ammonia made

percentage of ammonia made

pressure in atmospheres

350 °C
400 °C
450 °C

a How much ammonia is made at a temperature of 400 °C and a pressure of 300 atmospheres?

.... [1 mark]

D–C

b How does pressure affect the yield?

.... [1 mark]

B–A*

c Explain why the optimum temperature for the reaction is 450 °C.

....

....

.... [3 marks]

3 a Describe three different factors that affect the cost of making a new chemical.

....

.... [3 marks]

b How is the production of ammonia linked to world food production?

D–C

....

....

.... [3 marks]

c High pressure is used to increase the percentage yield and the reaction rate. Describe the difference between **reaction rate** and **percentage yield**.

....

.... [2 marks]

B–A*

d Why can a **low** percentage yield be acceptable for some processes?

....

.... [2 marks]

Acids and bases

1 a i What is an alkali?

.... [1 mark]

ii Finish this word equation for neutralisation.

.... + base → salt + ... [1 mark]

b Name the ion that all acids form in solution.

.... [1 mark]

c Explain how the concentration of hydrogen ions is linked to pH.

.... [1 mark]

d Write a **balanced symbol equation** to show neutralisation.

.... [2 marks]

2 a How does the pH of an acid change when an alkali is added to it?

.... [1 mark]

b **Universal indicator solution** can be used to measure the acidity of a solution. A few drops are added to the test solution and then the colour of the solution is compared to a standard colour chart.

Describe how the colour changes when a strong acid is added to the alkali to neutralise it.

....

....

.... [3 marks]

3 a Write the **word equation** for the reaction between copper carbonate and sulfuric acid.

.... [2 marks]

b Name the **salt** made when:

i sulfuric acid reacts with calcium carbonate .... [1 mark]

ii nitric acid reacts with potassium hydroxide .... [1 mark]

iii hydrochloric acid reacts with sodium carbonate .... [1 mark]

iv nitric acid reacts with copper oxide .... [1 mark]

c Using the four reactants in question 2b, **construct balanced symbol** equations for each one.

.... [2 marks]

.... [2 marks]

.... [2 marks]

.... [2 marks]

Fertilisers and crop yields

1 a Why do fertilisers need to be dissolved?

....

.... [1 mark]

D–C

b How do fertilisers increase crop yield?

....

....

.... [2 marks]

B–A*

2 a How is farming linked to eutrophication?

....

.... [1 mark]

D–C

b Describe the main stages of eutrophication.

....

....

....

....

....

....

....

.... [6 marks]

B–A*

3 a Fertilisers are made by reacting an acid and an alkali. Name the fertilisers which would be made by reacting:

i nitric acid and ammonia

.... [1 mark]

D–C

ii phosphoric acid and potassium hydroxide

.... [1 mark]

b Potassium nitrate fertiliser is made by adding an acid to an alkali.

i Suggest suitable reactants.

....

.... [2 marks]

B–A*

ii How can crystals be prepared from a solution of the fertiliser?

....

.... [2 marks]

1 a Name **two** methods of extracting salt from underground deposits found in Cheshire.

....

....

.... **[2 marks]**

D–C

b Suggest why salt mining can create problems.

....

.... **[2 marks]**

2 Electrolysis can be used to breakdown sodium chloride solution.

a What is formed:

i At the **anode**?

.... **[1 mark]**

ii At the **cathode**?

.... **[1 mark]**

iii In the **solution**?

.... **[1 mark]**

D–C

b Why are **inert electrodes** needed?

.... **[1 mark]**

c Give the formulae of the four ions present in sodium chloride solution.

.... **[2 marks]**

B–A*

d Describe, using a **half-equation**, how hydrogen is formed at the cathode.

.... **[2 marks]**

3 a Name the two chemicals which make household bleach.

....

.... **[2 marks]**

D–C

b Explain why the chlor-alkali industry is economically important to the UK.

....

....

....

.... **[3 marks]**

B–A*

C2 Extended response question

This equipment can be used to produce a fertiliser.

burette — acid

flask — alkali

Describe how a pure sample of ammonium phosphate could be produced. Include details of the chemicals you would use.

🛈 The quality of written communication will be assessed in your answer to this question.

....
....
....
....
....
....
....
....
....
....
....
....
....
....
....
....
....
....
....
....
....
....
....
....
....
....
.... **[6 marks]**

Heating houses

1 Kelly opens the front door on a very cold morning. Her mother complains that the house is getting cold. Use your ideas about energy flow to explain why the house gets cold.

....

....

.... **[2 marks]**

D–C

2 Explain the difference between temperature and heat.

....

....

.... **[2 marks]**

B–A*

3 A police helicopter uses a thermal imaging camera to take a picture at night. It is looking for a car that has recently been abandoned in a field after a high speed chase. Describe how the thermogram can help locate the car.

....

....

.... **[3 marks]**

D–C

4 a Finish the sentence.

The energy needed to raise the temperature of 1 kg of a material by 1 °C is known as the

.... **[1 mark]**

D–C

b i Jamie heats 500 g of seawater in a beaker from 20 °C to 90 °C. The specific heat capacity of seawater is 3900 J/kg °C. How much energy is needed to heat the sea water?

....

....

....

.... **[2 marks]**

B–A*

ii The energy supplied is greater than this. Suggest where this additional energy has been used.

....

.... **[1 mark]**

D–C

5 a What physical quantity is measured in units of J/kg?

.... **[1 mark]**

D–C

b When iron changes from solid to liquid, energy is transferred but there is no temperature change until all of the solid has changed into liquid. Use your ideas about molecular structure to explain why there is no change in temperature at iron's melting point.

....

....

....

....

....

.... **[2 marks]**

B–A*

Keeping homes warm

1 a 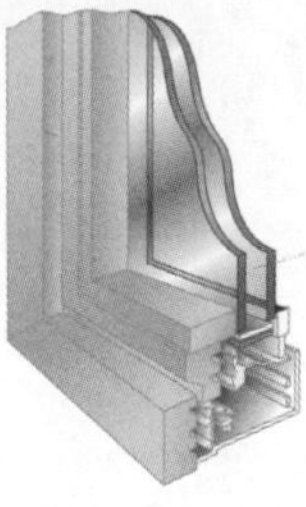 The diagram shows a section through a double glazed window. Michael says that it is just as effective to use a piece of glass twice the thickness. Use your ideas about energy transfer to explain why double glazing is better.

....

....

.... **[2 marks]**

b New homes are built with insulation blocks in the cavity between the inner and outer walls. The blocks have shiny foil on both sides.

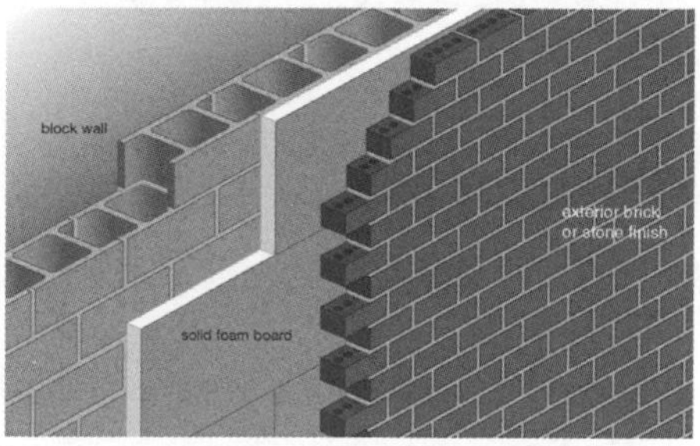

i Explain how the insulation blocks reduce energy transfer by conduction and convection.

....

....

....

.... **[3 marks]**

II Explain how the shiny foil helps to keep a home warmer in winter and cooler in summer.

....

....

....

D–C

2 a A brick in a wall is a better conductor of heat than air in the cavity between the walls. Explain why.

....

....

.... **[2 marks]**

b Hot air rises. Explain why.

....

....

.... **[2 marks]**

B–A*

3 Dan heats his house with coal fires. He is told that his fires are 32% efficient.

a Explain what is meant by 32% efficient.

....

.... **[1 mark]**

b Dan pays £9.50 for a 25 kg bag of coal. How much of that money is usefully used in heating his house?

....

....

.... **[2 marks]**

c Suggest why coal fires are so inefficient?

....

.... **[1 mark]**

D–C

A spectrum of waves

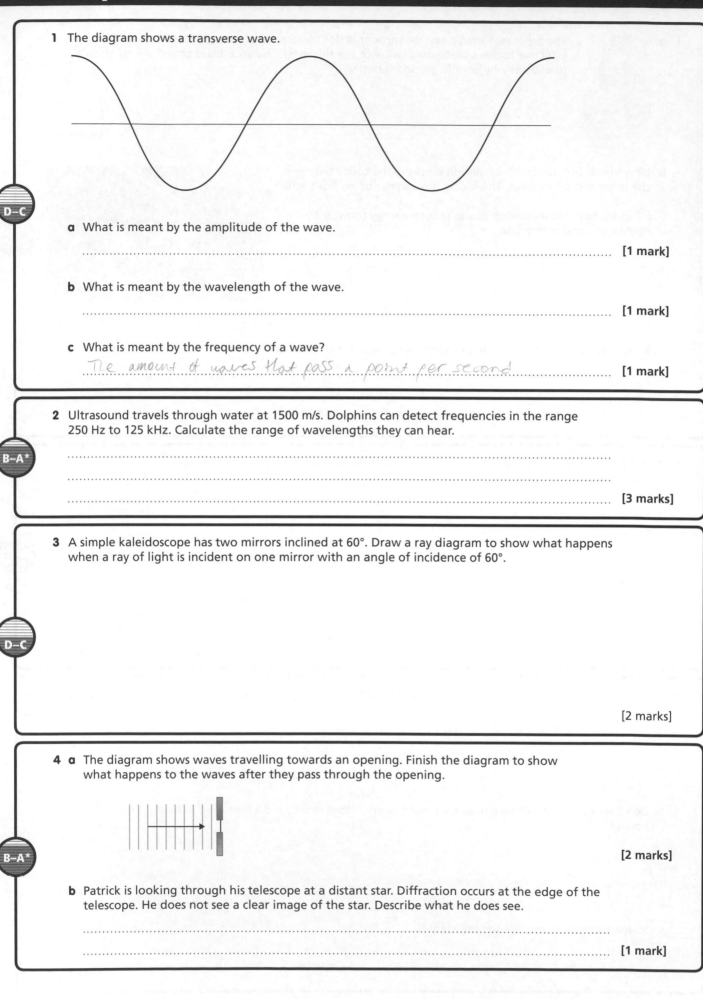

1 The diagram shows a transverse wave.

a What is meant by the amplitude of the wave.

.... [1 mark]

b What is meant by the wavelength of the wave.

.... [1 mark]

c What is meant by the frequency of a wave?

The amount of waves that pass a point per second .... [1 mark]

2 Ultrasound travels through water at 1500 m/s. Dolphins can detect frequencies in the range 250 Hz to 125 kHz. Calculate the range of wavelengths they can hear.

....

....

.... [3 marks]

3 A simple kaleidoscope has two mirrors inclined at 60°. Draw a ray diagram to show what happens when a ray of light is incident on one mirror with an angle of incidence of 60°.

[2 marks]

4 a The diagram shows waves travelling towards an opening. Finish the diagram to show what happens to the waves after they pass through the opening.

[2 marks]

b Patrick is looking through his telescope at a distant star. Diffraction occurs at the edge of the telescope. He does not see a clear image of the star. Describe what he does see.

....

.... [1 mark]

Light and lasers

1 Why is Morse code an example of a digital signal and not an analogue signal?

....

....

.... [2 marks]

D–C

2 Tina's science teacher shines a laser light onto a screen. It is brighter than the white light from the laboratory lights. Use your ideas about frequency to explain the difference between laser light and white light. You may draw diagrams to help you answer the question.

....

....

.... [3 marks]

B–A*

3 a The diagrams show three rays of light travelling from water into air. The three angles of incidence are (x) smaller than the critical angle (y) equal to the critical angle (z) larger than the critical angle.

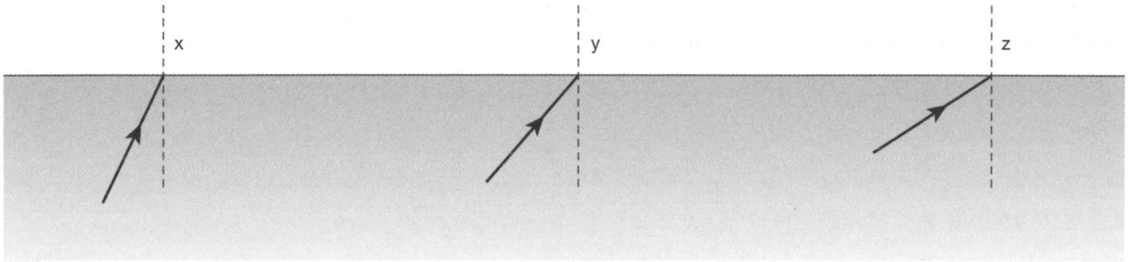

i Finish the diagrams to show what happens to the rays of light after they meet the water/air boundary. [4 marks]

ii Show clearly, on the correct diagram, the critical angle. Label it c. [1 mark]

D–C

b Sonia's doctor wants to look at the inside of her stomach. The doctor does so without using surgery. Describe how the doctor can see inside Sonia's stomach.

....

....

....

....

....

....

....

.... [3 marks]

B–A*

Cooking and communicating using waves

1 a Microwave ovens take less time to cook food than normal ovens. Suggest why.

....

.... **[1 mark]**

b Microwaves are suitable to communicate with space craft thousands of kilometres away, but mobile phones often cannot receive a signal just a few kilometres from the nearest transmitter. Why do microwave signals seem to work better in space than they do on Earth?

....

....

....

.... **[2 marks]**

D–C

2 The diagram shows the electromagnetic spectrum.

| radio | microwave | infrared | visible | ultraviolet | X-ray | gamma ray |

a Which part of the electromagnetic spectrum transfers the most energy?

.... **[1 mark]**

b An electric iron and an element from an electric fire both emit infrared radiation. The iron transfers less energy than the element. Explain how the wavelength of radiation from the iron differs from the wavelength of radiation from the element.

....

....

....

....

.... **[2 marks]**

B–A*

3 The diagram shows a transmitter on top of a hill. It transmits microwave mobile phone signals as well as radio and television signals.

The house, behind the other hill, can receive both radio and television signals but there is no mobile phone reception. Explain why.

....

....

....

....

.... **[2 marks]**

B–A*

Data transmission

1 a The change from analogue to digital transmission of television signals began in the United Kingdom in 2009. Write down one advantage of digital television.

.... [1 mark]

D–C

2 A ray of laser light is shone into one end of an optical fibre.

Finish the path of the ray as it passes into, through and out of the optical fibre. [2 marks]

D–C

3 A multinational company has thousands of computers which transmit data continually. The data is not transmitted using analogue signals. Explain the advantages of transmitting data digitally. You may use diagrams to illustrate your answer

....
....
....
....
....
....
....
....
....
....
....
....
....
.... [5 marks]

B–A*

Wireless signals

1 a Radio waves are refracted in the upper atmosphere. What happens to the amount of refraction if the frequency of the radio wave is decreased?

....

....

....

....

....

.... **[1 mark]**

b Why does the microwave beam sent by a transmitting aerial towards a satellite in orbit have to be **focused**?

....

....

....

....

....

.... **[1 mark]**

2 When Jenny is watching her television, she notices that there is a faint second picture slightly offset to the main picture.

Finish the sentence to explain why there is this 'ghost' picture.

Choose words from this list.

absorbed   dispersed   reflected   refracted

The aerial has received a direct signal from the transmitter and a signal that has been

.... **[1 mark]**

3 a Jenny listens to her favourite radio station. Every so often, she notices that she can hear a foreign radio station as well. Put ticks (✓) in the **two** boxes next to the statements that explain why this happens. **[1 mark]**

The foreign radio station is broadcasting on the same frequency. ☐

The foreign radio station is broadcasting with a more powerful transmitter. ☐

The radio waves travel further because of weather conditions. ☐

Jenny's radio needs new batteries. ☐

b Explain the benefits Jenny will get if she replaces her radio with a DAB radio?

....

....

....

....

....

.... **[1 mark]**

P1 Energy for the home

Stable Earth

1 a P waves and S waves are two of the waves which travel through the Earth after an earthquake. Finish the table by putting a tick (✓) in the correct box or boxes next to the description of the wave. The first one has been done for you. [3 marks]

description	P wave	S wave
pressure wave	✓	
transverse wave		
longitudinal wave		
travels through solid		
travels through liquid		

b How do scientists use the properties of P waves and S waves to find out the **size** of the Earth's core?

....

....

.... [2 marks]

c How do scientists use the properties of P waves and S waves to find out the **structure** of the Earth's core?

....

....

.... [2 marks]

2 Sandy wants to sunbathe and get a good tan.

a She is told that if she goes out in the Sun without sunscreen, she will burn in 15 minutes. How long can she safely sunbathe for if she uses a sunscreen with SPF 20?

....

.... [2 marks]

b The Earth's atmosphere contains a layer of ozone. This layer protects us from the effects of the Sun. When scientists first started measuring the thickness of the ozone layer, their results were unexpected. The layer was thinner than they thought it should be. They replaced all of their instruments. What should scientists do to confirm their results?

....

....

.... [1 mark]

3 Dave reads a newspaper article that claims CFCs are not responsible for the depletion of the ozone layer. He tells his friends at school that the science teachers have not been teaching them correctly about ozone depletion. Explain why Dave should not have believed what he read in the newspaper.

....

....

.... [2 marks]

P1 Extended response question

Many people now use mobile phones.

Jenny is walking in the hills of North Wales and her brother is walking around the Lake District. They keep in touch using their mobile phones. Explain reasons why the signal strength on both their phones changes, even as they walk only a short distance. Why do some people think it would be better if Jenny were to text her brother instead?

❗ The quality of written communication will be assessed in your answer to this question.

....

....

....

....

....

....

....

....

....

....

....

....

....

....

....

....

....

....

....

....

....

....

....

....

....

....

....

....

....

....

....

....

....

....

.... [6 marks]

Collecting energy from the Sun

1 a Write down **four** advantages of using photocells.

....

....

....

.... **[4 marks]**

D–C

b A p-n junction is made from two pieces of silicon. Explain how the two pieces of silicon are different and what causes the difference.

....

....

.... **[3 marks]**

B–A*

2 Name two methods of increasing the output from a photocell.

....

.... **[2 marks]**

3 a This is a question about passive solar heating.

During the day, short wavelength radiation from the Sun passes through the glass in the large window and warms the room.

Explain how the glass keeps the room warm during the night

.... **[1 mark]**

B–A*

b The diagram represents the electromagnetic spectrum.

X-rays	ultraviolet	visible light	infrared	radio

i Write the letter S on the diagram to show the wavelength of radiation from the Sun that is absorbed by plants in a greenhouse. **[1 mark]**

ii Write the letter P on the diagram to show the wavelength of radiation that is re-radiated from the plants in a greenhouse.

.... **[1 mark]**

4 a A wind turbine transfers the kinetic energy of the wind into electricity. What does the amount of electricity produced depend on?

.... **[1 mark]**

b Write down two disadvantages and two advantages of generating electricity using wind turbines.

Advantages

....

....

Disadvantages

....

.... **[4 marks]**

D–C

Generating electricity

1 a The diagram shows a model dynamo. When the coil is spun, a current is produced.

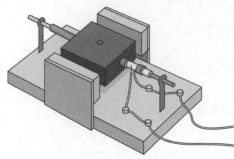

Write down two ways in which the size of the current can be increased.

....

.... **[2 marks]**

b A model generator consists of a coil of wire rotating between the poles of a magnet. How is the structure of a generator at a power station different from the model generator?

....

....

.... **[1 mark]**

c The oscilloscope trace shows the output from a generator. Each division of the timescale is 5 milliseconds.

i Mark on the diagram the period of the alternating voltage. **[1 mark]**

ii What is the peak voltage?

.... **[1 mark]**

2 A power station is about 30% efficient. If the fuel provides 200 MJ of energy, how much electrical energy will be produced?

....

....

....

....

....

....

.... **[1 mark]**

Global warming

1 a How do the greenhouse gases cause the temperature of the Earth to increase?

.... [1 mark]

b Which of the greenhouse gases is the most significant ?

.... [1 mark]

2 a How has deforestation affected the levels of carbon dioxide in the atmosphere?

.... [1 mark]

b Write down four natural sources of carbon dioxide.

.... [4 marks]

c Give three sources of methane gas caused by human activity.

.... [3 marks]

3 a Describe how electromagnetic radiation from the Sun can be trapped by gases in the Earth's atmosphere.

....

.... [3 marks]

b Dust in the atmosphere can cause either an increase or a decrease in the Earth's temperature. Explain how this is possible.

Increase in temperature is caused by:

.... [1 mark]

Decrease in temperature is caused by:

.... [1 mark]

4 a Most scientists agree that the average temperature of the Earth is increasing but what do they disagree on?

.... [1 mark]

b On what basis should governments make decisions on what action to take on global warming?

.... [1 mark]

c Describe two consequences of global warming.

....

.... [2 marks]

5 Say whether the following statements are based on scientific fact or opinion.

Average temperatures have risen by 0.8 degree Celsius around the world since 1880 according to NASA.

.... [1 mark]

The recent extreme weather, such as heat waves and tropical storms, is caused by climate change say some experts.

.... [1 mark]

As illustrated in Ice Core data from the Soviet Station Vostok in Antarctica, CO_2 concentrations in Earth's atmosphere has been rising for 18 000 years.

.... [1 mark]

D–C

B–A*

D–C

Fuels for power

1 a Each of the headlamp bulbs in Sammy's car is connected to a 12 V battery. When she switches on the headlamps, a current of 2 A passes through the bulb. Calculate the power rating of the bulb.

....

....

....

....

[2 marks]

b In her home, Sammy uses a 2.5 kW kettle for $\frac{1}{2}$ hour each day. Electricity costs 12p per kWh. How much does it cost Sammy each day to use her kettle?

....

....

....

[3 marks]

c Sammy sets her dishwasher to work overnight. Why is electricity cheaper during the night?

....

....

....

[1 mark]

d Sammy knows she needs to keep her electricity use down below 2 kWh per night. She has a 4 KW dryer, how long can she use it for?

....

....

[1 mark]

2 a List three factors which should be taken into consideration when deciding on which energy source to use in a particular situation.

....

....

....

[3 marks]

b Some people believe that the UK needs more nuclear power stations. Others believe that renewable fuel can provide all of our energy requirements. What information could scientists and engineers provide to help the government decide?

....

....

....

[3 marks]

3 The National Grid distributes electricity around the country at 400 000 V.

a Give **two** reasons why such high voltages are used.

....

[2 marks]

b How does using such a high voltage affect the current in the wires?

....

[1 mark]

c How would this help with energy losses?

....

[1 mark]

Nuclear radiations

1 a Answer true or false to each of the following statements about alpha, beta and gamma radiation.

Gamma radiation is stopped by paper.	
Alpha radiation has a range of a few centimetres in air.	
Beta radiation comes from the nucleus of an atom.	
Beta radiation can be absorbed by a thin sheet of paper.	

[4 marks]   D–C

b i Atoms are neutral because they contain the same number of positive protons and negative electrons. Explain how negative and positive ions are formed when an atom is exposed to radiation.

negative ions are formed by:

.... [1 mark]

positive ions are formed by:

.... [1 mark]

ii Describe the effects ionisation can have on the cells of the human body.

....

.... [2 marks]   B–A*

2 Gamma radiation is used to sterilise medical instruments. It has other medical uses as well.

a Write down one other medical use for gamma radiation.

.... [1 mark]

b What property of gamma radiation makes it suitable?

.... [1 mark]   D–C

3 A source of alpha radiation is used in a smoke alarm. Explain how a smoke alarm works.

....

....

.... [3 marks]

4 Radioactive waste must be stored securely for possibly thousands of years.

a Why must it be stored for so long?

....

.... [1 mark]   D–C

b Some people are worried that terrorists may make a nuclear bomb from nuclear waste. Discuss whether or not we should have any concerns about terrorists obtaining nuclear waste.

....

....

....

.... [3 marks]   B–A*

1 a There are many objects in our Solar System. Describe the following objects:

i a star

.... [1 mark]

ii a planet

.... [1 mark]

iii a meteor

.... [1 mark]

b Our Solar System is part of a galaxy and scientists think there may be a black hole at the centre of our galaxy.

i What is a galaxy?

.... [1 mark]

ii Describe a black hole.

.... [1 mark]

c The diagram represents the Moon in orbit around Earth.

i What is the name of the force that keeps the Moon in orbit around Earth?

.... [1 mark]

ii Add an arrow to the diagram to show the direction in which this force acts on the Moon.

[1 mark]

2 a When astronauts work outside a spacecraft, they have to wear special helmets with Sun visors. Why do the helmets need special visors?

.... [1 mark]

b NASA is planning to send a manned spacecraft to another planet after 2020. It is expected to cost £400 billion.

Which planet will the spacecraft go to? Explain the reason for your answer.

Planet:

.... [1 mark]

Explanation:

.... [1 mark]

3 Proxima Centauri is 4.22 light-years away from us.

Explain what is meant by the term light-year.

.... [1 mark]

Threats to Earth

1 a Most asteroids orbit the Sun in a belt between two planets. Which two planets?

.... [2 marks]   D–C

b Explain why asteroids have not joined together to form another planet.

....

.... [2 marks]   B–A*

c Describe two pieces of evidence scientists have discovered that support the theory that asteroids have collided with Earth in the past.

....

.... [2 marks]   D–C

2 There is evidence to suggest our Moon was a result of the collision between two planets. The iron core of the other planet melted and joined with the Earth's core.

Describe two pieces of evidence to support this theory.   B–A*

....

.... [2 marks]

3 The diagram shows the orbits of two bodies orbiting the Sun. One is a planet, the other is a comet.

a label the comet's orbit with the letter C. [1 mark]   D–C

b Write the letter X to show where on the orbit the comet is travelling at its fastest. [1 mark]   B–A*

c Why does a comet's tail always point away from the Sun?   D–C

.... [1 mark]

4 Scientists are constantly updating information on the paths of near-Earth objects (NEO)s.

a Why is it important to constantly monitor the paths of NEOs?   D–C

....

.... [2 marks]

b i Scientists may want to change the course of a NEO. Explain how this may be done.

....

.... [2 marks]

ii what would they need to take into consideration when planning to change the course of a NEO?   B–A*

....

.... [2 marks]

The Big Bang

1 a The Universe is expanding. Galaxies in the Universe are moving at different speeds. Which galaxies are moving the fastest?

.... **[1 mark]**

b Scientists know how fast galaxies are moving because they measure red shift.

Explain what is meant by red shift.

....

....

.... **[3 marks]**

c How does red shift provide information about the speed of a galaxy?

....

.... **[1 mark]**

2 A star starts its life as a swirling cloud of gas and dust.

a Describe what happens to this cloud to produce a glowing star.

....

....

....

.... **[4 marks]**

The end of a star's life depends on how big it is.

b What happens to a medium sized star, like our Sun, at the end of its life?

....

....

....

.... **[4 marks]**

3 Models of our universe have changed a lot over time.

a How did Galileo contribute to the changes?

....

.... **[1 mark]**

b Why did it take a long time for Galileo's ideas to be accepted?

....

.... **[1 mark]**

c What did Isaac Newton contribute to the model we have today?

....

.... **[1 mark]**

P2 Extended response question

Scientists have been making observations of our universe for a very long time. They have gathered a lot of information on the life cycle of the stars they observe.

Describe, in detail, the life cycle of a medium-sized star such as our Sun. Explain why scientists describe our Sun as a second generation star.

❗ The quality of written communication will be assessed in your answer to this question.

....
....
....
....
....
....
....
....
....
....
....
....
....
....
....
....
....
....
....
....
....
....
....
....
....
....
....
....
....
....
....
....
....
....
....
....

[6 marks]

Molecules of life

1 a Muscle cells contain many mitochondria. Explain why muscle cells need so many mitochondria.

....

.... **[2 marks]**

b Ribosomes are also found in cells but are smaller than mitochondria.

i Where are ribosomes found in cells?

.... **[1 mark]**

ii What is the job of ribosomes?

.... **[1 mark]**

2 a The diagram shows part of a DNA molecule.

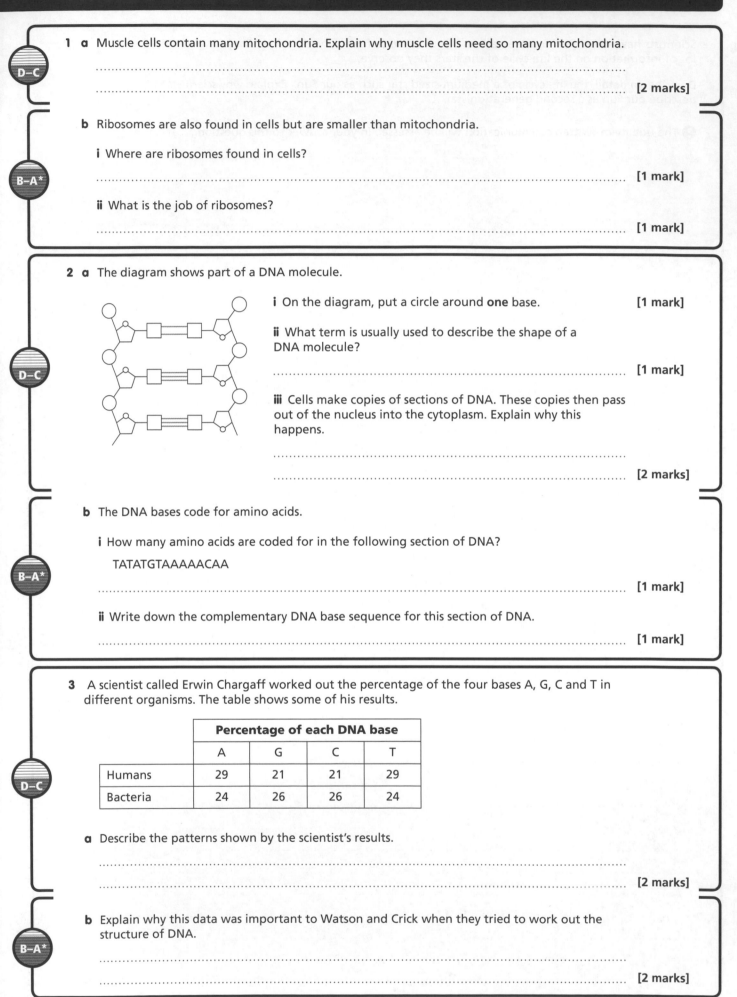

i On the diagram, put a circle around **one** base. **[1 mark]**

ii What term is usually used to describe the shape of a DNA molecule?

.... **[1 mark]**

iii Cells make copies of sections of DNA. These copies then pass out of the nucleus into the cytoplasm. Explain why this happens.

....

.... **[2 marks]**

b The DNA bases code for amino acids.

i How many amino acids are coded for in the following section of DNA?

TATATGTAAAAACAA

.... **[1 mark]**

ii Write down the complementary DNA base sequence for this section of DNA.

.... **[1 mark]**

3 A scientist called Erwin Chargaff worked out the percentage of the four bases A, G, C and T in different organisms. The table shows some of his results.

	Percentage of each DNA base			
	A	G	C	T
Humans	29	21	21	29
Bacteria	24	26	26	24

a Describe the patterns shown by the scientist's results.

....

.... **[2 marks]**

b Explain why this data was important to Watson and Crick when they tried to work out the structure of DNA.

....

.... **[2 marks]**

Proteins and mutations

1 a Draw straight lines to join each protein to its correct function.

Protein	Function
collagen	a carrier protein
haemoglobin	a hormone
insulin	a structural protein

[2 marks]

b Write down the name of the subunits that make up protein molecules.

.... [1 mark]

c Haemoglobin is called a globular protein because the molecules have a compact rounded shape. Collagen is a fibrous protein because the molecules are long and straight. Explain what causes these two proteins to have different shapes.

.... [1 mark]

2 a The enzyme amylase breaks down starch. Explain why amylase would not break down proteins.

.... [2 marks]

b Look at the graph. It shows the effect of temperature on the enzyme amylase.

i Describe the pattern shown in the graph.

....

....

.... [2 marks]

ii Write down the optimum temperature for this enzyme.

....

.... [1 mark]

iii Work out the Q_{10} for this enzyme between 10 and 20 °C. Answer = ... [2 marks]

iv Explain why the enzyme shows such an increase in rate between these temperatures.

....

.... [2 marks]

3 Changes to genes can happen spontaneously or are caused by factors in the environment.

a Write down one environmental factor that increases the chance of a change in a gene.

.... [1 mark]

b Why are changes to genes important for evolution?

.... [2 marks]

c The diagram shows how a white pigment is turned into a purple pigment in the petals of a flower using two enzymes. Explain how a change to the DNA could produce plants with red flowers.

Enzyme A   Enzyme B

white pigment ⟹ red pigment ⟹ purple pigment

....

.... [3 marks]

Respiration

1 a A horse is waiting to run in a race. During the race the horse will need to generate more ATP.

Explain the function of ATP.

....

.... **[2 marks]**

b The horse starts to run.

Complete the balanced symbol equation for aerobic respiration in the horse.

$C_6H_{12}O_6$ + ... $\rightarrow$ .... + ... **[2 marks]**

c The graph shows the lactic acid concentration in the horse's blood as it runs at different speeds.

i Describe the effect on lactic acid production as the horse runs faster.

....

.... **[2 marks]**

ii Identify what types of respiration are occurring at different points on the graph.

....

....

.... **[3 marks]**

iii Horses are more likely to damage their muscles when their blood lactic acid concentration is above 4 mmol per litre.

How fast can this horse run before this level is reached?

.... **[1 mark]**

d The horse stops running. Describe what happens to the lactic acid in the horse.

....

.... **[2 marks]**

2 a Suggest **one** way that the rate of respiration of an animal can be measured.

.... **[1 mark]**

b During a race a horse's temperature may increase by several degrees.

However, if it gets too high it can cause a fall in the rate of respiration in the horse's cells.

Suggest why this might be.

....

.... **[2 marks]**

Cell division

1 a Humans are multicellular organisms. There can be advantages to being multicellular rather than unicellular. Explain why.

....

.... [2 marks]

b The table shows the surface area and volume of different cubes.

i Finish the table by calculating the surface area to volume ratio of each cube. The first one has been done for you.

Cube	Surface area in cm²	Volume in cm³	Ratio
A	24	8	$\frac{24}{8} = 3.0$
B	54	27	
C	96	64	
D	150	125	

[3 marks]

ii Explain why larger, multicellular organisms need to develop special exchange surfaces, such as lungs. Use data from the table to help you.

....

.... [2 marks]

2 Look at the statements about cell division.

a Put a tick (✓) next to each statement that refers to the type of cell division that makes new body cells.

The new cells are diploid. ☐   The new cells show variation. ☐

Four new cells are made. ☐   Before cells divide, DNA replication takes place. ☐

The new cells contain 23 chromosomes. ☐   [2 marks]

b DNA replication is called 'semiconservative'. This is because two new DNA molecules are produced and each one has one new strand and one original strand of the DNA molecule.

Explain how DNA replication produces this result.

....

.... [2 marks]

3 a Scientists have discovered a mutation in the DNA of mice. They have found that this change makes the mice produce sperm without an acrosome.

The sperm that are produced without an acrosome **cannot** fertilise an egg. Explain why.

....

.... [2 marks]

b The diagram shows a cell dividing. The cell is shown during the first division of meiosis and during the second division.

Describe the differences between the two diagrams.

....

....

.... [3 marks]

The circulatory system

1 a Red blood cells are adapted to do their job. They are disc shaped and have no nucleus.

Explain how these adaptations help them do their job.

Disc shaped: .... [1 mark]

No nucleus: .... [1 mark]

b Write down the name of the chemical that makes red blood cells red.

.... [1 mark]

c Explain how this chemical transports oxygen around the body.

....

.... [3 marks]

2 Three different types of blood vessels transport blood around the body.

a Describe the role of these blood vessels in circulating blood around the body.

....

....

.... [3 marks]

b The diagrams show the three different types of blood vessels.

| A | B | C |

i Write the name of each type of blood vessel in the box under the correct diagram.

[2 marks]

ii Describe how blood vessel **C** is adapted for its function.

....

.... [2 marks]

3 Look at the diagram of the heart.

a On the diagram of the heart, label the bicuspid valve and the aorta. [2 marks]

b The left ventricle has a thicker wall than the right ventricle. Explain why.

....

.... [2 marks]

c The human heart is part of a double circulatory system. Write down one advantage of having a double circulatory system rather than a single system.

.... [1 mark]

D–C

B–A*

D–C

B–A*

D–C

B–A*

Growth and development

1 Look at the diagram of a bacterial cell.

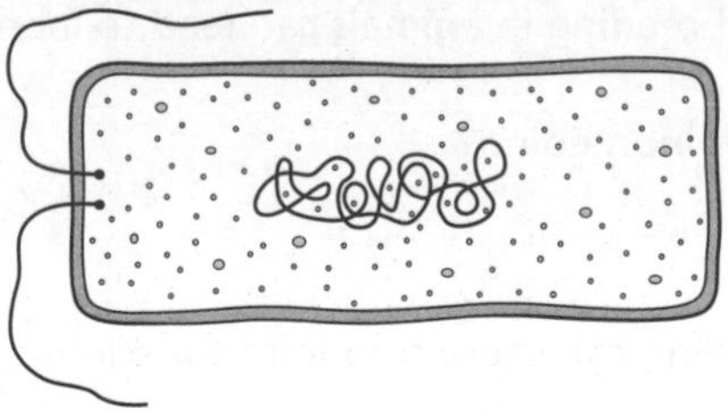

   a Apart from size, write down two differences
between the structure of a bacterial cell and
a human cheek cell.

....

....

[2 marks]

D–C

   b Describe how the shape of a bacterial chromosome is different from a human chromosome.

.... [1 mark]

B–A*

2 The graph shows the growth curve for males and females.

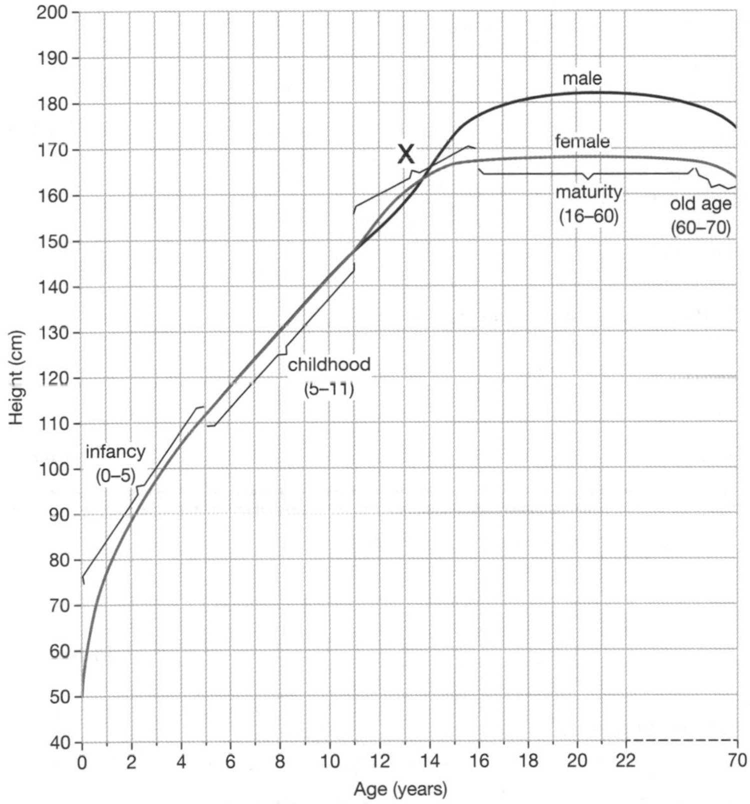

   a Which growth phase is
marked as **X** on the graph?

....

.... [1 mark]

D–C

   b Suggest the reason for the
difference in the height between
a girl and a boy at age 13.

....

....

....

....

.... [2 marks]

   c The graph uses height as a
measure of growth. What
are the advantages and
disadvantages of measuring
growth in this way?

....

....

....

B–A*

.... [2 marks]

3 For a fertilised egg to grow into an embryo the cells need to divide and change.

   a Write down the name that is given to cells in the embryo that can form specialised cells.

.... [1 mark]

D–C

   b Similar unspecialised cells exist in the body after birth. How do these unspecialised cells
differ from those found in the embryo?

....

B–A*

.... [2 marks]

4 What is a meristem?

.... [1 mark]

D–C

New genes for old

1 Selective breeding in animals can lead to inbreeding.

D–C

a What is inbreeding?

.... [1 mark]

B–A*

b Inbreeding can cause problems for scientists who are trying to mate endangered animals.

Explain what problems occur from inbreeding.

....

.... [2 marks]

2 Beta-carotene is found in carrots but not rice. The gene for beta-carotene can be transferred from carrots to rice.

D–C

a Suggest why this transfer might be useful.

....

.... [2 marks]

b Suggest one possible risk of genetic engineering.

.... [1 mark]

B–A*

c Outline the steps that would be needed in this genetic engineering of rice.

....

....

.... [3 marks]

3 Scientists hope to be able to transfer genes into people with genetic disorders.

a What name is given to this type of process?

.... [1 mark]

D–C

b Read this article about the development of this process.

> The treatment was daring. In 1990 a little American girl underwent the world's first experiment – an attempt to repair her immune system for the rest of her life. The treatment involved putting genes into her white blood cells, using a virus to inject the genes. The normal virus genes had first been removed.
>
> Today, she is alive and well. Without the treatment it is unlikely she could have survived. Her survival signalled the possibility to cure many life-threatening genetic disorders. However, after 5,000 patients had participated in 350 trials, things began to go wrong. First, in 1999, a young man died of a massive immune reaction to the gene treatment. Three years later, a French baby developed leukaemia as a by-product of the treatment. Experts say the virus that inserted the genes mistakenly turned on a cancer-causing gene.
>
> Scientists are now learning from these mistakes and further research into gene therapy is taking place.

Give arguments for and against the use of this treatment on people.

....

.... [3 marks]

Cloning

1 a Cows can be cloned using **nuclear transfer**.

What is meant by nuclear transfer?

....

.... [2 marks]

b Scientists are hoping to solve organ transplant problems by cloning pigs.

i Suggest how the cloning of pigs could help solve organ transplant problems.

....

.... [2 marks]

ii Suggest one other reason that people might want to clone pigs.

.... [1 marks]

D–C

2 The diagram shows a planned possible technique to produce a human cloned embryo for therapeutic use. The technique is similar to that used to produce Dolly the sheep.

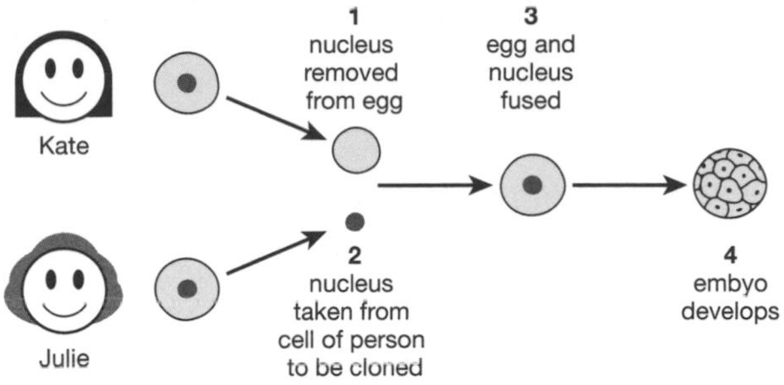

Kate

1
nucleus
removed
from egg

3
egg and
nucleus
fused

2
nucleus
taken from
cell of person
to be cloned

Julie

4
embyo
develops

B–A*

a What is done between steps 3 and 4 to make the cell divide?

.... [1 mark]

b If the embryo was allowed to grow into a baby, would it most resemble Kate, or Julie?

Explain your answer.

....

.... [2 marks]

3 Strawberry plants can reproduce sexually or asexually.

a Write down one advantage and one disadvantage to a gardener of reproducing strawberry plants asexually.

....

.... [2 marks]

D–C

b Plants like strawberries can be artificially reproduced by tissue culture.

Describe what is meant by the terms 'aseptic technique' and 'growth medium' in this process.

....

.... [2 marks]

B–A*

B3 Extended response question

Collagen is an important protein in the body.

Sometimes, disorders involving collagen cause problems in the skin and joints. They can also cause blood vessels to weaken and arteries may burst. These disorders can be inherited.

Explain how an inherited disorder can affect collagen and suggest why it causes these particular problems.

❗ The quality of written communication will be assessed in your answer to this question.

....
....
....
....
....
....
....
....
....
....
....
....
....
....
....
....
....
....
....
....
....
....
....
....
....
....
....
....

[6 marks]

Ecology in the local environment

1 Look at the kite diagram showing the distribution of different species of barnacles on rocks on a sea shore.

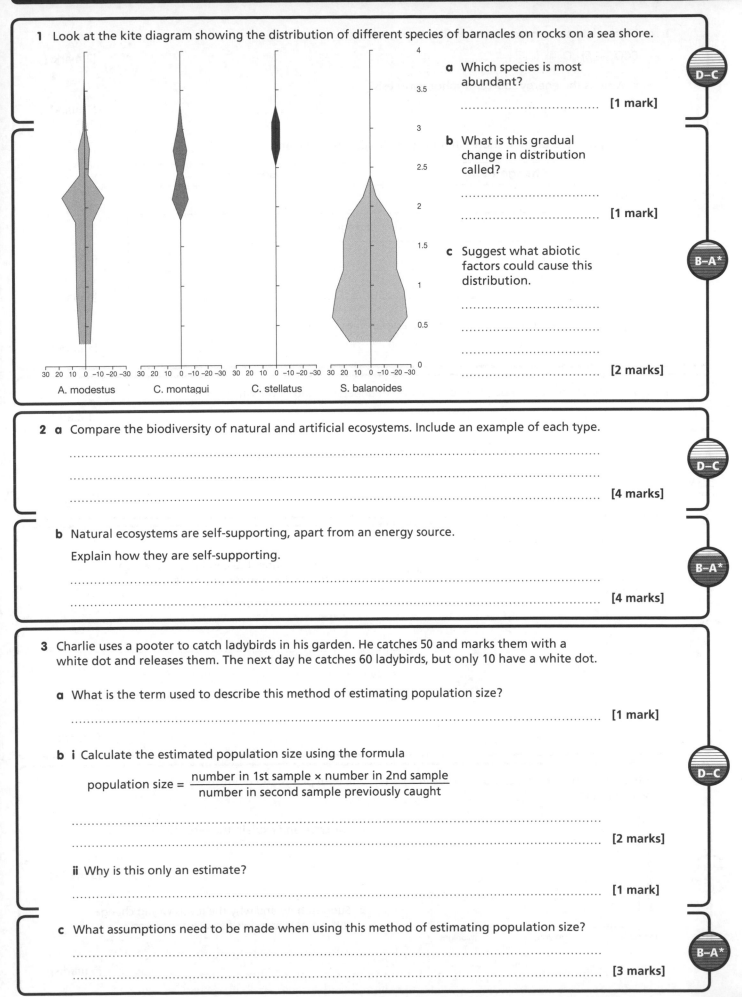

A. modestus   C. montagui   C. stellatus   S. balanoides

a Which species is most abundant?

.... [1 mark]

D–C

b What is this gradual change in distribution called?

....

.... [1 mark]

c Suggest what abiotic factors could cause this distribution.

....

....

....

.... [2 marks]

B–A*

---

2 a Compare the biodiversity of natural and artificial ecosystems. Include an example of each type.

....

....

.... [4 marks]

D–C

b Natural ecosystems are self-supporting, apart from an energy source.

Explain how they are self-supporting.

....

.... [4 marks]

B–A*

---

3 Charlie uses a pooter to catch ladybirds in his garden. He catches 50 and marks them with a white dot and releases them. The next day he catches 60 ladybirds, but only 10 have a white dot.

a What is the term used to describe this method of estimating population size?

.... [1 mark]

b i Calculate the estimated population size using the formula

$$\text{population size} = \frac{\text{number in 1st sample} \times \text{number in 2nd sample}}{\text{number in second sample previously caught}}$$

....

.... [2 marks]

D–C

ii Why is this only an estimate?

.... [1 mark]

c What assumptions need to be made when using this method of estimating population size?

....

.... [3 marks]

B–A*

Photosynthesis

1 a i Complete the symbol equation for photosynthesis.

$6CO_2 + 6H_2O \rightarrow$ ....

[2 marks]

ii What is the energy source for photosynthesis?

.... [1 mark]

b i Complete the table to show what happens to glucose in a plant.

Changed to:	Used for:
cellulose	
	storage
proteins	
fats and oils	

[4 marks]

ii If glucose is not changed to other substances, what is it used for?

.... [1 mark]

2 a What was Priestley's important contribution in understanding photosynthesis?

.... [1 mark]

b Explain how the use of isotopes changed our understanding of photosynthesis.

....

.... [3 marks]

3 Look at the diagram of a greenhouse.

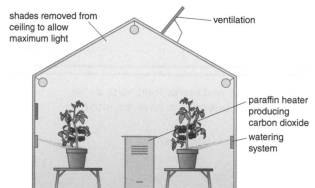

shades removed from ceiling to allow maximum light

ventilation

paraffin heater producing carbon dioxide

watering system

Describe how the conditions necessary for photosynthesis are achieved.

....

....

....

....

....

....

.... [3 marks]

4 The levels of oxygen and carbon dioxide in a water tank with plants and goldfish were recorded over 24 hours.

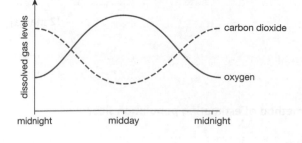

carbon dioxide

oxygen

dissolved gas levels

midnight   midday   midnight

a Describe and explain the results.

....

.... [3 marks]

b Suggest how and why the levels would change if the plants and goldfish were removed.

....

.... [3 marks]

Leaves and photosynthesis

1 a Complete the table showing leaf adaptations and their uses.

Adaptation:	Uses:
broad leaves	increase surface area to get more light
thin leaves	
	to absorb light from different parts of the spectrum
	support and transport
guard cells	

[4 marks]

b An insect called a leaf miner burrows through the inside of leaves and eats the cells.

i Name two types of cells it will eat.

.... [2 marks]

II Explain why the insect will cause colourless lines on the leaves.

.... [2 marks]

2 Look at the graph. It shows the absorption spectrum of two plant pigments.

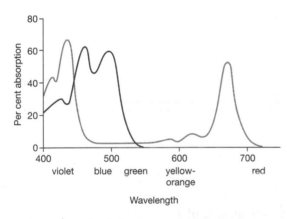

Per cent absorption

400 500 600 700
violet blue green yellow- red
orange

Wavelength

a Name two photosynthetic pigments.

....

.... [1 mark]

b What do you understand by the term 'absorption spectrum'?

....

....

.... [1 mark]

c How does the information in the graph show that the leaves maximise the use of energy from the Sun?

....

.... [2 marks]

d Which part of the light spectrum is not used in photosynthesis?

.... [1 mark]

e If you wanted the maximum growth in your plants, which wavelengths of light would you shine on them?

.... [2 marks]

f The longer wavelengths of light are absorbed or reflected by the sea. Suggest what effect this will have on plants growing on the sea bed.

....

....

.... [3 marks]

Diffusion and osmosis

1 Ruby opens a bottle of perfume. The perfume evaporates. After a few seconds, Ruby smells the perfume.

She draws a diagram to explain this.

molecules of air

perfume molecules

perfume

a Using the same symbols, draw in the second box the position of molecules where you expect them to be after a few seconds. **[2 marks]**

b Name the process involved in these changes.

.... **[1 mark]**

c Explain how differences in concentration cause these changes.

....

.... **[2 marks]**

d Describe three ways of increasing the rate of these changes.

1 ....

2 ....

3 .... **[3 marks]**

2 Complete the definition of osmosis.

Osmosis is the movement of .... across a

.... membrane.

It takes place from an area of a .... solution to an area of a

.... solution.

The movement is a consequence of the .... movement of individual particles. **[5 marks]**

3 When placed in very salty water, plant cells become plasmolysed.

a Describe what happens to the contents of the plant cells.

....

.... **[2 marks]**

b What are the possible consequences of plasmolysis in plant cells?

....

.... **[2 marks]**

c Animal cells behave differently when placed in very salty water. Explain why.

....

.... **[2 marks]**

Transport in plants

1 a Complete the sentences about transport of water and food in plants.

Water and food travel through plants inside .... bundles.

Water travels inside .... cells.

Food travels inside .... cells. **[3 marks]**

D–C

b Describe how xylem vessels are different from phloem cells.

....

.... **[3 marks]**

B–A*

---

2 Look at the diagram of the lower leaf surface of two leaves kept in different conditions.

stoma closed   stoma open

leaf A   leaf B

a Count the number of open stomata in Leaf A and Leaf B.

Open stomata in Leaf A: ....

Open stomata in Leaf B: .... **[1 mark]**

D–C

b Which leaf has been kept in dark conditions? Explain your answer.

....

.... **[2 marks]**

c Explain how the plant can change its stomatal apertures.

....

....

.... **[3 marks]**

B–A*

---

3 Four leaves from the same plant were smeared with petroleum jelly in different ways, weighed and suspended from a string. The experiment was left for two hours and the leaves were reweighed.

Leaf	Treatment	Initial weight g	Final weight g	Weight change g
1	No petroleum jelly used	10.9	7.2	3.7
2	Petroleum jelly on both leaf surfaces	9.8	9.2	0.6
3	Petroleum jelly on lower surface	10.1	9.1	1.0
4	Petroleum jelly on upper surface	9.9	7.2	2.7

D–C

a What was the main cause of weight loss?

.... **[1 mark]**

b Use the data to show that most stomata are in the lower surface.

....

.... **[2 marks]**

c Suggest and explain the expected results if Leaf 1 was kept at a slightly higher temperature.

....

....

.... **[3 marks]**

B–A*

Plants need minerals

1 Look at the diagram of plants grown in conditions lacking certain minerals.

poor growth, yellow leaves — A
poor root growth, discoloured leaves — B
poor fruits and flowers, discoloured leaves — C
yellow leaves — D

D–C

a Write down the mineral that each plant is lacking.

Plant A: ....   Plant B: ....

Plant C: ....   Plant D: ....

[4 marks]

b Plants lacking nitrogen are usually smaller than usual. Explain why.

.... [2 marks]

---

2 Complete the table about the use of some minerals.

B–A*

Element	Use of element
nitrogen	make proteins
phosphorus	carries genetic information
magnesium	

[3 marks]

---

3 Look at the diagram showing how minerals are absorbed by root hairs.

carrier
+ energy
minerals
+ energy
outside cell   cell membrane   inside cell

B–A*

a What is the process called?

.... [1 mark]

b Describe how this process is different from diffusion.

....

....

.... [4 marks]

Decay

1 Ali wants to make garden compost. He wants his garden waste to decay quickly.

   a What conditions should Ali provide?

      Put a tick (✓) in the box next to each correct answer.

      A temperature of 25 °C   ☐

      A temperature of 75 °C   ☐

      A large amount of water   ☐

      Plenty of carbon dioxide   ☐

      Plenty of oxygen   ☐   **[2 marks]**

   b Name **three** useful animals he would expect to find in his compost heap.

      1 ....

      2 ....

      3 ....   **[3 marks]**

   c Suggest different views people may have on having compost heaps in their garden.

      ....

      ....   **[2 marks]**

   d **i** Explain why detritivores are important in decay.

      ....

      ....   **[2 marks]**

     ii Explain how saprophytes are important in decay.

      ....

      ....   **[2 marks]**

   e Explain how and why temperature affects the rate of decay.

      ....

      ....

      ....   **[4 marks]**

2 Ali uses a refrigerator and freezer for his garden produce.

   Explain how putting foods in:

   a a refrigerator at 5 °C

   b a deep freeze at −22 °C

   will help to keep them for longer.

   ....

   ....

   ....

   ....   **[4 marks]**

D–C

B–A*

D–C

Farming

1 Plants can be grown without the use of soil.

   a What is this system of intensive farming called?

   .... **[1 mark]**

   b i Explain the **advantages** of this system.

   ....

   .... **[2 marks]**

   ii Explain the **disadvantages** of this system.

   ....

   .... **[2 marks]**

2 In 1935 large cane toads from South America were introduced into Queensland in Australia to control insect pests that attack sugar cane plants.

The toads had little effect on the insect pest population. They have now spread into nearly all areas of Australia and have no natural predators. They are a pest, eating the smaller native toad and causing a great deal of damage. Because of their large size and poisonous skin, they have few predators.

The insect pests are now controlled by insecticides.

   a This is an example of introducing an organism into an ecosystem to control another organism. What is this method called?

   .... **[1 mark]**

   b Suggest why the introduction of cane toads was thought to be better than using insecticides.

   ....

   .... **[2 marks]**

   c Suggest why the introduction of cane toads was not a success.

   ....

   .... **[2 marks]**

3 The use of crop rotation is important in farming.

   a Some farmers use a crop rotation system. Explain why.

   ....

   .... **[2 marks]**

   b Some organic farmers vary the seed planting time of their crops. Explain why.

   ....

   .... **[2 marks]**

B4 Extended response question

Materials can enter and leave cells.

The diagram represents a plant cell in water.

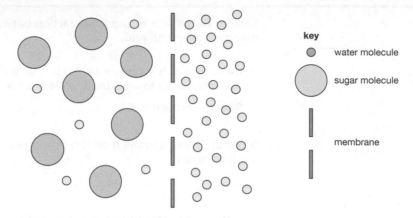

key

water molecule

sugar molecule

membrane

Explain your predicted movement of molecules and the importance of any movement.

❗ The quality of written communication will be assessed in your answer to this question.

....

....

....

....

....

....

....

....

....

....

....

....

....

....

....

....

....

....

....

....

....

....

....

....

....

.... [6 marks]

Rate of reaction (1)

1 a What is the rate of reaction?

.... [1 mark]

2

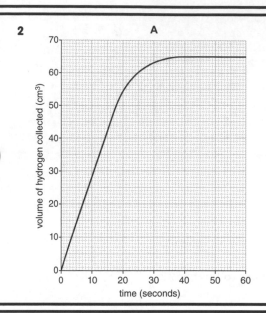

Look at **Graph A**. It records the reaction between magnesium and an acid.

a If the rate of reaction measured in the first 10 seconds is X, what are the units used to measure X?

Rate of reaction = X .... [1 mark]

b What happens to the rate of reaction after 30 seconds?

.... [1 mark]

c If the acid is not used up at the end of the reaction, which is the limiting reactant?

.... [1 mark]

3

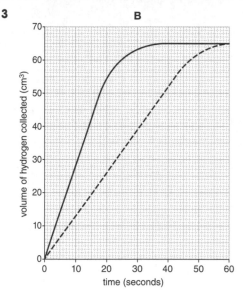

Look at **Graph B**.

a Does the second line represent a reaction with a faster or slower rate of reaction?

.... [1 mark]

b Calculate the rate of reaction of the second reaction between 20 to 40 seconds. Show your method and include the units.

....

....

.... [3 marks]

c Predict how much hydrogen will be given off at 35 seconds.

.... [1 mark]

4

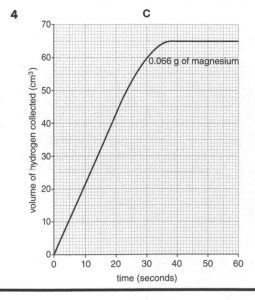

0.066 g of magnesium

Look at **Graph C**. This graph shows the second reaction from Graph B.

a If half the amount of magnesium is used in a third reaction, sketch the result you would expect to find on to the graph, if the path of this reaction was measured. [2 marks]

b Explain your answer in terms of how the particles react together.

....

....

.... [2 marks]

Rate of reaction (2)

1 Look at Graph A. It shows the reaction between magnesium and an acid at 30 °C.

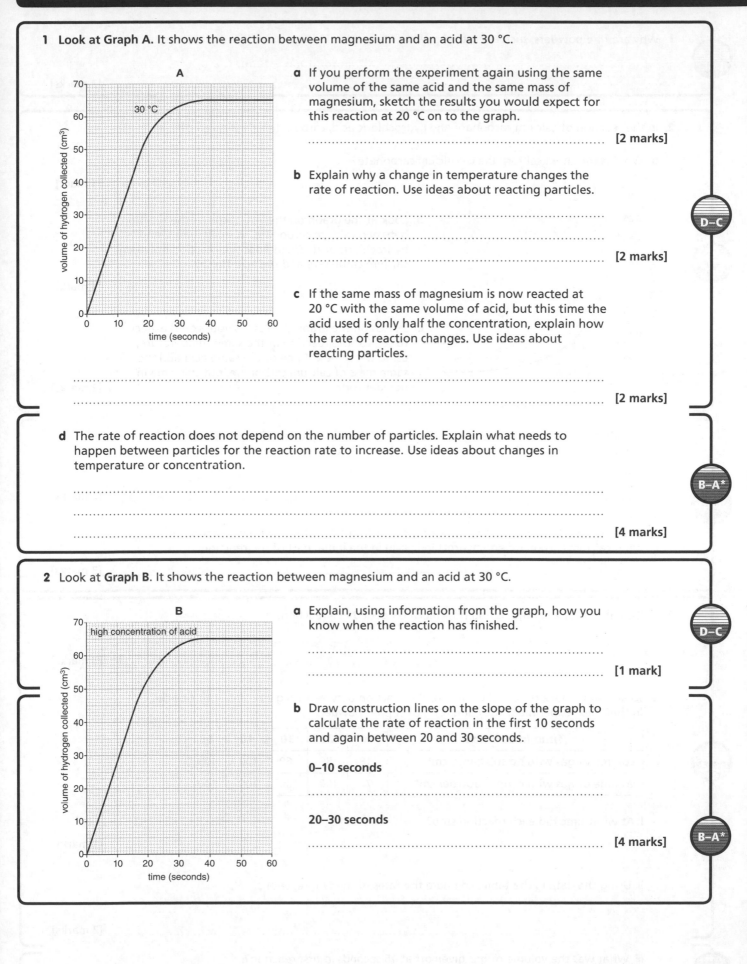

a If you perform the experiment again using the same volume of the same acid and the same mass of magnesium, sketch the results you would expect for this reaction at 20 °C on to the graph.

.... [2 marks]

b Explain why a change in temperature changes the rate of reaction. Use ideas about reacting particles.

....

....

.... [2 marks]

D–C

c If the same mass of magnesium is now reacted at 20 °C with the same volume of acid, but this time the acid used is only half the concentration, explain how the rate of reaction changes. Use ideas about reacting particles.

....

.... [2 marks]

d The rate of reaction does not depend on the number of particles. Explain what needs to happen between particles for the reaction rate to increase. Use ideas about changes in temperature or concentration.

....

....

.... [4 marks]

B–A*

2 Look at **Graph B**. It shows the reaction between magnesium and an acid at 30 °C.

a Explain, using information from the graph, how you know when the reaction has finished.

....

.... [1 mark]

D–C

b Draw construction lines on the slope of the graph to calculate the rate of reaction in the first 10 seconds and again between 20 and 30 seconds.

0–10 seconds

....

20–30 seconds

.... [4 marks]

B–A*

D–C 1 Why are fine powders, such as custard powder, dangerous in large quantities in factories?

....

.... [3 marks]

2 In the reaction of calcium carbonate and hydrochloric acid, carbon dioxide is given off.

   a What is the chemical formula of calcium carbonate?

.... [1 mark]

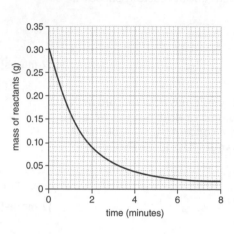

   b Look at the graph on the left. It shows the reaction between calcium carbonate, in small lumps, and hydrochloric acid. When does the reaction between calcium carbonate and the acid finish?

.... [1 mark]

   c Sketch on the graph the results you would expect to see for this reaction at the same temperature, using the same volume of the same acid and the same mass of calcium carbonate, but this time in powder form. [2 marks]

   d Explain the difference in the rate of reaction between lump calcium carbonate and its powdered form, with hydrochloric acid. Use ideas about reacting particles.

....

.... [2 marks]

B–A*   **e** Draw construction lines on the slope of the graph to calculate the rate of reaction of the lump calcium carbonate and hydrochloric acid in the initial part of the reaction.

.... [3 marks]

3 **a** Describe what a catalyst is and how much is needed in a reaction.

....

.... [3 marks]

   b Jo and Akira have collected results for the reaction between 1.5 g zinc and sulfuric acid. In their next experiment they add 0.1 g of copper powder.

D–C

Time in seconds	10	20	30	40	50
volume of gas with no substance cm^3	18	30	60	100	100
volume of gas with copper powder cm^3	70	100	100	100	100

   i At what time did each reaction stop?

.... [1 mark]

   ii Using the data in the table, compare the rates of the two reactions.

....

.... [2 marks]

B–A*   **iii** What was the volume of gas given off at 15 seconds in first reaction?

.... [1 mark]

Reacting masses

1 a How many atoms are there in the formula of calcium hydroxide, $Ca(OH)_2$?

.... [1 mark]

b Work out the relative formula mass of $Ca(OH)_2$.

A_r (Ca = 40, O = 16, H = 1)

.... [1 mark]

c Show that when sodium hydroxide reacts with nitric acid to make sodium nitrate and water that mass is conserved.

$$NaOH + HNO_3 \rightarrow NaNO_3 + H_2O$$

A_r (Na = 23, O = 16, H = 1, N = 14)

....

.... [2 marks]

d Use the balanced symbol equation to show that mass is conserved in this reaction.

$$CaCO_3 + 2HNO_3 \rightarrow Ca(NO_3)_2 + H_2O + CO_2$$

A_r (Ca = 40, C = 12, O = 16, H = 1, N = 14)

....

....

.... [3 marks]

2 a Zinc reacts with hydrochloric acid to make zinc chloride.

$$Zn + 2HCl_2 \rightarrow ZnCl_2 + H_2$$

Jo needs to make a certain amount of zinc chloride. How will she make sure that the amount of zinc she uses is the limiting factor?

....

.... [3 marks]

b Zinc carbonate, $ZnCO_3$, decomposes on heating to give zinc oxide (ZnO) and carbon dioxide.

i Write a balanced symbol equation for this reaction.

.... [2 marks]

ii Calculate how much ZnO is made when 12.50 g of $ZnCO_3$ decomposes.

A_r (Zn = 65, O = 16, C = 12)

....

....

....

.... [3 marks]

iii How much $ZnCO_3$ would you need to decompose to get 22 g of CO_2?

....

.... [2 marks]

Percentage yield and atom economy

1 Leo and Lesley have made some crystals of magnesium sulfate, although they have not made as much as they had hoped to. They wanted to make 42 g but only made 28 g.

D–C

a What was their **actual yield**?

.... [1 mark]

b What was their **predicted yield**?

.... [1 mark]

c i How could they calculate their **percentage yield**?

....

.... [1 mark]

ii Complete their calculation.

....

.... [1 mark]

B–A*

2 An industrial company makes sulfuric acid on a very large scale. Explain why the company wants the percentage yield of the process to be as high as possible.

....

.... [2 marks]

---

3 **a** Sodium nitrate is made using sodium hydroxide and nitric acid (water is an undesired product).

$$NaOH + HNO_3 \rightarrow NaNO_3 + H_2O$$

$$A_r (Na = 23, O = 16, H = 1, N = 14)$$

D–C

i Calculate the relative formula mass of both products.

....

.... [2 marks]

ii Using the formula for atom economy, calculate the atom economy for sodium nitrate.

....

....

....

.... [3 marks]

b Calculate the atom economy for calcium nitrate from the symbol equation and relative formula masses provided (water is an undesired product).

$$Ca(OH)_2 + 2HNO_3 \rightarrow Ca(NO_3)_2 + 2H_2O$$

$$A_r (Ca = 40, O = 16, H = 1, N = 14)$$

B–A*

....

....

.... [3 marks]

c Explain why a company wants the atom economy of a process to be as high as possible.

....

....

.... [3 marks]

C3 Chemical economics

Energy

1 Choose the best correct word from the list to complete sentence (a).

Choose another correct word to complete sentence (b).

catalytic   exothermic   endothermic   relative

a Bond breaking is a process that is ....   [1 mark]

b Bond making is a process that is ....   [1 mark]

c The reaction between nitrogen and hydrogen to make ammonia is an exothermic reaction. Explain why. Include ideas about bond breaking and bond making in your answer.

....

....

....   [3 marks]

2 a Todd and Terri want to compare the energy transferred by two fuels, A and B. Describe how they can do this by using the apparatus in the diagram.

100 g water

spirit burner

....

....

....

....

....

....

....   [5 marks]

b **Fuel A** heats the water (100 g) from 20 °C to 50 °C. They use the formula:

Energy transferred (J) = mass of water × specific heat capacity × temperature change

Work out how much energy Fuel A has transferred (specific heat capacity of water is 4.2 J/g °C)

....

....

....

....   [3 marks]

c If **Fuel B** releases 16 800 J using 1.4 g of fuel, calculate the energy released per gram of Fuel B.

....

....   [3 marks]

d If 3.0 g of **Fuel A** was used to heat the water in the experiment in part (b), work out which fuel gives out the most energy per gram.

....

....

....   [2 marks]

Batch or continuous?

1 a Why are pharmaceuticals made in batch processes?

....

.... **[2 marks]**

b Why are some chemicals, like ammonia, made using continuous processes?

....

.... **[2 marks]**

c Explain **one** advantage of batch processes.

.... **[1 mark]**

d Explain the advantages of continuous processes.

....

.... **[2 marks]**

e Explain the disadvantages of batch processes.

.... **[1 mark]**

2 a Explain why it is often expensive to make and develop new pharmaceutical drugs.

....

....

.... **[3 marks]**

b Explain why it is difficult to test and develop new pharmaceutical drugs that are safe to use.

....

.... **[2 marks]**

3 a Anya and Mark are researchers who have returned with a newly discovered plant which they think may contain an important chemical Z. Describe how they could extract the chemical from the plant and compare it to two other chemicals, P and R.

....

....

....

.... **[4 marks]**

b Anya and Mark test chemical Z for purity. They produce these results.

Chemical	Melting point °C	Boiling point °C	Chromatographic shift Rf
Z (new)	63.5 °C	124 °C	2 cm in 10 mins
P (pure old)	64.1 °C	122 °C	2 cm in 10 mins
R (pure old)	63.0 °C	126 °C	1.5 cm in 10 mins

Explain why they think that chemical Z may be a sample of impure P and not R.

....

....

.... **[3 marks]**

C3 Chemical economics

Allotropes of carbon and nanochemistry

1 a Explain why diamond, graphite and fullerenes are allotropes of carbon.

.... [1 mark]

b Explain why fullerenes can be used in new drug delivery systems.

.... [1 mark]

2 a Explain why diamond is used in cutting tools and jewellery.

....

.... [2 marks]

b Explain why graphite is used in pencil leads and lubricants.

....

.... [2 marks]

c Use ideas about structure and bonding to answer these questions.

i Explain why diamond does not conduct electricity.

....

.... [2 marks]

ii Explain why graphite conducts electricity.

....

.... [2 marks]

iii Explain why graphite is slippery.

....

.... [2 marks]

iv Explain why both diamond and graphite have high melting points.

....

.... [2 marks]

3 a Explain why diamond and graphite have a giant molecular structure.

....

.... [2 marks]

b Substances that have giant molecular structures have particular properties. Describe some of these properties and explain why some substances have them.

....

.... [2 marks]

c Explain how the structure of nanotubes enables them to be used as catalysts.

.... [1 mark]

D–C

D–C

B–A*

D–C

B–A*

C3 Extended response question

Tom and Leah compare the energy per gram from two fuels, **A** and **B**. Describe the experiments they carry out and explain the calculations they need to do.

❶ The quality of written communication will be assessed in your answer to this question.

....
....
....
....
....
....
....
....
....
....
....
....
....
....
....
....
....
....
....
....
....
....
....
....
....
....
....
....
....
....
....
....

[6 marks]

Atomic structure

You can refer to the periodic table (page 248) to answer the questions in this section.

1 a What are the particles of the nucleus of an atom?

.... [1 mark]

b Finish the table to show the relative mass and charge of the particles of an atom.

	Relative charge	Relative mass
electron		0.0005 (zero)
proton	+1	
neutron		

[2 marks]

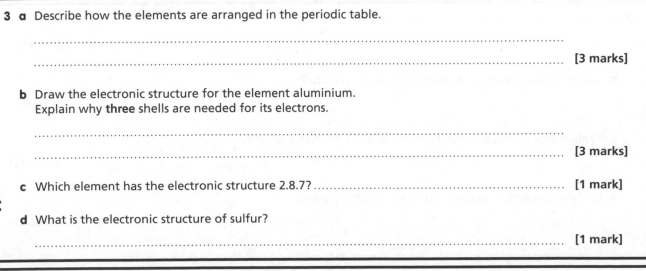

D–C

c What is the total number of protons, electrons and neutrons in an atom of potassium? Complete the table.

Atomic number	Mass number	Number of protons	Number of electrons	Number of neutrons
19	39			

[3 marks]

d Explain why atoms are neutral.

.... [1 mark]

B–A*

e What is the total number of protons, electrons and neutrons in Mg^{2+}? Complete the table.

Atomic number	Mass number	Number of protons	Number of electrons	Number of neutrons
12	24			

[3 marks]

2 a What is an isotope? .... [1 mark]

b Isotopes of an element have different numbers of neutrons in their atoms. Complete the table.

Isotope	Electrons	Protons	Neutrons
$^{12}_{6}C$	6		
$^{14}_{6}C$			

[2 marks]

D–C

c Which isotope has 17 protons, 17 electrons and 18 neutrons?

.... [1 mark]

B–A*

3 a Describe how the elements are arranged in the periodic table.

....
.... [3 marks]

b Draw the electronic structure for the element aluminium. Explain why **three** shells are needed for its electrons.

....
.... [3 marks]

D–C

c Which element has the electronic structure 2.8.7? .... [1 mark]

d What is the electronic structure of sulfur?

.... [1 mark]

B–A*

4 Explain why John Dalton's theory of atomic structure was **provisional**.

.... [1 mark]

D–C

Ionic bonding

1 **a** Put **M** next to the box that describes a **metal atom**, and **N** next to the box that describes a **non-metal**.

An atom that has extra electrons in its outer shell and needs to **lose** them to be stable. ☐

An atom that has 'spaces' in its outer shell and needs to **gain** them to be stable. ☐ [1 mark]

b Draw a diagram to show how an atom of lithium transfers an electron to an atom of fluorine.

[2 marks]

c Explain why lithium forms a **positive** ion.

.... [1 mark]

d Explain why fluorine forms a **negative** ion.

.... [1 mark]

e Explain how the ionic bonding in lithium fluoride means that it becomes a solid.

....

.... [2 marks]

f What is the formula of calcium chloride if the calcium ion is Ca^{2+} and the chloride ion is Cl^-?

.... [1 mark]

g Draw a 'dot and cross' model to show the bonding in potassium chloride (outer shell only).

[2 marks]

h Draw a 'dot and cross' model to show the bonding in potassium oxide (outer shell only).

[2 marks]

2 **a** Describe the structure of magnesium oxide.

.... [1 mark]

b When is magnesium oxide able to conduct electricity?

.... [1 mark]

c Explain why a solution of sodium chloride can conduct electricity.

.... [2 marks]

d Explain why sodium chloride has a high melting point.

.... [1 mark]

e Explain why magnesium oxide has an even higher melting point than sodium chloride.

....

.... [2 marks]

C4 The periodic table

The periodic table and covalent bonding

1 a Non-metals combine by covalent bonding. What happens in covalent bonding?

.... [1 mark]

b How do you describe water and carbon dioxide? Underline the correct answer.

giant ionic lattice   simple molecules with weak intermolecular force

simple molecules with strong intermolecular force   small ionic lattices   [1 mark]

D–C

2 a Draw a 'dot and cross' model to show the bonding in hydrogen fluoride (outer shell only).

[2 marks]

b Draw a 'dot and cross' model to show the bonding in carbon dioxide (outer shell only).

B–A*

[2 marks]

c Explain why sulfur dioxide has a low melting point and does not conduct electricity.

....

.... [2 marks]

3 a Explain why magnesium belongs to group 2 in the periodic table.

.... [1 mark]

b Explain why fluorine belongs to the same group as chlorine in the periodic table.

.... [1 mark]

c Explain why neon is in period 2 but potassium is in period 4 in the periodic table.

....

.... [2 marks]

D–C

d In which group is boron?

.... [1 mark]

e In which period is phosphorus?

.... [1 mark]

4 a What did both Newlands and Mendeleev notice about the behaviour of elements that helped them to start to classify them?

.... [1 mark]

D–C

b Explain two ways in which Mendeleev's ideas were shown to be correct when more evidence became available.

....

.... [2 marks]

B–A*

The group 1 elements

1 a Predict what will happen when rubidium is added to water.

....

....

.... [3 marks]

b Which of rubidium or caesium will react more vigorously? Explain your answer.

....

.... [2 marks]

c Construct the balanced symbol equation for the reaction of potassium with water.

[2 marks]

d Use the information in the table to predict the properties of rubidium. Complete the table. You do not need to give exact figures, just the range of values between which the boiling point and atomic radius will fall.

Element	Melting point in °C	Boiling point in °C	Atomic radius in nm
Na	98	892	0.185
K	64	774	0.227
Rb	39		
Cs	28	671	0.265

[2 marks]

2 a Explain why group 1 elements have similar properties.

.... [1 mark]

b Explain how group 1 elements form ions with stable electronic structures, and construct an ionic equation to demonstrate this.

....

....

.... [3 marks]

c Explain why this equation shows that the process is an oxidation reaction.

.... [1 mark]

d Explain how the ease of losing an electron relates to the trend of reactivity of group 1 elements with water.

....

.... [3 marks]

3 Akira and Jo have been asked to demonstrate to a junior class how group 1 chemicals are used in fireworks to give different colours. How will they demonstrate this to show three different flames? You may use a diagram to help explain your answer.

....

....

....

.... [4 marks]

The group 7 elements

1 a Write in the table the physical appearance of the group 7 elements at room temperature.

	Physical appearance	**Melting point in °C**	**Boiling point in °C**
fluorine			
chlorine		−101	−35
bromine		−7	59
iodine		114	184
astatine			

[2 marks]

b Write in the table your prediction for the temperatures above or below which the melting points and boiling points for fluorine and astatine will fall. You are not expected to know the actual figures for the melting points and boiling points.   [2 marks]

2 a Construct the word equation for the reaction between sodium and bromine.

.... [1 mark]

b Construct the balanced symbol equation for the reaction between potassium (K) and chlorine.

.... [2 marks]

c Construct the balanced symbol equation for the reaction between rubidium and iodine.

.... [2 marks]

3 a What would you see if chlorine was bubbled through potassium iodide?

.... [1 mark]

b What type of reaction is this? .... [1 mark]

c Construct the word equation for the reaction between potassium iodide and chlorine.

.... [1 mark]

d Explain why there is no reaction when bromine is mixed with lithium fluoride.

.... [1 mark]

e Explain why, in theory, astatine would be made if chlorine was bubbled into potassium astatide.

.... [1 mark]

f Explain why group 7 elements have similar properties. Use ideas about ions.

.... [1 mark]

g Construct an equation to show the formation of an iodide ion from an iodine molecule. Explain why this is reduction.

....
.... [2 marks]

h Explain the trend of reactivity in the halogens group in terms of electron transfer.

....
.... [2 marks]

Transition elements

1 a A compound that contains a transition element is often coloured. Match the colours to the compounds.

copper compounds		often pale green
iron(II) compounds		often orange/brown
iron(III) compounds		often blue

[1 mark]

b Transition elements or their compounds are often catalysts. What is a catalyst?

.... [1 mark]

D–C

c Give two examples of transition elements as catalysts.

....

.... [2 marks]

d What two substances are made when manganese carbonate is heated?

.... [1 mark]

e Construct the word equation for the thermal decomposition of copper carbonate.

.... [1 mark]

f What would you see when copper carbonate decomposes?

.... [1 mark]

B–A*

g Write a balanced symbol equation of the thermal decomposition of zinc carbonate.

.... [2 marks]

2 a Describe how you would use a solution of sodium hydroxide to identify the presence of the transition metal ions: Cu^{2+} Fe^{2+} Fe^{3+}

Describe what you would do and what you would **see**.

....

....

....

....

.... [5 marks]

D–C

b Write a balanced symbol equation for the reaction between Fe^{2+} ions and hydroxide ions.

.... [2 marks]

B–A*

c Write a balanced symbol equation for the reaction between Fe^{3+} ions and hydroxide ions.

.... [2 marks]

Metal structure and properties

1   Metals have specific properties that make them suitable for different uses.

Property	Iron	Copper	Aluminium	Tin	Gold	Silver	Lead
hardness in mohs	4	3	2.8	1.5	2.5	2.5	1.5
density in g/cm³	7.9	8.9	2.7	7.3	19.2	10.5	11.3
electrical conductivity × 10⁷ in siemen/cm	1.00	5.96	3.50	0.917	4.52	6.30	0.455
melting point in °C	1080	1358	933	505	1338	1234	601

a   Explain why you would make electrical cables out of silver or copper rather than lead or iron.

.... [1 mark]

b   Choose which metal you would use to make a model aircraft. Justify your answer.

....

.... [2 marks]

c   What property of tin makes it more useful in solder than copper?

.... [1 mark]

2

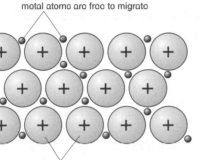

electrons from outer shells of metal atoms are free to migrate

metal ions

a   Explain why metals have high melting points and high boiling points.

....

.... [1 mark]

b   Describe metallic bonding and explain how metals conduct electricity.

....

....

....

.... [3 marks]

B–A*

3   a   Describe what is meant by a superconductor.

.... [1 mark]

b   What are the potential benefits of superconductors?

....

....

.... [3 marks]

c   Explain some of the drawbacks of superconductors.

....

.... [2 marks]

Purifying and testing water

1 The water in a river is cloudy and often not fit to drink. To turn it into the clean water in taps it is passed through a **water purification** works.

a Label the three main stages in water purification. **[1 mark]**

b Explain the three processes in detail.

....

....

.... **[3 marks]**

c How do some pollutants get into the water before or after purification?

....

.... **[2 marks]**

d Some soluble substances are not removed. Name two examples.

.... **[1 mark]**

e Sea water is undrinkable. Techniques such as distillation remove the dissolved substances. What is the disadvantage of using distillation?

.... **[1 mark]**

2 a Three samples of water, A, B and C, have the following reactions with test reagents. What type of sodium salts do the samples contain? Complete the boxes beneath the results table to identify the sodium salt in A, B and C.

Test reagent	A	B	C
silver nitrate	white solid	yellow solid	no solid
barium chloride	no solid	no solid	white solid
sodium salt contained			

[3 marks]

b Construct the word equation for the reaction between barium chloride and magnesium sulfate.

.... **[1 mark]**

c What kind of reaction is this?

.... **[1 mark]**

d Construct the word equation of a reaction between silver nitrate and a sodium salt that will produce a cream solid.

.... **[2 marks]**

e Construct the balanced symbol equation for the reaction in (2a) between $BaCl_2$ and $MgSO_4$.

.... **[2 marks]**

f Construct the balanced symbol equation for the reaction between potassium chloride (KCl) and silver nitrate ($AgNO_3$).

.... **[2 marks]**

C4 Extended response question

Jo and Sam have been given six salts; A, B, C, D, E and F that need identifying. They know they contain Na^+, K^+, Cu^{2+}, Fe^{2+}, Li^+ and Fe^{3+} ions.

Describe the two experiments that they can do and explain how they would identify which ion each salt contained.

❶ The quality of written communication will be assessed in your answer to this question.

....

....

....

....

....

....

....

....

....

....

....

....

....

....

....

....

....

....

....

....

....

....

....

....

....

....

....

....

....

....

....

....

....

[6 marks]

Speed

1 The graph shows Ashna's walk to her local shop and home again.

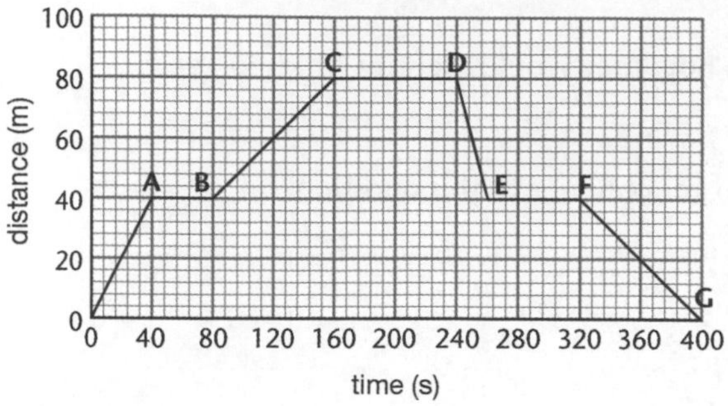

a How long did she spend in the shop?

.... **[1 mark]**

b How far is the shop from her home?

.... **[1 mark]**

c During which part of her journey did she walk fastest?

.... **[1 mark]**

d Calculate Ashna's speed between 0 and A.

....

....

.... **[2 marks]**

e Calculate Ashna's speed between F and G.

....

....

.... **[2 marks]**

2 A cycle track is 500 m long. Imran completes 10 laps. He cycles at an average speed of 45 km/h.

a How long did he take to complete ten laps?

....

....

.... **[3 marks]**

b Imran put on a spurt in the last lap, completing it in 35 s. What was his average speed, in m/s, for the last lap?

....

....

.... **[2 marks]**

c Calculate Imran's average speed for the first nine laps.

....

....

.... **[3 marks]**

D–C

B–A*

B–A*

Changing speed

1 Darren is riding his bicycle along a road. The speed–time graph shows how his speed changed during the first minute of his journey.

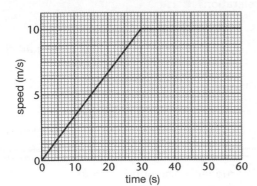

a Darren makes the same journey the next day but:
  – increases his speed at a steady rate for the first 20 s, reaching a speed of 10 m/s
  – travels at a constant speed for 10 s
  – slows down at a steady rate for 15 s to a speed of 5 m/s
  – travels at a constant speed of 5 m/s.

  Plot the graph of this journey on the same axes. [4 marks]

b How could you calculate the distance Darren travelled in the first minute of his original journey?

.... [1 mark]

2 The way in which the speed of a car changes over a 60 s period is shown in the table.

a Plot a speed–time graph for the car using the axes given.

time in s	speed in m/s
0	0
5	5
10	10
15	15
20	15
25	15
30	15
35	15
40	15
45	11
50	7
55	3
60	0

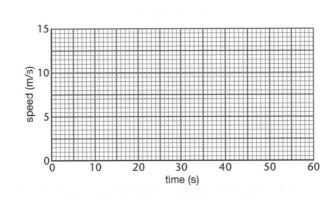

[4 marks]

b Calculate the deceleration between 40 and 55 seconds.

....
....
....
.... [3 marks]

c Use the graph to calculate the distance travelled by the car.

....
.... [2 marks]

3 A cheetah accelerates at 6 m/s². How fast is it moving after 5 seconds, starting from rest?

....
.... [2 marks]

Forces and motion

1 A car of mass 500 kg accelerates steadily from 0 to 40 m/s in 20 s.

D–C

a What is the size of the resultant force which produces this acceleration?

....

.... [2 marks]

b Calculate the acceleration of the car if the resultant force is increased to 1250 N.

....

.... [2 marks]

2 Helen is driving her car on a busy road when the car in front brakes suddenly. She puts her foot firmly on her brake pedal and just manages to stop without hitting the car in front.

a Write down two things, apart from speed, that could increase a driver's thinking distance.

....

.... [2 marks]

D–C

b Later that day Helen is driving at high speed on a motorway.

i How will Helen's thinking distance change?

.... [1 mark]

ii Explain why.

....

.... [2 marks]

B–A*

c Explain the effect worn tyres have on braking distance.

.... [2 marks]

3 Finish the graphs to show how thinking distance and braking distance change with speed. Some points have been plotted for you.

[3 marks]

B–A*

─┼─ thinking

─✕─ braking

Work and power

1 Hilary lifts a parcel of weight 80 N onto a shelf 2 m above the ground.

    a Calculate the amount of work Hilary does.

    ....

    .... [2 marks]

    b i How high could Hilary lift a parcel of weight 60 N if she did the same amount of work in 1.5 s?

    ....

    .... [2 marks]

    ii What is Hilary's power output now?

    ....

    .... [2 marks]

D–C

2 Chris and Abi both have a mass of 60 kg. They both run up a flight of stairs 3 m high. Chris takes 8 s and Abi takes 12 s.

    a What can you say about the power of Chris and Abi?

    .... [1 mark]

    b Calculate Chris' weight. (Take g = 10 N/kg.)

    ....

    .... [2 marks]

D–C

    c Calculate Chris' power.

    ....

    .... [2 marks]

    d Calculate Abi's power.

    ....

    .... [2 marks]

B–A*

    e Chris tells Abi that on Jupiter they would each weigh 1560 N. What is the gravitational field strength on Jupiter?

    .... [1 mark]

3 The table shows the fuel consumption of three cars in miles per gallon (mpg).

Car fuel	Consumption (mpg)
A	48
B	34
C	28

    a Which car has the best fuel consumption?

    .... [1 mark]

    b Which car is likely to be most powerful?

    .... [1 mark]

D–C

    c We should keep our fuel consumption to a minimum to protect the environment. Why?

    ....

    ....

    .... [3 marks]

Energy on the move

1 Use the data about fuel consumption to answer the following questions.

Car	Fuel	Engine size in litres	Miles per gallon	
			urban	non-urban
Renault Megane	petrol	2.0	25	32
Land Rover	petrol	4.2	14	24

D–C

a Suggest why fuel consumption is better in non-urban conditions.

....

.... [2 marks]

b How many gallons of petrol would a Land Rover use on a non-urban journey of 96 miles?

.... [2 marks]

c Suggest why a Renault Megane uses less fuel for the same journey?

.... [1 mark]

B–A*

d Explain how driving style can affect fuel consumption.

....

....

.... [3 marks]

2 a How do battery-powered cars pollute the environment?

.... [1 mark]

D–C

b Give one advantage and one disadvantage of solar-powered cars compared to battery-powered cars.

....

.... [2 marks]

3 Sam is driving a car of mass 1200 kg at a speed of 20 m/s.

a Calculate the kinetic energy of the car.

....

....

.... [2 marks]

B–A*

b When Sam suddenly applies the brakes, the car travels 32 m before it stops. Sam suggests that the braking distance would be 16 m if he was travelling at 10 m/s. Explain why Sam is not correct.

....

.... [2 marks]

Crumple zones

1. Kevin was involved in an accident on the M1 motorway. Luckily he was not seriously hurt but his car was badly damaged. Kevin's car had crumple zones, seat belts, airbags and ABS brakes.

   a Finish the table by describing how each feature helps to reduce Kevin's injuries.

Safety feature	How it works
seat belt	
crumple zones	
air bag	

   [3 marks]

   D–C

   b Describe how ABS brakes work.

   ....

   ....

   ....   [2 marks]

   B–A*

2. Marie was a passenger in a car travelling at 25 m/s when the driver braked sharply to avoid a dog that had run into the road. Marie was wearing her seat belt and this brought her to a stop in 0.5 s. Marie's mass is 55 kg.

   a Calculate the average force the seat belt exerted on her body.

   ....

   ....

   ....   [2 marks]

   D–C

   b If Marie had not been wearing her seat belt, she would have hit the windscreen which would have brought her head to a stop in 0.002 s. Calculate what the average force on her head would have been.

   ....

   ....

   ....   [2 marks]

3. To minimise injury the forces acting on the people in a car during a car accident must be as small as possible.

   a Explain why this means safety features must reduce the deceleration of the car on impact.

   ....

   ....   [2 marks]

   B–A*

   b Explain how one safety feature is designed to reduce the deceleration of the car on impact.

   ....

   ....   [2 marks]

Falling safely

1 Charlie drops a golf ball and a ping pong ball from a height of 30 cm above the ground.
Both balls hit the ground together although their masses are different.

D–C

   a Why do they both hit the ground at the same time?

   ....

   .... [1 mark]

2 Sarah is a sky diver. She has a mass of 60 kg.

D–C

   a On the diagram mark and name the forces acting on Sarah as she falls. [2 marks]

   b As Sarah leaves the aircraft, she starts to accelerate.

   Describe any difference in the size of the forces acting on her just after leaving the aircraft.

   .... [1 mark]

   c Sarah's acceleration decreases as she falls. Explain why.

   .... [1 mark]

   d Eventually she is travelling at terminal speed. Describe any difference in the size of the forces acting on her when she is falling at a constant speed.

   .... [1 mark]

B–A*

   e Sarah opens her parachute. Explain what happens to each of the forces acting on her now?

   ....

   .... [2 marks]

B–A*

3 We often use the value 10 m/s^2 as the value for acceleration due to gravity on Earth. Describe how the acceleration due to gravity changes depending on where you are.

   ....

   ....

   .... [3 marks]

The energy of games and theme rides

1 Finish the sentences. Choose words from this list.

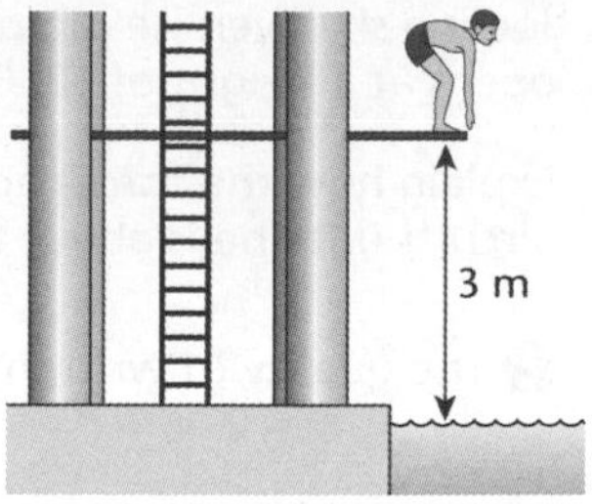

gravitational potential energy (GPE)   kinetic energy (KE)

Rob is about to dive into the swimming pool.

He has .... As he dives ....

changes to .... Rob climbs to the 10 m board.

He has more .... than before.   **[4 marks]**

D–C

2 The diagram shows a roller coaster. The carriages are pulled up to B by a motor and then released.

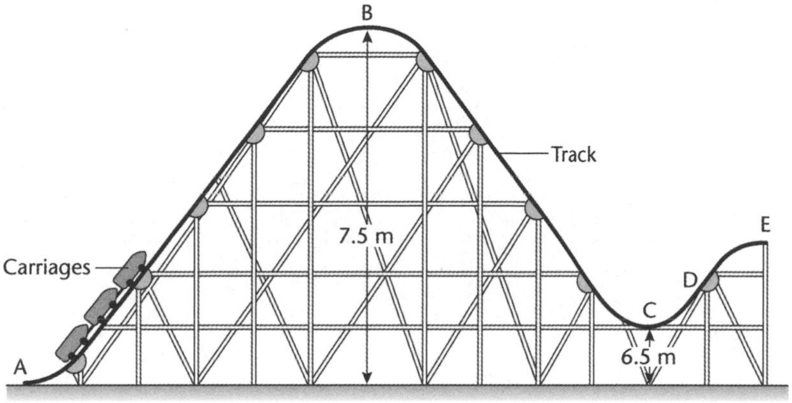

a At which point, A, B, C, D or E do the carriages have the greatest kinetic energy?

.... **[1 mark]**

D–C

b Describe the main energy change as the carriages move from B towards C.

.... **[1 mark]**

c Why must the height of the next peak at E be less than that at B?

....

.... **[1 mark]**

d The theme park decides to build a faster roller coaster. Suggest how they could modify the design to achieve this.

....

....

....

.... **[3 marks]**

3 49 000 years ago, an asteroid struck Earth and formed the Barringer crater in the Arizona desert. It is estimated the asteroid was travelling at 11 km/s when it struck. What is the equivalent height the asteroid fell from assuming it did not reach terminal speed and the effect of the Earth's atmosphere was negligible.

B–A*

....

....

.... **[2 marks]**

P3 Extended response question

Joe is a skydiver. He jumps from a plane at a height of 10 000m and free falls until his parachute opens at a height of 3000 m.

Explain how the forces acting on him affect the speed of his descent from the time he leaves the aircraft until he is about to land on the ground.

❗ The quality of written communication will be assessed in your answer to this question.

....
....
....
....
....
....
....
....
....
....
....
....
....
....
....
....
....
....
....
....
....
....
....
....
....
....
....
....
....
....
....
....

[6 marks]

Sparks

1 a Sally stands on an insulating mat. She puts her hands on the dome of an uncharged Van de Graaff generator. The Van de Graaff generator is switched on and Sally's hair starts to stand on end.

i Why does Sally stand on an insulating mat?

.... [1 mark]

ii Why must the Van de Graaff generator be uncharged when Sally puts her hands on it?

.... [1 mark]

iii What happens to Sally when the Van de Graaff generator is switched on?

.... [1 mark]

iv Why does this make Sally's hair stand on end?

.... [2 marks]

b Donna hangs two small, light plastic balls on nylon threads side by side.

She touches each ball with a charged polythene strip.

Why do the balls repel each other?

....

.... [2 marks]

c Paul does the same experiment but this time he only touches one ball with a charged polythene strip. The balls move towards each other but do not touch.

Give as much detail as you can to explain why this happens.

....

....

.... [2 marks]

2 a Use your knowledge of electrostatics to explain the following.

i You sometimes get an electric shock on closing a car door after a journey.

....

.... [2 marks]

ii You should never shelter under a tree during a thunderstorm.

.... [1 mark]

iii Cling film often sticks to itself as it is unrolled.

.... [1 mark]

b A factory uses machinery with moving parts. How does the machinery become charged?

....

.... [1 mark]

c Why do the operators at the factory stand on rubber mats?

....

.... [3 marks]

D–C

B–A*

B–A*

Uses of electrostatics

1 An electrostatic precipitator is placed in the chimney of a power station. It contains some wires and plates which are connected to a high voltage supply.

D–C

a If the wires are given a negative charge, what charge will be gained by the plates?

.... [1 mark]

b What charge do the soot particles gain as they pass close to the wires?

.... [1 mark]

B–A*

c Explain what has happened to the soot particles in terms of electrons.

.... [1 mark]

d What force is acting between the soot particles and the plates?

.... [1 mark]

2 In a bicycle factory the frames are painted using an electrostatic sprayer. The paint is positively charged. The frames are given the opposite charge to the paint.

D–C

nozzle is
charged up
positively

object to be painted
is negatively charged

a Why does the paint spread out on leaving the sprayer?

....

.... [2 marks]

b Why are the bicycle frames given the opposite charge to the paint?

....

.... [2 marks]

c Lana decides to repaint her bicycle frame. She uses a paint spray that charges the paint positively but she does not charge the frame.

i What charge, if any, does Lana's bicycle frame acquire?

.... [1 mark]

B–A*

ii What problem does this cause?

.... [1 mark]

3 A defibrillator delivers an electric shock through the chest wall to the heart.

a What does the electric shock do to the heart?

.... [1 mark]

b The paddles of a defibrillator, charged from a high voltage supply, are placed on the patient's chest. Why should the chest be clean shaven and dry?

D–C

.... [2 marks]

c Why does the operator call out 'clear' before using the paddles?

.... [1 mark]

d A current of about 50 A passes through the patient for about 4 ms (0.004 s).

In general, such a large current would be fatal. Why can it be used in this situation?

.... [1 mark]

Safe electricals

1 a Adeela sets up the circuit shown. She then moves the variable resistor to include a greater length of wire into the circuit.

variable
resistor

   i What effect will this have on the brightness of the lamp?

.... **[1 mark]**

   ii Add an ammeter and voltmeter to Adeela's circuit to allow her to measure the current through the lamp and the potential difference across the lamp.

[2 marks]

   iii If the voltmeter reads 6 V and the ammeter 0.25 A, calculate the resistance of the lamp.

....

.... **[3 marks]**

D–C

b Adeela adjusts the variable resistor to 30 W and the current increases to 0.4 A.

   i What is the voltage across the lamp now?

....

.... **[3 marks]**

   ii Explain the change in resistance in terms of electron movement.

....

....

.... **[3 marks]**

B–A*

2 A battery has positive and negative terminals. Give two differences between the voltage from a battery and mains voltage.

....

.... **[2 marks]**

D–C

3 a i Label the live (L), neutral (N) and earth (E) wires in the plug shown. **[3 marks]**

               ii Which wire, live, neutral or earth, is a safety wire?

               .... **[1 mark]**

               iii How does it work?

               ....

               .... **[2 marks]**

b i An electric kettle passes a current of 10.5 A when working normally. Should the plug contain a 5 A or 13 A fuse?

.... **[1 mark]**

D–C

   ii Why are fuses always connected in the live wire?

.... **[2 marks]**

   iii The kettle is made from metal. How do the fuse and earth wire stop a person receiving an electric shock if they touch the kettle when it is faulty?

....

.... **[4 marks]**

Ultrasound

1 a Sound is a longitudinal wave.

  i Explain how sound travels through the air to reach your ear.

  ....

  .... **[2 marks]**

  ii How does the frequency of a note change if its pitch increases?

  .... **[1 mark]**

b What is 'ultrasound'?

  .... **[1 mark]**

c Light is an example of a transverse wave. What is the difference between a longitudinal and a transverse wave?

  ....

  .... **[2 marks]**

D–C

---

2 High-powered ultrasound is used to treat a patient with kidney stones.

a How does ultrasound do this?

  ....

  .... **[3 marks]**

b Why must high-powered ultrasound be used?

  .... **[2 marks]**

D–C

---

3 a Finish the sentences about an ultrasound scan.

  A .... of ultrasound is sent into a patient's body. At each boundary

  between different .... some ultrasound is ....

  and the rest is transmitted. The returning .... are used to build up

  an .... of the internal structure. A .... is placed

  on the patient's body between the ultrasound .... and their

  .... Without it nearly all the .... would be

  .... by the .... **[6 marks]**

b Give two factors that affect the amount of ultrasound sent back to the detector at each interface within the body.

  ....

  .... **[2 marks]**

c Air has a density of 1.3 kg/m^3. Soft tissue has an average density of 1060 kg/m^3.

  i Explain why the gel used should have a density of about 1060 kg/m^3.

  .... **[2 marks]**

  ii The time delay for an echo from ultrasound in soft tissue at a depth of 0.16 m was 0.2 ms (0.0002 s). Calculate the speed of ultrasound in soft tissue.

  ....

  .... **[4 marks]**

B–A*

What is radioactivity?

1 a Complete the table showing the three types of nuclear radiation.

Type of radiation	Charge	What it is	Particle or wave
alpha			
beta			
gamma			

[3 marks]

b Name the type of nuclear radiation that:

i does not change the composition of the nucleus .... [1 mark]

ii travels at about one-tenth the speed of light .... [1 mark]

2 a What is meant by 'half-life'? .... [1 mark]

b

The graph shows how the activity of cobalt-60 changes with time. Use the graph to find the half-life of cobalt-60.

Show clearly, on the graph, how you got your answer. [2 marks]

3 Radon, $^{220}_{86}$Rn, is radioactive. It decays to an isotope of polonium, $^{216}_{84}$Po with a half-life of 52 s.

a How many protons are there in a radon nucleus?

.... [1 mark]

b How many neutrons are there in a radon nucleus?

.... [1 mark]

c What is the name of the particle emitted in this decay?

.... [1 mark]

d i Write a nuclear equation to describe the decay of a radon nucleus.

.... [3 marks]

ii Show that the atomic mass and mass number are conserved in this decay.

.... [2 marks]

e When a beta particle is emitted from a nucleus the mass number is unchanged but the atomic number increases by one. Explain how this is possible.

.... [2 marks]

f Finish the nuclear equation for the decay of iodine-131, emitting a beta particle.

$^{131}_{53}$I $\rightarrow$ $^{x}_{y}$Xe + $^{0}_{-1}\beta$ x = ...   y = ... [2 marks]

Uses of radioisotopes

1 **a** Suggest two natural sources of background radiation.

.... **[2 marks]**

b Suggest two sources of background radiation that arise from human activity.

.... **[2 marks]**

2 Andrew works for an oil company. A leak has been reported in an underground pipe. He decides to locate the leak by introducing a small amount of radioisotope into the pipe.

a **i** What sort of radiation should the radioisotope emit?

.... **[1 mark]**

ii Explain your choice. ....

.... **[2 marks]**

b What radiation detector could Andrew use? .... **[1 mark]**

c How will Andrew tell the site of the leak from his results?

.... **[1 mark]**

3 Smoke alarms use a source of alpha radiation in a small chamber.

oppositely
charged
plates

americium-241
source of alpha particles

a Why is alpha radiation more suitable than either gamma or beta radiation for use in a smoke alarm?

....

.... **[1 mark]**

b Explain how the smoke alarm works.

....

....

.... **[4 marks]**

4

activity (B1)
250
200
150
100
50
0
0   2   4   6   8   10   12
time (thousands of years)

Trees contain carbon-14 which is radioactive. The graph shows how the activity of 1 kg of wood changes after a tree has died. This can be used to estimate the age of objects that were once alive.

a This method cannot be used to date a wooden bowl made from a tree that died less than 200 years ago.

Explain why.

.... **[1 mark]**

b Why can carbon-14 not be used to date rocks?

.... **[1 mark]**

c 1 kg of wood found in an archaeological dig had an activity of 150 Bq. Use the graph to estimate its age. Show your answer on the graph.

....

.... **[2 marks]**

D–C

D–C

B–A*

D–C

D–C

B–A*

Treatment

1 a Why are x-rays and gamma rays suitable to treat cancer patients?

....
.... [2 marks]

b Why are alpha and beta particles not suitable to treat cancer patients?

....
.... [1 mark]

c Explain why nuclear radiation has to be used on patients very carefully.

....
.... [3 marks]

d How are radioisotopes for medical use produced?

.... [1 mark]

2 a What is a radioactive tracer? ....
.... [2 marks]

b Why is it used? ....
.... [2 marks]

c What sort of radiation should a tracer emit?

.... [1 mark]

d X-rays and gamma rays have similar properties. Why are gamma rays suitable to use as tracers but x-rays are not?

....
.... [2 marks]

3 Three similar sources of radiation are used to destroy a brain tumour.

source of radiation

tumour

source of radiation

source of radiation

a i What type of radiation would be most suitable?

.... [1 mark]

ii Give a reason for your choice.

....
....
.... [2 marks]

b Why are there three sources of radiation arranged as shown?

....
....
.... [2 marks]

c Describe an alternative technique to destroy a brain tumour that uses only one source of radiation.

....
.... [2 marks]

Fission and fusion

1 a Complete the following sentences explaining how a power station works.

The .... provides heat to boil the ....

to produce ....

The pressure of the .... turns the ....

which turns the .... making electricity. **[3 marks]**

b What is meant by 'fission'?

.... **[2 marks]**

c Enriched uranium is used as the fuel in a nuclear power station.

What is 'enriched uranium'?

.... **[2 marks]**

d A slow neutron can be captured by a uranium-235 nucleus and the 'new' nucleus then becomes unstable.

i Describe what happens after the capture of the neutron by the uranium nucleus.

....

.... **[3 marks]**

ii How can this one event become a continuous chain reaction?

.... **[1 mark]**

2 In one type of nuclear power station graphite is used as a moderator and boron rods are used to control the number of nuclear fissions in a given time.

a What does a 'moderator' do?

.... **[1 mark]**

b Why is a moderator necessary?

.... **[1 mark]**

c How do the boron rods control the number of fissions?

....

.... **[2 marks]**

3 a What is nuclear fusion?

.... **[1 mark]**

b Under what conditions does nuclear fusion take place in stars?

.... **[1 mark]**

c Why would fusion be preferable to fission as an energy source for the future?

....

.... **[2 marks]**

d Why are the claims that cold fusion is possible not widely accepted?

.... **[1 mark]**

P4 Extended response question

Doctors sometimes diagnose or treat a patient's illness without using surgery. They can use a number of different methods including x-rays, ultrasound or radioactive isotopes.

The doctor suspects that a female patient, who is pregnant, may have a tumour in her lung.

Suggest how the doctors should diagnose her condition and explain the reasons why this method is more suitable than the others.

❗ The quality of written communication will be assessed in your answer to this question.

....
....
....
....
....
....
....
....
....
....
....
....
....
....
....
....
....
....
....
....
....
....
....
....
....
....
....
....
....
....
....
.... [6 marks]

B1 and B2 Grade booster checklist

B1	
I can describe the factors that increase or decrease blood pressure.	
I can explain how a diet containing high levels of saturated fats and salt can increase the risk of heart disease.	
I can explain why protein deficiency is common in developing countries.	
I can calculate a person's BMI and their EAR for protein.	
I can explain the difference between active and passive immunity.	
I can describe changes in lifestyle that may reduce the risk of some cancers.	
I can describe the path taken by a spinal reflex.	
I can explain the differences between monocular and binocular vision.	
I can explain why damage to ciliated cells can lead to a 'smoker's cough'.	
I can interpret data on the alcohol content, measured in units of alcohol, of different alcoholic drinks.	
I can describe the dangerous effects of high and low temperatures on the body.	
I can explain how Type 2 diabetes can be controlled by diet but Type 1 diabetes needs insulin doses.	
I can describe shoots as being positively phototropic and negatively geotropic.	
I can describe the commercial uses of plant hormones.	
I can describe how sex chromosomes, XX in female and XY in male, determine the sex of an individual.	
I can explain the causes of genetic variation.	
I am working at grades D–C	

I can explain the possible consequences of having high or low blood pressure.	
I can explain how narrowed coronary arteries, together with a thrombosis, increase the risk of a heart attack.	
I can describe the storage of carbohydrates (as glycogen) in the liver and fats (as adipose tissue) in the skin.	
I can describe the differences between first and second class proteins.	
I can interpret data on types of cancer and their survival/mortality rates.	
I can describe the benefits and risks of being immunised.	
I can explain how the eye accommodates (focuses light) for distant and close objects.	
I can explain how a nerve impulse is transmitted across a synapse.	
I can explain the action of depressant and stimulant drugs on synapses.	
I can interpret information on reaction times, accident statistics and alcohol levels.	
I can explain how negative feedback mechanisms are used in homeostasis.	
I can explain how insulin helps to regulate blood sugar levels by converting excess glucose into glycogen.	
I can use information on auxins to interpret data from phototropism experiments.	
I can explain how different levels of auxin cause shoot curvature towards light.	
I can use and explain a monohybrid cross involving dominant and recessive alleles.	
I can use genetic diagrams to predict the possibilities of inherited disorders in the next generation.	
I am working at grades B–A*	

B2	
I can name organisms using the binomial system.	
I can recall the definition of a species.	
I can describe how energy is lost from food chains.	
I can explain why pyramids of energy and pyramids of biomass can be different shapes.	
I can describe how plants obtain nitrogen from the soil.	
I can describe the role of decomposers in the carbon and nitrogen cycles.	
I can explain the difference between parasitism and mutualism.	
I can explain the reasons for changes in predator and prey numbers.	
I can explain some of the adaptations that animals have for living in cold conditions.	
I can explain some of the adaptations that organisms have for living in dry conditions.	
I can use Darwin's theory of natural selection to explain how evolution occurs.	
I can state why Darwin's theory was not well accepted by many people.	
I can describe how indicator species can show levels of pollution.	
I can explain the causes and consequences of global warming.	
I can explain why it is considered important to set up conservation programmes.	
I can explain what sustainable development means.	
I am working at grades D–C	

I can describe the difference between an artificial and a natural classification system.	
I can explain that organisms can be similar because they may be closely related or because they may have evolved to live in similar environments.	
I can describe the problems involved in constructing pyramids of biomass.	
I can explain why food chains always have a limited number of trophic levels.	
I can recall the roles of all four types of bacteria in the nitrogen cycle.	
I can describe how oceans can act as carbon sinks.	
I can explain why the peak in numbers of predators and prey do not coincide.	
I can explain what is meant by an ecological niche.	
I can explain why animals in colder climates tend to be larger.	
I can describe the differences between specialists and generalists.	
I can describe the importance of reproductive isolation in speciation.	
I can explain the difference between Lamarck's theory and Darwin's theory.	
I can recall the advantages and disadvantages of using indicator species to judge pollution levels.	
I can explain why people in different countries have different carbon footprints.	
I can explain the importance of genetic variation in species.	
I can recognise some of the difficulties in trying to protect whales.	
I am working at grades B–A*	

C1 and C2 Grade booster checklist

C1

I can describe how fractional distillation separates crude oil in a column with a temperature gradient.	
I can interpret data about the supply and demand of crude oil fractions and describe the cracking of alkanes.	
I can suggest key factors that need to be considered when choosing fuels and interpret data to pick the best.	
I can describe an experiment to show the products of combustion of a hydrocarbon in a good supply of air.	
I can describe a simple carbon cycle involving the processes of photosynthesis, combustion and respiration.	
I can describe how the present day atmosphere evolved from gases from inside the Earth and photosynthesis.	
I can recognise and interpret information on the displayed formulae of alkanes, alkenes and polymers.	
I can describe how bromine is used to test for an alkene, because orange bromine water decolourises.	
I can compare GORE-TEX® to nylon and say its breathable properties make it useful to active outdoor people.	
I can explain the environmental and economic issues related to uses and disposal of polymers.	
I can recall that protein molecules change shape when they are heated and that this is called denaturing.	
I can construct the balanced symbol equation for the decomposition of sodium hydrogencarbonate given some or all of the formulae.	
I can recall that alcohols react with acids to make an ester and water, and describe this as an experiment.	
I can explain why a perfume needs certain properties, e.g. being non-toxic and able to easily evaporate.	
I can describe paint as a colloid where particles are dispersed with particles of a liquid but are not dissolved.	
I can explain when thermochromic pigments are suitable and how phosphorescent pigments glow in the dark.	
I am working at grades D–C	

I can explain why crude oil separates by fractional distillation due to molecular size and intermolecular forces.	
I can understand that during boiling intermolecular forces break but covalent bonds within molecules do not.	
I can explain why the amount of fossil fuels being burnt is increasing due to increasing world population.	
I can construct the balanced symbol equation for the incomplete combustion of a given hydrocarbon fuel.	
I can evaluate the effects of human influences on air composition by deforestation and population increase.	
I can describe one theory of the evolution of the present day atmosphere from volcano degassing.	
I can explain reactions of bromine and alkenes as addition reactions forming colourless di-bromo compounds.	
I can draw the displayed formula of an addition polymer given the displayed formula of its monomer.	
I can relate the stretching of a plastic to the weak intermolecular forces between its polymer molecules.	
I can explain why the size of the holes in the PTFE membrane of Gore-Tex® materials makes them breathable.	
I can explain how starch grains swell up, burst and spread out to lose the rigid structure of the cell walls when potatoes are cooked.	
I can explain that the hydrophilic ends of an emulsifier bond to water and the hydrophobic ends bond to oil.	
I can explain the volatility of perfumes by the weak attraction between particles of the liquid perfume.	
I can explain that the particle attraction in nail varnish is stronger than between water and nail varnish.	
I can explain how oil paints dry by the solvent evaporating then the oil oxidising in atmospheric oxygen.	
I can recall that phosphorescent pigments are much safer than the alternative radioactive substances.	
I am working at grades B–A★	

C2

I know the lithosphere is the Earth's crust and outer part of the mantle, and that it is made of tectonic plates.	
I know that the type of volcanic eruption depends on the type of magma with thick lava making it explosive.	
I know that limestone decomposes on heating to make calcium oxide and can write the word equation.	
I know that bricks are made from clay and cement is made when limestone and clay are heated together.	
I know how to purify copper by electrolysis and can explain the advantages and disadvantages of recycling it.	
I know that brass is an alloy of copper and zinc, that solder is made from lead and tin and can give some uses.	
I know that rusting is an oxidation reaction of iron and the word equation to make hydrated iron(III) oxide.	
I know that a car body made from aluminium will corrode less and will be lighter than one made from steel.	
I can write an equation for the Haber process and know that high pressure and an iron catalyst are needed.	
I know how factors affect costs in making substances, e.g. higher temperatures lead to higher energy costs.	
I know that acids neutralise bases to make salt and water and can show these changes in pH using indicators.	
I know how to make a salt and predict its name, e.g. sodium hydroxide and nitric acid makes sodium nitrate.	
I can identify the argument for the use of fertilisers as the world population is rising so more food is needed.	
I know that eutrophication and pollution of water supplies can result from the excessive use of fertilisers.	
I know that chlorine is made at the anode and hydrogen at the cathode in the electrolysis of brine.	
I can describe how sodium hydroxide and chlorine are used to make household bleach.	
I am working at grades D–C	

I can explain plate tectonic theory including how convection currents cause slow movement of the plates.	
I know that iron-rich basalt is formed from runny lava and silica-rich rhyolite is formed from thick lava.	
I can explain why rocks such as granite (igneous), marble (metamorphic) and limestone (sedimentary) have different hardness.	
I can write a balanced symbol equation (with no formula given) for the decomposition of $CaCO_3$.	
I can describe that impure copper is used as an anode in $CuSO_4$ solution when purifying copper by electrolysis.	
I can explain how the equation $Cu^{2+} + 2e^- \rightleftharpoons Cu$ shows that Cu^{2+} ions gain electrons to form Cu atoms (reduction).	
I can explain the advantages of using aluminium for car bodies being better fuel economy and longer lifetime.	
I can evaluate information on materials used to manufacture cars showing their use and ability to be recycled.	
I can explain that high temperature decreases the yield of NH_3 and 450 °C is chosen as optimum temperature.	
I can explain that optimum conditions are used to give lowest cost, not the fastest reaction or highest % yield.	
I can explain that acids contain H^+ ions, alkalis contain OH^- ions and neutralisation involves $H^+ + OH^- \rightleftharpoons H_2O$	
I can construct the balanced symbol equations for neutralisation, e.g. $2KOH + H_2SO_4 \rightleftharpoons K_2SO_4 + 2H_2O$	
I can explain how fertilisers give extra essential elements and the process of eutrophication (if overused).	
I can describe which reactants to use and the experimental method to make a named synthetic fertiliser.	
I can explain why the electrolysis of NaCl solution involves both oxidation and reduction, using equations.	
I can explain the economic importance of the chlor-alkali industry in producing the bulk chemicals, Cl_2 and NaOH.	
I am working at grades B–A★	

P1 and P2 Grade booster checklist

P1

I can interpret data on rate of cooling and understand the consequences of the direction of energy flow.	
I can understand the concepts of specific heat capacity and specific latent heat and use the equations.	
I can explain how energy is transferred and how losses to the atmosphere can be reduced.	
I can interpret data on energy saving strategies and understand the importance of energy efficiency.	
I can describe the features of a transverse wave and determine wavelength and frequency from a diagram.	
I can understand that refraction is due to a wave travelling at different speeds in different materials.	
I can describe how light and the Morse code have been used for communication.	
I can describe how light behaves when its angle of incidence is below, above and equal to the critical angle.	
I can provide reasons for poor mobile phone reception.	
I can explain how scientists check each other's results by publishing their findings.	
I can describe how infrared signals carry information to control electrical and electronic devices.	
I can recall how the properties of digital signals led to the digital switchover for television and radio.	
I can describe how reflection and refraction of radio waves can be an advantage and can be a disadvantage.	
I can state the advantages and disadvantages of DAB broadcasts.	
I can recall the properties of P waves and S waves from an earthquake.	
I can interpret data about how sunscreen and skin tone can protect the skin from damage when sunbathing.	

I am working at grades D–C

I can explain the difference between temperature and heat.	
I can explain why temperature does not change during a change of state.	
I can explain why the trapped air in cavity wall insulation further limits energy transfer through the walls.	
I can relate the design features of a house to the reduction in energy loss.	
I can describe the diffraction pattern for waves and say how this depends on the size of the gap.	
I can describe diffraction effects in telescopes and other optical instruments.	
I can explain advantages and disadvantages of using different electromagnetic waves for communication.	
I can explain how a laser beam is used in a CD player.	
I can explain how microwaves and infrared radiation transfer energy to materials.	
I can provide reasons for signal loss and describe how this loss can be reduced.	
I can explain how the signal from an infrared remote device uses digital signals to control electronic devices.	
I can describe the advantages of using digital signals and optical fibres for the rapid transmission of data.	
I can explain how long distance communication is achieved.	
I can explain why there is no interference when listening to digital radio.	
I can explain how seismic waves can be used to provide evidence for the structure of the Earth.	
I can describe how the ozone layer protects the Earth and the effects of its depletion.	

I am working at grades B–A*

P2

I can describe advantages and disadvantages of using photocells to produce electricity.	
I can describe advantages and disadvantages of wind turbines.	
I can describe how a simple alternating current generator works and how to increase the output.	
I can calculate the efficiency of a power station.	
I can explain how human activity and natural phenomena both have effects on weather patterns.	
I can distinguish between opinion and evidence based statements in the global warming debate.	
I can calculate the amount of electricity used in kilowatt hours and use this to find its cost.	
I can explain that transformers are used in the National Grid to reduce energy waste and costs.	
I can describe the relative penetrating power of alpha, beta and gamma radiations.	
I can describe some methods of disposing of radioactive waste.	
I can recall the relative sizes and nature of planets, stars, comets, meteors, galaxies and black holes.	
I can recall some difficulties of manned space travel and explain how information from space can be sent back to Earth.	
I can describe some of the evidence for past large asteroid collisions.	
I can describe how a collision between two planets could have resulted in the Earth-Moon system.	
I can recall that all galaxies are moving apart and that microwave radiation is received from all parts of the Universe.	
I can describe the end of the life cycle of small and large stars.	

I am working at grades D–C

I can describe how light produces energy in a photocell and how to increase the current output.	
I can explain how passive solar heating works.	
I can recall that an efficient solar collector must track the position of the Sun in the sky.	
I can rearrange the formula to calculate efficiency in a power station.	
I can explain the greenhouse effect in detail.	
I can explain how scientists agree on the greenhouse effect but disagree on its causes.	
I can explain describe the advantages and disadvantages of using off peak electricity.	
I can explain how increasing the voltage in electricity transmission reduces energy waste through heat.	
I can describe experiments to show the relative penetrating powers of alpha, beta and gamma radiation.	
I can describe the advantages and disadvantages of nuclear power and explain the problems of dealing with radioactive waste.	
I can explain that gravitational attraction provides the centripetal force for orbital motion in our Solar System.	
I can explain why a light-year is a useful unit for measuring distances in space.	
I can explain why the speed of a comet changes as it approaches a star.	
I can suggest possible action that could be taken to reduce the threat of near-Earth objects.	
I can explain how the Big Bang theory accounts for red shift and how the age of the Universe can be estimated.	
I can explain the properties of a black hole.	

I am working at grades B–A*

B3 and B4 Grade booster checklist

B3

I can explain why liver and muscle cells have large numbers of mitochondria.	
I can describe the shape of a DNA molecule.	
I can explain why enzymes are specific.	
I can describe the effect of temperature on enzyme action.	
I can recall the symbol equation for aerobic respiration.	
I can explain why anaerobic respiration occurs during exercise.	
I can explain some of the advantages of being multicellular.	
I can recall the uses of mitosis and meiosis in mammals.	
I can explain how a red blood cell is adapted to its function.	
I can label the main parts of the mammalian heart.	
I can identify simple differences between bacterial cells and plant and animal cells.	
I can recall the function of stem cells.	
I can recognise that selective breeding may lead to inbreeding.	
I can explain some potential benefits and risks of genetic engineering.	
I can describe some possible uses of cloning.	
I can describe what is meant by nuclear transfer.	
I am working at grades D–C	

I can describe the position and function of ribosomes in a cell.	
I can explain how the bases in DNA code for proteins.	
I can explain why changing temperatures affect enzyme action.	
I can explain how changes to genes can alter proteins.	
I can recall the function of ATP.	
I can explain what is meant by oxygen debt.	
I can describe the main stages in mitosis.	
I can explain how meiosis produces haploid cells.	
I can explain how haemoglobin transports oxygen.	
I can explain how the main types of blood vessels are adapted for their functions.	
I can explain the advantages and disadvantages of different measures of growth.	
I can explain the difference between adult and embryonic stem cells.	
I can explain some of the consequences of inbreeding.	
I can describe the main stages in genetic engineering.	
I can describe the cloning technique used to produce Dolly the sheep.	
I can explain why cloning plants is easier than cloning animals.	
I am working at grades B–A*	

B4

I can calculate an estimate of a population size.	
I can compare the biodiversity of natural and artificial ecosystems.	
I can recall and use the balanced symbol equation for photosynthesis.	
I can explain why plants carry out respiration at all times.	
I can name and locate the main parts of a leaf.	
I can explain how leaves are adapted for photosynthesis.	
I can explain the net movement of particles in diffusion.	
I can describe the process of osmosis.	
I can recall that transpiration is the evaporation and diffusion of water from inside leaves	
I can interpret data on transpiration rates.	
I can explain why plants need nitrates, phosphates, potassium and magnesium.	
I can relate mineral deficiencies to their symptoms.	
I can describe the effects of temperature, oxygen and water on the rate of decay.	
I can explain how food preservation methods reduce the rate of decay.	
I can describe how plants can be grown without soil (hydroponics).	
I can explain the advantages and disadvantages of biological control.	
I am working at grades D–C	

I can explain what it means for an ecosystem to be self-supporting.	
I can describe zonation of species across a habitat.	
I can explain how isotopes have increased our understanding of photosynthesis.	
I can explain the effects of limiting factors on the rate of photosynthesis.	
I can explain how the cellular leaf structure is adapted for efficient photosynthesis.	
I can interpret data on the absorption of light by photosynthetic pigments.	
I can explain how the rate of diffusion can be increased.	
I can predict the direction of water movement in osmosis.	
I can describe the structure of xylem and phloem.	
I can explain how the cellular leaf structure is adapted to reduce water loss.	
I can describe how mineral elements are used to produce useful plant compounds.	
I can explain how minerals are taken up by root hairs using active transport.	
I can explain why changing temperature, oxygen and water affect the rate of decay.	
I can explain how saprophytic fungi use extracellular digestion.	
I can explain the advantages and disadvantages of hydroponics.	
I can explain how intensive food production improves the efficiency of energy transfer.	
I am working at grades B–A*	

C3 and C4 Grade booster checklist

C3

I know that the rate of reaction measures how much product is formed in a fixed time period in g/s or cm³/s.	
I know that the amount of product formed is directly proportional to the amount of limiting reactant used.	
I know that the rate of reaction depends on the number of collisions between reacting particles.	
I can draw sketch graphs to show the effects of changing temperature or concentration on the rate of reaction.	
I know that catalysts are specific to particular reactions and only small amounts are needed.	
I can draw sketch graphs to show the effects of surface area on the reaction rate and the amount of product formed.	
I can calculate the M_r of substances like $Ca(OH)_2$ that have formulae with brackets if given the A_r.	
I can show that mass is conserved during a reaction using a simple equation and the relative formula masses.	
I know and can use the formula: $\text{percentage yield} = \dfrac{\text{actual yield} \times 100}{\text{predicted yield}}$	
I know and can use the formula: $\text{atom economy} = \dfrac{M_r \text{ of desired products} \times 100}{\text{sum of } M_r \text{ of all products}}$	
I know that bond breaking is an endothermic process and bond making is an exothermic process.	
I know how to measure the energy given out by fuel, using a calorimeter, and can describe the experiment.	
I can explain why most pharmaceuticals are produced by batch and other bulk chemicals by a continuous process.	
I know that chemicals are extracted from plants by crushing, boiling, dissolving and using chromatography.	
I know that diamond, graphite and fullerenes are different structures of the same element, so are allotropes.	
I know that graphite can be used in lubricants, as it is slippery, and in pencil leads, as it is slippery and black.	

I am working at grades D–C

I know how to interpret data from a table, graph or in writing to calculate the rate of reaction and use the correct units.	
I can explain why the amount of product made is directly proportional to the amount of limiting reactant used.	
I can explain that reacting particles make more successful collisions between particles at higher temperatures.	
I can explain that reacting particles have higher collision frequencies between particles at higher pressures.	
I can explain that reacting particles have more collisions in a powdered reactant than a lump of reactant.	
I can extrapolate a graph showing that the rate of a reaction changes when the reactant surface area changes.	
I can show that mass is conserved in a reaction, given a symbol equation and the relative formula masses.	
I can use a balanced symbol equation to calculate the mass of product formed from a given mass of reactant.	
I can explain why industrial processes need to have high percentage yields to reduce the reactants wasted.	
I can explain why an industrial process needs to have high atom economy to reduce the unwanted products.	
I know if the energy taken in bond breaking is more than given out in bond making, the reaction is endothermic.	
I know and can use the formula: $\text{energy per gram} = \dfrac{\text{energy released (in J)}}{\text{mass of fuel burnt (in g)}}$	
I can explain advantages and disadvantages of batch and continuous manufacturing processes from data given to me.	
I can explain that it is difficult to test new drugs as they need to have clinical trials and be safe under the law.	
I can explain that diamond and graphite have high melting points due to strong covalent directional bonds.	
I can explain that nanotubes can be used as catalysts as they have a high surface area for reactions to happen.	

I am working at grades B–A*

C4

I know that the nucleus is made of protons of mass 1 and charge of +1, and neutrons of mass 1 and charge 0.	
I know that isotopes are atoms of elements that have the same atomic number but different mass numbers.	
I can deduce the formula of an ionic compound from the formula of the positive and negative ions.	
I know that solid sodium chloride is a giant ionic lattice of positive ions strongly attracted to negative ions.	
I know that non-metals combine together by sharing electron pairs and that this is called covalent bonding.	
I can recognise the period to which an element belongs by the number of occupied electron shells in its atom.	
I can explain that group 1 elements have similar properties as they all have one electron in their outer shell.	
I can describe how to use a flame test to identify the presence of lithium, sodium and potassium compounds.	
I can identify the metal halide formed when a group 1 element reacts with a group 7 element.	
I know that when chlorine reacts with solutions of metal halides, it displaces bromides as Br_2 and iodides as I_2.	
I know that carbonates of transition metals decompose on heating to make the metal oxide and carbon dioxide.	
I know how using sodium hydroxide solution identifies Cu^{2+}, Fe^{2+} and Fe^{3+} as different coloured precipitates.	
I know that metals have high melting points and boiling points due to strong metallic bonds.	
I know potential benefits of superconductors include super-fast electronic circuits and powerful electromagnets.	
I know that the water purification process includes filtration, sedimentation and chlorination to kill microbes.	
I know that the reaction of barium chloride with sulfates is an example of a precipitation reaction.	

I am working at grades D–C

I can explain that an atom is neutral as it has the same number of positive protons as negative electrons.	
I can deduce the electronic structure of the first twenty elements in the periodic table when I know the atomic number, mass number and the charge on the particle.	
I can explain the ionic bonding in simple compounds, using the 'dot and cross' model.	
I can use ideas about structure and bonding to explain the melting point and electrical conductivity of sodium chloride.	
I can use a 'dot and cross' model to explain the bonding in molecules containing single or double covalent bonds.	
I can use ideas about structure and bonding to explain why carbon dioxide and water have low melting points.	
I can construct a balanced symbol equation for the reaction between any group 1 element and water.	
I know that oxidation is loss of electrons and can explain why a process is oxidation from its ionic equation.	
I can predict the properties of fluorine and astatine, given the properties of the other group 7 elements.	
I know that reduction is gain of electrons and can explain why a process is reduction from its ionic equation.	
I can construct balanced symbol equations for the thermal decomposition of $FeCO_3$, $CuCO_3$, $MnCO_3$ and $ZnCO_3$.	
I can construct balanced symbol equations for the reactions between Cu^{2+}, Fe^{2+} and Fe^{3+}, and OH^- ions.	
I know metallic bonds are attractions between a 'sea' of delocalised electrons and close-packed positive ions.	
I can explain some drawbacks of superconductors, including the fact that they only operate at very low temperatures.	
I can explain the disadvantages of using the distillation of sea water to make fresh water.	
I can construct balanced symbol equations for the reactions of barium chloride with sulfates, and silver nitrate with halides.	

I am working at grades B–A*

P3 and P4 Grade booster checklist

P3	
I can interpret the relationship between initial speed, final speed, distance and time.	
I can interpret the gradient of a speed–time graph.	
I can draw and interpret the shapes of speed–time graphs.	
I can describe the significance of positive and negative acceleration and calculate acceleration.	
I can describe and interpret the relationship between force, mass and acceleration.	
I can explain the factors that affect thinking distance, braking distance and their implications on road safety.	
I can calculate weight from knowledge of mass and gravitational field strength.	
I can calculate power knowing the time it takes to perform work.	
I can use and apply the equation $KE = \frac{1}{2}mv^2$.	
I can explain how electrically powered cars do cause pollution but not at their point of use.	
I can describe the relationships between momentum, mass, velocity, force and time.	
I can explain how and why crumple zones, seat belts and air bags reduce injuries.	
I can explain the motion of a falling object in terms of balanced and unbalanced forces.	
I can recognise that all objects fall with the same acceleration at a point on the Earth's surface.	
I can use the equation $GPE = mgh$ and interpret examples of energy transfer from GPE to KE.	
I can interpret the energy transfers of a roller-coaster and the effects of mass and speed on kinetic energy.	
I am working at grades D–C	

I can interpret the relationship between speed, distance and time to include the effect of changing quantities.	
I can draw and interpret distance–time graphs.	
I can calculate acceleration and distance travelled from speed–time graphs.	
I can interpret the relationship between acceleration, speed change and time.	
I can use and manipulate the equation linking force, mass and acceleration.	
I can interpret and explain the shapes of graphs for thinking and braking distance against speed.	
I can perform a two-stage calculation linking work done and power.	
I can explain the derivation of the equation power = force × speed.	
I can apply the ideas of kinetic energy to everyday situations such as braking distance.	
I can explain and evaluate the factors upon which fuel consumption depends.	
I can use Newton's second law of motion to explain the relationship between force, mass and acceleration.	
I can evaluate car safety features and explain how ABS brakes work.	
I can explain why objects reach a terminal speed.	
I can understand that gravitational field strength varies with height and position on the Earth's surface.	
I can understand the energy transfers when a body is falling at a terminal speed.	
I can use and apply the relationship $mgh = \frac{1}{2}mv^2$.	
I am working at grades B–A*	

P4	
I can state that like charges repel and opposite charges attract.	
I can describe electrostatic phenomena in terms of transfer of electrons which have a negative charge.	
I can describe problems and dangers caused by static electrical charge.	
I can explain how static electrical charge can be useful in dust precipitators, defibrillators, crop spraying and paint spraying.	
I can calculate resistance and describe how the resistance of a wire can be changed.	
I can describe the functions of the three wires and the fuse in a three pin plug.	
I can describe the features of a longitudinal wave.	
I can recall what ultrasound is and describe how it is used to scan the body and to break down kidney and other stones.	
I can explain and use the concept of half-life of radioactive isotopes.	
I can describe radioactivity as naturally occurring radiation from the nucleus of an unstable atom and give examples of sources of background radiation.	
I can recall the natures of alpha, beta and gamma radiation.	
I can describe how radioactive tracers are used in industry and in hospitals.	
I can describe other uses of radioactive isotopes such as in smoke detectors or radioactive dating.	
I can describe how x-ray images are produced.	
I can describe how nuclear fission is used in power stations and how the reaction is controlled.	
I can describe nuclear fusion and explain why it is difficult to produce power from it.	
I am working at grades D–C	

I can explain how problems with static electricity, including electric shocks, can be reduced.	
I can explain uses of electrostatics in terms of electron movement.	
I can rearrange the formula for resistance.	
I can explain the reasons for using fuses and circuit breakers in circuits.	
I can compare the motion of particles in longitudinal and transverse waves.	
I can explain how ultrasound is used in medicine.	
I can explain why ultrasound is used instead of x-rays for some scans.	
I can interpret graphical data on radioactive decay and half-life.	
I can construct and balance nuclear equations for alpha and beta decay.	
I can explain why alpha particles are so strongly ionising.	
I can evaluate the relative significance of sources of background radiation.	
I can explain how radiocarbon dating finds the age of old materials.	
I can explain how x-rays and gamma rays are produced.	
I can explain how radioactive sources are used to treat cancer.	
I can explain what causes a chain reaction.	
I can explain why 'cold fusion' is still not accepted by most scientists.	
I am working at grades B–A*	

Modern periodic table

Group 1	2											3	4	5	6	7	0
																	1 H hydrogen (1)
7 Li lithium (3)	9 Be beryllium (4)											11 B boron (5)	12 C carbon (6)	14 N nitrogen (7)	16 O oxygen (8)	19 F fluorine (9)	4 He helium (2)
23 Na sodium (11)	24 Mg magnesium (12)											27 Al aluminium (13)	28 Si silicon (14)	31 P phosphorus (15)	32 S sulfur (16)	35 Cl chlorine (17)	20 Ne neon (10)
39 K potassium (19)	40 Ca calcium (20)	45 Sc scandium (21)	48 Ti titanium (22)	51 V vanadium (23)	52 Cr chromium (24)	55 Mn manganese (25)	56 Fe iron (26)	59 Co cobalt (27)	59 Ni nickel (28)	64 Cu copper (29)	65 Zn zinc (30)	70 Ga gallium (31)	73 Ge germanium (32)	75 As arsenic (33)	79 Se selenium (34)	80 Br bromine (35)	40 Ar argon (18)
85 Rb rubidium (37)	88 Sr strontium (38)	89 Y yttrium (39)	91 Zr zirconium (40)	93 Nb niobium (41)	96 Mo molybdenum (42)	99 Tc technetium (43)	101 Ru ruthenium (44)	103 Rh rhodium (45)	106 Pd palladium (46)	108 Ag silver (47)	112 Cd cadmium (48)	115 In indium (49)	119 Sn tin (50)	122 Sb antimony (51)	128 Te tellurium (52)	127 I iodine (53)	84 Kr krypton (36)
133 Cs caesium (55)	137 Ba barium (56)	139 La lanthanum (57)	178 Hf hafnium (72)	181 Ta tantalum (73)	184 W tungsten (74)	186 Re rhenium (75)	190 Os osmium (76)	192 Ir iridium (77)	195 Pt platinum (78)	197 Au gold (79)	201 Hg mercury (80)	204 Tl thallium (81)	207 Pb lead (82)	209 Bi bismuth (83)	210 Po polonium (84)	210 At astatine (85)	131 Xe xenon (54)
223 Fr francium (87)	226 Ra radium (88)	227 Ac actinium (89)															222 Rn radon (86)

Modern periodic table. You need to remember the symbols for the highlighted elements.

B1 Answers

B1 Understanding organisms

Page 134 Fitness and health

1 a Take more regular exercise; eat a healthier diet; reduce salt intake; drink less alcohol; avoid stress; don't smoke; maintain a healthy weight
(Any 2)

   b With high blood pressure, increased risk of small blood vessels bursting; if in brain a stroke happens; if in kidney, kidney damage occurs

2 a Risk factors of hypertension, smoking and high cholesterol show overall decrease; risk factor of being overweight has increased

   b Three out of four risk factors reduced, so expect reduction in incidence of heart disease; however, being overweight shows a large increase, so overall effect may mean little change; change in risk factors will take many years to show effect
(Any 2)

3 Being fit is the ability to do exercise – this will not stop/kill pathogens

4 Carbon monoxide reduces the oxygen capacity of red blood cells; it combines with the haemoglobin in red blood cells; preventing combination with oxygen/oxy-haemoglobin; so the blood contains less oxygen; can be fatal
(Any 4)

Page 135 Human health and diet

1 a Kwashiorkor

   b Overpopulation; limited investment in agriculture

2 a 40 000 g = 40 kg; EAR = 40 × 0.6 = 24 g

   b May be vegetarian or vegan and not ensuring adequate nutrition; may have poor self-image

   c Plant proteins are second-class proteins, i.e. *one* plant source alone does not contain all the essential amino acids; the body cannot make its own essential amino acids; so Simon will need to eat a combination of plant foods that make a first-class protein, e.g. beans on toast
(Any 2)

3 a Carbohydrates

   b Amino acids

   c Extra glycogen is stored in liver; excess protein cannot be stored

Page 136 Staying healthy

1 a Pathogens; toxins

   b Antigens; antibodies

2 a Parasite: mosquito/*Plasmodium*; host: human

   b Drain stagnant water to kill larvae of mosquito/vector; put oil on water to prevent larvae of mosquito/vector breathing; spray insecticide to kill adult mosquito/vector; take medication to kill *Plasmodium* inside body
(Any 2)

3 a i At 20 days the level of immunity from passive immunity is twice that of/much higher than active immunity

     ii After 60 days there is no immunity from passive immunity, active immunity level is still high

   b In passive immunity the body receives antibodies; this results in a high immunity level being quickly reached; in active immunity the body makes its own antibodies continually; so the level of antibodies and immunity remains high

   c Passive immunity (no mark) because immunity response is quicker

   d Antibiotics are used to kill bacteria; bacteria show differences, some will be resistant to the antibiotic; these resistant strains will survive due to less competition; so new antibiotics will need to be developed since existing antibiotics will not work
(Any 3)

Page 137 The nervous system

1 a Reflected should be refracted; optic nerve should be retina; spinal cord should be brain

   b Narrower field of view; poorer judgement of distance

   c Explanation of accommodation, i.e. ability to change focus; for distant objects the lens is made thinner; by the suspensory ligaments becoming tighter; and the ciliary muscles relaxing; accept explanation for near objects

2 a A: cell body; B: axon; C: sheath

   b Axon

   c C

   d Nerve impulse triggers release of neurotransmitter substance; which diffuses; across synapse; and binds with receptor molecules; in membrane of next neurone; triggering new impulse
(Any 5)

Page 138 Drugs and you

1 Depressant linked to alcohol; painkiller linked to paracetamol; stimulant linked to caffeine; hallucinogen linked to LSD

2 a i 251

     ii Fewer men smoke; people now know risks/publicity/warnings on cigarette packets/ban in public areas

   b Cause more neurotransmitter substances; to diffuse across synapse; more nerve impulses transmitted

3 a Reading, texting and having alcohol all increase reaction distances; at both 35 and 70 mph; greater effect at 70 mph; greatest effect from reading, then texting, then alcohol; using alcohol roughly doubles reaction distances
(Any 3)

   b Not reliable since only one person tested; condition of car/tyres/road not known

Page 139 Staying in balance

1 a Maintaining a constant internal environment

   b Optimum temperature for action of many enzymes

   c Sweat more; therefore lose too much water

   d i A negative feedback mechanism cancels out a change; it returns it to normal

     ii Give an example, e.g. body too hot so blood temperature rises; hypothalamus in brain detects this change; triggers sweating and vasodilation

2 a Type 2 diabetes caused by too little insulin or by cells not reacting to it; amounts of carbohydrates eaten can be altered; to suit activity

   b i Glucose enters blood

     ii Insulin; converts excess glucose into glycogen

     iii No insulin; so glucose remains in blood

B2 Answers

Page 140 Controlling plant growth

1 a Makes roots grow; because rooting powder contains auxins/plant hormones

   b i Because they kill only some specific weeds

   ii Contains plant hormones; which makes plant/roots grow too fast

2 a Positively; gravity; geotropic; auxins

   b i Auxins

   ii Tip

   iii Shoot curves/grows away; from the side which has the agar block; agar block contains auxin; which diffuses into the left side of the shoot; higher concentration of auxin on left side; causes more cell elongation on left side

   (Any 5)

Page 141 Variation and inheritance

1 a Sudden change in gene/chromosome

   b In gamete formation; genes mixed up; in fertilisation; recombination of genes from two parents

   (Any 2)

2 the same; 23; a different

3 a Correct lengths of chromosomes; correct letters; correct gender

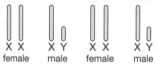

X X   X Y   X X   X Y
female male   female   male

   b Two sex chromosomes X and Y; these separate in gamete formation; male is XY, female is XX; ratio from cross is 1:1

   (Any 3)

   c i Correct gametes; correct combinations (as shown in diagram)

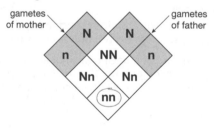

gametes of mother   gametes of father

N   N

n   NN   n

Nn   Nn

nn

   ii nn ringed (as shown in diagram)

Page 142 B1 Extended response question

5–6 marks

Uses chart to work out that Rick has an ideal blood pressure, Shabeena is pre-high blood pressure while Alan and Toni have high blood pressure. Applies knowledge of how a poor life style (being overweight, stress, high alcohol intake, smoking) increases blood pressure and a healthy lifestyle (regular exercise, balanced diet) lowers blood pressure to ideal levels. Realises that Alan, Shabeena and Toni have high risk of stroke/kidney damage. Links other information on the risks of smoking (lung cancer) and poor diet (diabetes) to the data. All information in answer is relevant, clear, organised and presented in a structured and coherent format. Specialist terms are used appropriately. There are few, if any, errors in grammar, punctuation and spelling.

3–4 marks

Correctly applies data to blood pressure chart. Gives a limited description of some life style factors and consequence but links to actual people not clear. Links to smoking risks and poor diet brief or missing. For the most part the information is relevant and presented in a structured and coherent format. Specialist terms are used for the most part appropriately. There are occasional errors in grammar, punctuation and spelling.

1–2 marks

Errors in applying data to the blood pressure chart. An incomplete/confused description is given of reasons and consequences for data. Answer may be simplistic. There may be limited use of specialist terms. Errors of grammar, punctuation and spelling prevent communication of the science.

0 marks

Insufficient or irrelevant science. Answer not worthy of credit.

B2 Understanding our environment

Page 143 Classification

1 a Phylum; class; order; genus; species

   b DNA analysis; to see how similar it is

2 a A group of organisms that can interbreed; to produce fertile offspring

   b i Bobcat and ocelot

   ii Both belong to the same genus

3 It has some features of a reptile and some of a bird; an organism that has features of more than group (like this one, which has feathers and also has teeth) will be difficult to classify

4 a Hybrid

   b It is infertile/has some of the characteristics of a horse and some of a zebra

5 They share a fairly recent common ancestor; therefore have similar genes; both live in similar habitats; have developed similar adaptations

   (Any 3)

Page 144 Energy flow

1 a i the dry mass of living material; at each trophic level

   ii one rose bush can feed many beetles; pyramid of numbers does not take into account the size of the organism

   b Measuring biomass involves removing all the water; destructive process; need to collect all the parts of the rose bush

   (Any 2)

2 a Respiration

   b i Keep the cows in warm conditions; restrict their movement

   ii have more energy available for growth

   c i 3056 – (1909 + 1022); = 125kJ

   ii 125/3056 × 100; = 4.1%

   iii So much energy is lost between each trophic level; if we ate crops there would only be one energy transfer; less energy would be lost

   (Any 2)

B2 Answers

Page 145 Recycling

1 a Combustion; respiration; photosynthesis

  b Less oxygen in waterlogged soils; decomposers cannot respire

  c Turn into limestone; limestone weathered by acid rain; chemical reaction releases carbon dioxide

2 a To make proteins

  b i Ammonia

    ii Nitrifying bacteria

  c Nitrogen-fixing bacteria; in soils; or in root nodules of legumes; convert nitrogen to ammonia or nitrates; action of lightning; combines oxygen and nitrogen

(Any 4)

Page 146 Interdependence

1 a Mates

  b i Where it lives; its role in the community

    ii They both live in the same part of the forest; both eat acorns

2 a The numbers go up when there are fewer owls because fewer lemmings are eaten; when the owl numbers increase the lemming numbers drop as more are eaten

  b When there are lots of lemmings, then owls can survive and breed; it takes a while for owl numbers to increase

3 Mutualism

4 a Mites

  b Bees gain nectar; flowers get pollinated

  c Pea plant gives the bacteria some sugars made by photosynthesis; the bacteria fix nitrogen into compounds that the plant can use

Page 147 Adaptations

1 a Thick fur for insulation; layer of blubber under the skin; wide feet stop it sinking into the snow; claws to hold on to the ice; white colour for camouflage; small ears lose less heat; large body to retain heat

(Any 3)

  b Hibernation; body reactions slow down; conserve food reserves; less food is available in winter

(Any 3)

  c Polar bears live in colder climates; smaller ears/larger body means smaller surface area to volume ratio; so less heat loss

2 a Less water loss through leaves; protect cacti from being eaten

  b Sun basking/absorb Sun's heat; to warm body up

3 If one food is in short supply they can change to another; different foods are available in different areas

Page 148 Natural selection

1 a Can digest the acorns more easily

  b In the genes; which are passed on in the sex cells

  c Some of the red squirrels would have a mutation; this would allow them to digest acorns; they are more likely to survive; pass on this gene; over many generations the population will all be able to digest

(Any 4)

2 Some bacteria are resistant to antibiotics; they can survive and reproduce

3 a Worried that people would disagree with him; most people believed that God created all organisms as they are now

  b Organisms on small islands do not grow as fast; pass this on to their offspring

  c Acquired characteristics are not inherited

Page 149 Population and pollution

1 a Increase in temperatures/global warming/ greenhouse effect; rising sea levels/flooding; drought

(Any 2)

  b i CFCs

    ii UV light causes skin cancer

2 a Exponential

  b Burn less fossil fuels; produce less waste that cannot be recycled; eat less/local food; less travel

(Any 2)

3 a Indicator species

  b Mussels, damselfly larvae and bloodworms all survive in polluted water; if it was clean you would also find mayfly and stonefly larvae

  c Advantage: more accurate result at any one time; Disadvantage: more expensive; does not monitor over a very long period

Page 150 Sustainability

1 a More tourism; better transport; other resources

  b To see how closely related they are; so prevent inbreeding/maintain genetic variation

2 a For: provides food/jobs; Against: may lead to their extinction

  b Areas too large to police

3 a Taking enough resource from the environment to supply an increasing population but leaving enough behind to ensure a supply for the future and prevent permanent damage

  b Restricting numbers stops too many being killed; preventing killing small fish allows them to mature to breed

  c It is difficult to dispose of the increased waste produced without causing pollution; the people need more food which needs to be provided without destroying habitats; increased demand for energy which could cause pollution if generated from fossil fuels

Page 151 B2 Extended response question

5–6 marks

A description of how energy is lost from a food chain and some realisation that efficiency of transfer is different between different trophic levels. Includes an appreciation that vegetarianism involves fewer transfers of energy and so less energy loss. Answers are backed up by at least one calculation of efficiency. All information in answer is relevant, clear, organised and presented in a structured and coherent format. Specialist terms are used appropriately. Few, if any, errors in grammar, punctuation and spelling.

3–4 marks

Descriptions of how energy is lost from food chains and an appreciation that vegetarianism would provide humans with more energy. Figures are quoted but not processed. For the most part the information is relevant and presented in a structured and coherent format. Specialist terms are used for the most part appropriately. There are occasional errors in grammar, punctuation and spelling.

1–2 marks

An incomplete description, naming some processes by which energy is lost but no analysis of figures or appreciation of the importance of the number of trophic levels. Answer may be simplistic. There may be limited use of specialist terms. Errors of grammar, punctuation and spelling prevent communication of the science.

0 marks

Insufficient or irrelevant science. Answer not worthy of credit.

C1 Carbon Chemistry

Page 152 Making crude oil useful

1 a A molecule containing carbon and hydrogen; only

b

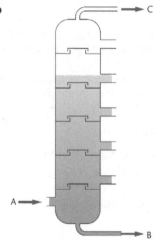

   i A At the bottom of the tower (left-hand side)
   ii B It 'exits' through the bottom of the tower
   iii C At the top of the tower
   iv The fraction with a low boiling point, LPG

c The forces between molecules are called intermolecular forces; these forces are broken during boiling/the molecules of a liquid separate from each other as molecules of gas; the molecules in different fractions have different-length chains; this means that the forces between the molecules are different
Either: molecules such as those that make up bitumen and heavy oil are very long; there are strong forces of attraction between the molecules; the strong forces mean that the molecules are difficult to separate/a lot of energy is needed to pull one molecule away from another; these fractions have high boiling points
Or: Molecules such as petrol have short molecules; each molecule does not have very strong attractive forces; these molecules are easily separated/less energy is needed to pull the short molecules apart; short molecules have very low boiling points
(Answer should be logical and well-constructed, any 1 from introduction, any 2 from either option below it)

2 Oil slicks can harm animals, pollute beaches and destroy unique habitats for long periods of time; clean-up operations are extremely expensive and the detergents used cause problems to wildlife

3 a An alkene has a double bond between carbon atoms
  b Polymers
  c 32%
  d An industrial cracking plant near the distillation refinery; the heavy oil/large hydrocarbon molecules in surplus supply could be cracked; petrol/smaller more useful molecules would be produced in the process to support demand

Page 153 Using carbon fuels

1 a i Coal is bulky and dirty whereas petrol is volatile; coal produces acid fumes but petrol produces less acid fumes
    ii High energy value; good availability
  b There is greater demand because the population is increasing/more fossil fuels are being consumed; greater consumption produces more carbon dioxide; carbon dioxide is a greenhouse gas and contributes to global warming
(Any 2)

2 a i Hydrocarbon fuel + oxygen → carbon dioxide + water
(1 mark for reactants, 1 mark for products)
    ii The candle is burnt; and the products are passed through test reagents (if they are positive, then carbon dioxide and water are formed)
  b Less soot is made; more heat is released; toxic carbon monoxide gas is not produced
  c $C_5H_{12} + 8O_2 \rightarrow 5CO_2 + 6H_2O$
(1 mark for all 4 symbols, 1 mark for balancing the equation)
  d $C_3H_8 + 2O_2 \rightarrow 3C + 4H_2O$
(1 mark for all 4 symbols, 1 mark for balancing the equation)

Page 154 Clean air

1 a Nitrogen (largest portion); oxygen; carbon dioxide (smallest portion)
  b 78%; 21%; 0.035%
  c Plants photosynthesise using up carbon dioxide; <u>photosynthesis</u> produces oxygen; animals respire using up oxygen; animal <u>respiration</u> produces carbon dioxide; fuels undergo <u>combustion</u> using up oxygen; fuel combustion produces carbon dioxide
(Answer must include underlined terms)

2 a Increasing population: energy use has increased so levels of carbon dioxide have increased; deforestation: levels of carbon dioxide have increased/oxygen levels have not increased
  b Carbon dioxide and water vapour were released from early volcanoes; nitrogen is unreactive so levels built up over time; photosynthetic organisms developed that used up carbon dioxide and released oxygen

3 a When there is incomplete combustion in petrol-powered cars, it produces poisonous gases such as carbon monoxide; oxides of nitrogen can contribute to photochemical smog and acid rain; if there is sulfur in the fuel, acid rain can be made
  b Carbon monoxide
  c Oxygen and nitrogen can combine at the very high temperatures on the surface
  d They are both found in clean air/they are non-toxic

C1 Answers

Page 155 Making polymers

1 a It contains an oxygen atom

b It contains a double bond

c The bromine water is orange; when the alkene is present, the water is decolourised/it turns colourless

2 a A compound with at least one double bond between carbon atoms

b Bromine solution reacts with the alkene in the unsaturated compound which undergoes an 'addition reaction'; it turns from orange to colourless; a new di-bromo compound is formed

3 a C

b High pressure; catalyst

c A double bond between carbon atoms

d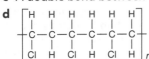

Illustration must include: 4 or 6 carbon atoms and alternate H and Cl atoms on the bottom; brackets and bonds through either end and n at end

e
$$
\begin{array}{c}
\text{H} \quad \text{CH}_3 \\
| \qquad | \\
\text{C}=\text{C} \\
| \qquad | \\
\text{H} \quad \text{H}
\end{array}
$$

Illustration must include: Two carbon atoms joined by a double bond and CH_3 on top right-hand side; only 4 other atoms/groups joined to two carbon atoms

f The reaction needs high pressure and a catalyst; the reaction causes the double bond in the monomer to break/each of the two carbon atoms forms a new bond; the reaction continues until it is stopped/it makes a long molecule

Page 156 Designer polymers

1 a Any reasonable example of use; and suitability. For example:

Use	Suitability
Contact lens	flexible
Drain pipe	rigid
Wound dressing	waterproof

b i Waterproof; flexible; tough; lightweight; keeps UV light out

(Any 2)

ii It keeps water vapour from body sweat in

c i The membrane has pores which are larger than a water vapour molecule and therefore moisture from sweat passes through; the membrane pores are small enough to prevent water droplets from passing inwards

ii It is fragile so would not be strong enough for the purpose of the fabric in outdoor clothing

2 a i Wastes valuable land

ii Creates toxic gases

iii Difficult to sort different polymers

b They do not have to be disposed of in landfill sites or burned; they can decay by bacterial action or be dissolved

3 a i Strong covalent bond label

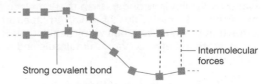

ii Intermolecular forces label

b Some plastics have weak intermolecular forces of attraction between the polymer molecules; these polymer molecules can slide over one another/ separate easily; other plastics form intermolecular chemical bonds/cross-linking bridges between polymer molecules; intermolecular bonds are strong so the polymer molecules cannot slide over one another so these polymers are rigid/the chains cannot easily separate

Page 157 Cooking and food additives

1 a Change shape (permanently); the process is called denaturing

b Starch grains swell and spread out; the cell wall ruptures making the texture softer

c Sodium hydrogencarbonate → sodium carbonate + carbon dioxide + water

d $2NaHCO_3 \rightarrow Na_2CO_3 + CO_2 + H_2O$

(1 mark for all 4 symbols, 1 mark for balancing the equation)

2 a The hydrophobic end is a fat-loving end; the hydrophilic end is a water-loving end; the molecule is an emulsifier

b The hydrophobic end is attracted to fat; the hydrophilic end is attracted to water; the emulsifier holds the two types of molecules together (in close proximity)

Page 158 Smells

1 a Acid + alcohol → ester + water

b Put some alcohol; in a test tube; add an equal amount of ethanoic acid; heat in a boiling water bath; diagram of test tube in a heated beaker of water (accept also flask and condenser as a reflux apparatus)

(Any 3 for 3 marks)

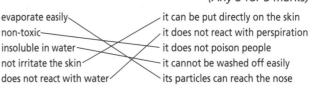

(4 marks – 5 correct, 3 marks – 4 correct, 2 marks – 3 correct, 1 mark – 2 correct)

3 Solution

4 Particles of a liquid are weakly attracted to each other; when some particles of a liquid increase their kinetic energy the force of attraction between the particles is overcome; when the attraction is broken, the particles escape through the surface of the liquid into the surroundings (this is evaporation); if the evaporation happens easily the liquid is said to be volatile and the substance can reach the nose

5 Perfumes and solvents

C2 Answers

6 This is because the force of attraction between two water molecules is stronger; than that between a water molecule and a molecule of nail varnish; the force of attraction between two nail varnish molecules is stronger; than between varnish molecule and water molecule

Page 159 Paints and pigments

1 Particles/solute are dispersed through a liquid/solvent

2 Although the particles are not small enough to dissolve, they are not large enough to settle

3 The solvent evaporates and the paint dries to form a continuous film

4 The solvent evaporates; the oil is oxidised by atmospheric oxygen

5 Answers could include: a cup, so that it can change colour if it is too hot to touch; a baby's bath toy so that it can warn if the water is too hot

6 a Combine them with acrylic paints of different colours

b The paint would contain a blue thermochromic pigment; the pigment would have been mixed with yellow acrylic paint (to make green); when heated the blue pigment would have turned colourless leaving the paint looking yellow

7 They absorb energy; the energy is released as light over a long period of time

8 Radioactive materials; phosphorescent pigments are safer

Page 160 C1 Extended response question

5–6 marks
At least two properties of perfumes with reasons for why these are relevant. A complete explanation of evaporation using kinetic theory.
All information in the answer should be relevant, clear, organised and presented in a structured and coherent format. Specialist terms should be used appropriately. There should be few, if any, errors in grammar, punctuation and spelling.

3–4 marks
At least two properties of perfumes with reasons why these are relevant. Some attempt at the explanation of evaporation using kinetic theory.
The information should generally be relevant and presented in a structured and coherent format. Specialist terms should mostly be used appropriately. There may be occasional errors in grammar, punctuation and spelling.

1–2 marks
One or two properties of perfumes with an attempt at an explanation of evaporation but without reference to kinetic theory. Answers may be simplistic. There will be limited use of specialist terms. Errors of grammar, punctuation and spelling are likely to prevent effective communication of the science.

0 marks
Insufficient or irrelevant science such as repeating the question. Answer not worthy of credit.

C2 Chemical resources

Page 161 The structure of the Earth

1 a Crust; top part of the mantle

b Crust too thick to drill through; need to rely on infrequent seismic waves or man-made explosions

c Convection currents

d Oceanic plates are denser than continental plates; when plates meet the oceanic plate sinks; when continental plates collide they buckle to form mountains; deep in the mantle the oceanic plate melts; the magma forces its way up; volcanoes form and erupt, recycling rock

2 a The theory explains the evidence; it has been discussed/tested by many scientists

b Wegener's evidence could be interpreted in different ways; little actual evidence had been collected; there was a reluctance to give up current theories

(Any 2)

3 a To predict eruptions/to save lives; to find out information about the Earth's structure

b The composition of the lava varies

c The type of eruption depends on the type of lava; iron-rich basalt is runny and flows out; silica-rich rhyolite explodes to create a more violent eruption

Page 162 Construction materials

1 a Limestone, marble, granite

b

Building material	aluminium	brick	glass
Raw material	bauxite (aluminium) ore	clay	sand

c Limestone and clay are heated together

d i Granite is an <u>igneous</u> rock, made of <u>interlocking crystals</u> which <u>solidifies</u>.

ii Marble is a <u>metamorphic</u> rock, formed from <u>limestone</u> by <u>heat</u> and <u>pressure</u>.
(1 mark for each underlined word)

2 a Calcium carbonate → calcium oxide + carbon dioxide

b Concrete is made by mixing cement, sand and small stones with water; it can be strengthened by adding steel bars/grids

c Concrete is strong under compression (squashing force); concrete is weak under tension (pulling force)/labelled diagram illustrating these points; reinforced concrete has steel bars making it harder and more flexible/labelled diagram illustrating these points

Page 163 Metals and alloys

1 a Label to left (negative) side

b Impure copper dissolves (into the electrolyte)/impure copper acts as an anode; copper moves from the anode to the cathode; copper from the electrolyte is plated on to the cathode/cathode gains mass because copper is deposited; electrical currents/electrons flows around the circuit

c Advantages: the cost to melt it is low; people earn money from selling it; less mining needed, so there is less noise pollution and dust, and fewer trucks; the cost of copper is kept down

(Any 2)

C2 Answers

Disadvantages: it leads to fewer jobs in mining; some items are hard to separate; pure and less pure copper must not be mixed which can be difficult to check; some items contain hardly any copper so it is not worthwhile; the separation process can create pollution; it is hard to persuade people to recycle

(Any 2)

d $Cu \rightarrow Cu^{2+} + 2e^- \: / \: Cu - 2e^- \rightarrow Cu^{2+}$

(1 mark for all 3 symbols, 1 mark for balancing the equation)

e The loss of electrons (at the anode) is oxidation; the gain in electrons (at the cathode) is reduction; the reaction can be represented by the equation $Cu^{2+} + 2e^- \rightarrow Cu$

2 a

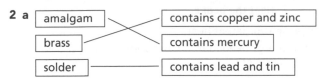

b Smart alloys are useful because they remember their original shape; they are a good material to use for spectacle frames because if the frame gets bent, it can return to the correct shape rather than being thrown away

Page 164 Making cars

1 a Salt water accelerates rusting in steel and so car bodies made from steel rust more quickly when salt is put on the roads

b Aluminium has a protective oxide layer

c Iron + oxygen + water → hydrated iron(III) oxide
(1 mark for reactants, 1 mark for products)

2 a i A mixture containing at least one metal

ii Steel is harder; steel is less likely to corrode

b not very strong; higher cost

c Transparent; strong; does not shatter
(Any 2)

d lightweight; better fuel economy/corrosion-resistant; car has a longer life
(Any 2)

3 a Moist acidic air

b Aluminium; the other materials are affected more by the atmosphere, according to the table

4 It is the law that some parts should be recycled; there are valuable resources in the cars; disposal of valuable resources requires land
(Any 2)

Page 165 Manufacturing chemicals – making ammonia

1 a High pressure; 450 °C temperature; (iron) catalyst; recycling unreacted starting materials
(Any 3)

b $N_2 + 3H_2 \rightleftharpoons 2NH_3$
(1 mark for all 3 symbols, 1 mark for balancing the equation)

2 a 50%

b The higher the pressure, the higher the yield

c Higher temperatures break the ammonia down, so reducing the yield; lower temperatures reduce the reaction rate; 450 °C is the optimum temperature which gives a suitable yield at a reasonable rate

3 a Pressure; temperature; catalysts; recycling materials; automation
(Any 3)

b Ammonia is used to make nitrogen fertilisers; fertilisers are needed to increase crop yield; as world population increases more fertiliser is needed; the amount of land available for growth decreases, so the existing land needs to produce more crops
(Any 3)

c Reaction rate is about how fast the product is made; percentage yield is about the final amount obtained

d If the reaction can be repeated many times; if a high percentage of the starting materials can be recycled

Page 166 Acids and bases

1 a i An alkali is a base which dissolves in water

ii Acid, water

b Hydrogen ion/H^+ ion

c The greater the concentration of H^+ ions, the lower the pH

d $H^+ + OH^- \rightleftharpoons H_2O$
(1 mark for all 3 symbols, 1 mark for balancing the equation)

2 a The pH at the start is low, as alkali is added to the acid the pH rises

b The colour starts as purple showing it is acidic; as the acid neutralises the alkali the pH falls and the colour changes to blue; when neutral, the pH = 7 and the colour is green

3 a Copper carbonate + sulfuric acid → copper sulfate + water + carbon dioxide
(1 mark for reactants, 1 mark for products)

b i Calcium sulfate

ii Potassium nitrate

iii Sodium chloride

iv Copper nitrate

c i $H_2SO_4 + CaCO_3 \rightarrow CaSO_4 + H_2O + CO_2$
(1 mark for all 5 symbols, 1 mark for balancing the equation)

ii $HNO_3 + KOH \rightarrow KNO_3 + H_2O$
(1 mark for all 4 symbols, 1 mark for balancing the equation)

iii $2HCl + Na_2CO_3 \rightarrow 2NaCl + H_2O + CO_2$
(1 mark for all 5 symbols, 1 mark for balancing the equation)

iv $2HNO_3 + CuO \rightarrow Cu(NO_3)_2 + H_2O$
(1 mark for all 4 symbols, 1 mark for balancing the equation)

Page 167 Fertilisers and crop yields

1 a So substances are small enough to get into the roots

b They replace/provide essential elements/nutrients; they provide nitrogen needed to make plant proteins resulting in increased growth

2 a Farming uses fertilisers that wash off fields into rivers

b Fertilisers are washed off fields; fertilisers lead to increased nitrate and phosphate levels in the water; algae grow quickly on the surface because of the chemicals; the algae block the sunlight to other oxygen-producing plants; aerobic bacteria deplete the oxygen levels in the water; most living organisms die
(Answer should be logical and well-constructed)

P1 Answers

3 a i Ammonium nitrate
 ii Potassium phosphate
 b i Nitric acid; potassium hydroxide/potassium oxide
 ii Evaporate most of the water; remove from the heat

Page 168 Chemicals from the sea – the chemistry of sodium chloride

1 a Mining rock salt from the ground; solution mining
 b Subsidence of houses; brine can escape and affect habitats; noise; dust; trucks
(Any 2)

2 a i Chlorine
 ii Hydrogen
 iii Sodium hydroxide
 b To prevent unwanted reactions/so the electrodes do not dissolve
 c H^+ and Na^+; Cl^- and OH^-
 d $2H^+ + 2e^- \rightarrow H_2$
(1 mark for all 3 symbols, 1 mark for balancing the equation)

3 a Sodium hydroxide; chlorine
 b The industry produces a range of useful products; it produces raw materials for other industrial processes; products from the industry can be exported; the industry reduces need for importing
(Any 3)

Page 169 C2 Extended response question

5–6 marks
The chemicals used should be named (and suitable). For example, phosphoric acid and ammonium hydroxide/ammonia solution. A suitable indicator should be named. The method should include: repeat readings, use of the average to make the final fertiliser without indicator, evaporating the solution to form crystals.
All information in the answer should be relevant, clear, organised and presented in a structured and coherent format. Specialist terms should be used appropriately. There should be few, if any, errors in grammar, punctuation and spelling.

3–4 marks
The indicator used should be named and most of the method described, although some stages may be missing. The information should generally be relevant and presented in a structured and coherent format. Specialist terms should mostly be used appropriately. There may be occasional errors in grammar, punctuation and spelling.

1–2 marks
The description will be incomplete, with little detail given. Answers may be simplistic. There will be limited use of specialist terms. Errors of grammar, punctuation and spelling are likely to prevent effective communication of the science.

0 marks
Insufficient or irrelevant science. Answer not worthy of credit.

P1 Energy for the home

Page 170 Heating houses

1 a Energy flows from a warm to a cooler body; temperature of the warmer body falls

2 Temperature is a measure of hotness on an arbitrary scale; heat is a form of energy on an absolute scale

3 A thermogram uses colours to represent different temperatures; the car engine/tyres/exhaust will be hot; a thermogram will show colours representing high temperature against the cold field OR
Thermal imaging cameras detect IR radiation; hot objects such as recently used car engines give out IR; the hotter the object the brighter (whiter) the image produced

4 a Specific heat capacity
 b i Energy needed = mass × specific heat capacity × temperature change = 0.5 × 3900 × 70; = 136 500 J
 ii Heat the beaker

5 a Specific latent heat
 b Energy needed to break bonds; holding molecules together

Page 171 Keeping homes warm

1 a Particles in solid close together/particles in gas far apart/no particles in vacuum; gap between glass filled with gas or vacuum/more difficult to transfer energy than in solid
 b i Air in foam is a good insulator/reduces energy transfer by conduction; air is trapped/unable to move; reduces energy transfer by convection
 ii Energy from the room is reflected back in winter; energy from the Sun is reflected back in summer

2 a Particles are in constant motion and transfer kinetic energy – conduction; particles in solid close together so transfer energy easily/air is a gas so particles are far apart and it is more difficult to transfer energy
 b Air expands when heated; density = mass + volume, so increased volume means less density

3 a Only 32% of energy input is useful as energy
 b Output = 0.32 × 9.5; = £3.04
 c Energy is lost up the chimney

Page 172 A spectrum of waves

1 a The maximum displacement of a particle from its rest position/allow clear labelling of diagram from axis to peak or trough
 b The distance between two successive points having the same displacement and moving in the same direction/allow clear labelling of diagram showing this
 c Number of complete waves passing a point each second

2 Wavelength = speed ÷ frequency; $1500 \div 250000 = 6 \times 10^{-3}$ m; $1500 \div 125000 = 12 \times 10^{-3}$ m

3 Reflections from both mirrors; reflected ray returns along incident path

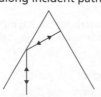

4 a Curved waves centres on gap; same wavelength
 b Spikes or rings around the star

P2 Answers

Page 173 Light and lasers

1 Dots and dashes series of on and off signals; not continuously variable signal

2 White light – many colours, different frequencies; laser light – one colour single frequency; laser light in phase

3 a i x angle of refraction greater than angle of incidence; y ray glancing along water/air boundary; z ray reflected back into water; angle of reflection = angle of reflection (by eye)

   ii c marked as angle between ray and normal in diagram y

   b Light down one set of optical fibres; reflected from internal organs; up a second set of fibres viewed by eyepiece or camera

Page 174 Cooking and communicating using waves

1 a Microwave radiation is more penetrating than infrared

   b Microwaves need line of sight; there are no obstructions in space

2 a Gamma rays

   b Wavelength of radiation from iron is longer than wavelength of radiation from element

(2 marks)

3 Radio waves diffracted around hill; short wavelength/ microwaves do not show much diffraction

Page 175 Data transmission

1 Greater choice of programmes; interact with programmes; information services

(Any 2)

2 3 – 5 reflections along length of fibre; ray reflected from surface with equal angles (by eye)

3 Interference on digital signals not evident; because signal is either high or low; many signals transmitted simultaneously/multiplexing; signals each divided up into short segments; transmitted signal takes segments from each, then recombined at end

Page 176 Wireless signals

1 a Less refraction at higher frequencies

   b Because it is aimed at a very small object

2 Reflected

3 a The foreign radio station is broadcasting on the same frequency; the radio waves travel further because of weather conditions

   b There will be no interference from other stations

Page 177 Stable Earth

1 a Transverse – S and longitudinal – P; travels through solid – P and S; travels through liquid – P

   b Waves refracted by core; cause shadow on opposite side of Earth

   c S waves do not pass through liquid; not detected on opposite side of Earth

2 a 20 × 15; 300 minutes /5 hours

   b Repeat their readings; consult other scientists about their findings

3 Newspaper may not be reliable source; based on only one piece of evidence; no peer review

Page 178 P1 Extended response question

5–6 marks

A detailed description of how microwave radiation requires line of sight and how it is affected by hills (little diffraction) and water (scattering the signals) A discussion of possible dangers from use of mobile phone near the head and that children are more at risk because their bodies are still developing.

All information in answer is relevant, clear, organised and presented in a structured and coherent format. Specialist terms are used appropriately. Few, if any, errors in grammar, punctuation and spelling.

3–4 marks

A limited description of some of the details of how microwave signals are transmitted and affected by hills and water. A mention of possible dangers from microwaves and children more at risk.

For the most part the information is relevant and presented in a structured and coherent format. Specialist terms are used for the most part appropriately. There are occasional errors in grammar, punctuation and spelling.

1–2 marks

An incomplete description, stating that microwave radiation is used and signal strength is affected by hills and water. A mention of possible dangers from microwaves.

Answer may be simplistic. There may be limited use of specialist terms. Errors of grammar, punctuation and spelling prevent communication of the science.

0 marks

Insufficient or irrelevant science. Answer not worthy of credit.

P2 Living for the future

Page 179 Collecting energy from the Sun

1 a Photocells do not need fuel; do not need cables; need little maintenance; use a renewable energy source; the operation of photocells cause no pollution or global warming

(Any 4)

   b n-type silicon has an impurity added to produce an excess of electrons and p-type silicon has a different impurity to produce an absence of free electrons

(3 marks)

2 a Increase the light intensity; increase the surface area exposed to light; decrease the distance to the light source

(Any 2)

3 a During the night the walls and floor radiate infrared at a longer wavelength and this is reflected by the glass so it stays inside the room

   b i S at left end of infrared

      ii P at right end of infrared

4 a The speed of the wind

   b Advantages – renewable; no pollution or global warming
      Disadvantages – unreliable; noisy; spoil the landscape

(Any 2)

P2 Answers

Page 180 Generating electricity

1 a Use a stronger magnet; increase the number of turns in the coil; spin the coil faster

(Any 2)

b In a power station the turbine spins the generator, electromagnets provide field, electromagnets turn inside coil where current is induced

c i Indicates one complete cycle on time axis

ii 300 V

2 a Turbine turns generator

b 60 MJ

Page 181 Global warming

1 a By absorbing the longer wavelength infrared radiation

b Water vapour

2 a Increased it

b Natural forest fires; volcanic eruptions; decay of dead plant and animal matter; escape from the oceans; respiration

(Any 4)

c The mining and burning of fossil fuels; cattle farming; rice paddies; the burying of waste in landfills

(Any 3)

3 a Sun's radiation has a short wavelength which is absorbed by the Earth. The Earth then re-radiates but with a longer wavelength. Longer wavelength radiation is absorbed by the greenhouse gases

(3 marks)

b Increase – The smoke from factories reflects radiation from the towns back to Earth
Decrease – Ash clouds from volcanoes reflect radiation from the Sun back into space

(1 mark each)

4 a They disagree on how much humans are contributing to global warming

b On the basis of scientific evidence

c Polar ice caps melting; rising sea levels

5 a Fact

b Opinion

c Fact

Page 182 Fuels for power

1 a 24 W

(2 marks)

b 15p

(3 marks)

c Less demand but supply still available

d 30 minutes

2 a Availability; ease of extraction; effect on the environment; associated risks

(Any 3)

b The advantages and disadvantages of each type of power station; the availability and ease of obtaining each type of fuel; the energy output of each type; the costs of building and maintaining each type; any sensible suggestion

(Any 3)

3 a Reduces energy loss; reduced cost

b The current goes down as the voltage goes up

c This leads to less heating in the wires and reduced energy losses

Page 183 Nuclear radiations

1 a False; true; true; false

b i Negative – radiation causes atom to gain one or more electrons

(2 marks)

Positive – radiation causes atom to lose one or more electrons

(2 marks)

ii Ionisation can lead to changes/mutations in DNA/ protein molecules may change shape; this mutation can lead to cancer

2 a Radioactive tracer; radiotherapy

(Any 1)

b Penetrates the body easily; can kill mutated cells

(Any 1)

3 The radiation ionises the oxygen and nitrogen atoms in air; this causes a very small electric current that is detected; when smoke fills the detector in the alarm during a fire the air is not so ionised; the current is less and the alarm sounds

4 a Waste can remain radioactive for thousands of years

b Radioactive waste is not suitable for making nuclear bombs; it could be used by terrorists to contaminate water supplies; or areas of land; or to frighten the public

(Any 3)

Page 184 Exploring our Solar System

1 a i A star is a ball of hot glowing gas giving out energy from fusion

ii A planet is a spherical object orbiting a star, which has cleared the neighbourhood of its orbit

iii A meteor is made from grains of dust that burn up as they pass through the Earth's atmosphere

b i A galaxy is a collection of stars

ii A black hole is are formed when a large star dies, you cannot see a black hole because light cannot escape from it

c i The centripetal force caused by gravity

ii Arrow on Moon towards centre of circle

2 a To prevent astronaut being the blinded by the Sun's glare

b Planet – Mars
Explanation – Mars is the only planet near enough to the Earth for humans to travel to

3 The distance light travels in one year

Page 185 Threats to Earth

1 a Mars; Jupiter

b The gravitational field of Jupiter is too strong and prevents this

(2 marks)

c Unusual metals have been found near craters; fossils are found below the metal layer but not above; fossil layers have been disturbed by tsunamis

(Any 2)

B3 Answers

2 The average density of Earth is 5500 kg/m³ while that of the Moon is only 3300 kg/m³; there is no iron in the Moon; the Moon has exactly the same oxygen composition as the Earth

(Any 2)

3 a C on elliptical orbit

 b X nearest the Sun

 c Solar wind pushes it away from Sun

4 a So that we could track their position accurately; and deal with any that came too near Earth

 b i Send a rocket with explosives near the NEO; and detonate the explosives to alter its path

 ii Be careful not to split asteroid into pieces which may still hit Earth; be careful that explosion is far enough away not to disrupt Earth

Page 186 The Big Bang

1 a The ones furthest away

 b Light spectra from distant galaxies are shifted to the red end of the spectrum

(3 marks)

 c The greater the red shift the faster the galaxy is moving

2 a Cloud pulled together by gravity; star becomes smaller, hotter and brighter; the core temperature is hot enough for nuclear fusion to take place; fusion releases energy

 b First becomes a red giant; gas shells, called planetary nebula, are thrown out; the core becomes a white dwarf; then cools to become a black dwarf

3 a He observed the planets' orbits with a telescope

 b Because the Church disagreed with the ideas

 c Newton's law of gravitation/ Newton used gravity to explain the orbit of the planets

Page 187 P2 Extended response question

5–6 marks
A detailed description of how the Sun started as a cloud of gas and dust pulled together by gravitational forces. When the temperature becomes high enough thermonuclear fusion starts to join hydrogen nuclei together to form helium nuclei. This main sequence phase continues for about 10 billion years until the hydrogen runs out and the star cools and expands to form a red giant. Next the outer layers called planetary nebulae are ejected and the core shrinks to form a white dwarf and then cools to a black dwarf. Includes the idea that our Sun was formed from the remains of a supernova remnant.
All information in answer is relevant, clear, organised and presented in a structured and coherent format. Specialist terms are used appropriately. Few, if any, errors in grammar, punctuation and spelling.

3–4 marks
A limited description of some of the details of our Sun's life cycle. Includes the idea that our Sun was formed from the remains of a supernova remnant.
For the most part the information is relevant and presented in a structured and coherent format. Specialist terms are used for the most part appropriately. There are occasional errors in grammar, punctuation and spelling.

1–2 marks
An incomplete description, may simply list the stages. Mentions the fact that the Sun was formed from the death of another star.

Answer may be simplistic. There may be limited use of specialist terms. Errors of grammar, punctuation and spelling prevent communication of the science.

0 marks
Insufficient or irrelevant science. Answer not worthy of credit.

B3 Living and growing

Page 188 Molecules of life

1 a To provide lots of energy; for contraction

 b In the cytoplasm

 c They are the site of protein synthesis

2 a i A circle around any one of the squares on the diagram

 ii Double helix

 iii Proteins are made in the cytoplasm; DNA cannot leave the nucleus

 b i 5

 ii ATATACATTTTTGTT

3 a C always equals G; T always equals A

 b They realised that C always bonds with G and T with A; this holds the two chains together

Page 189 Proteins and mutations

1 a Collagen – a structural protein, haemoglobin – a carrier protein, insulin – a hormone

 b Amino acids

 c They have a different order of amino acids; so the chains fold up differently

(Either)

2 a Enzymes are specific; it would be the wrong shape to fit the active site

 b i As the temperature increases the reaction is faster; after a certain temperature any increase will decrease the rate

 ii 41–42 °C

 iii $Q_{10} = \dfrac{4}{2} = 2$

 iv Increased movement of the molecules; which leads to more collisions

3 a Radiation; certain chemicals

 b Sometimes they produce an advantage; this is more likely to be passed on

 c Could be a change in the gene coding for enzyme B; a different base sequence might lead to a different order of amino acids; may stop enzyme B working so the red pigment is not converted to purple

Page 190 Respiration

1 a ATP traps the energy released by respiration; it passes it on to other processes that need it

 b $6O_2 \rightarrow 6CO_2 + 6H_2O$

 c i As the horse runs faster there is an increase in lactic acid levels; this is slow at first but increases more rapidly after about 8 m/s

 ii At slower speeds respiration is mainly aerobic; at high speeds there is more anaerobic respiration and so more lactic acid is made

 iii About 9 m/s

 d Gets sent to the liver; broken down with the use of oxygen

2 a Measure the rate of oxygen consumption/rate of carbon dioxide production

b Respiration is controlled by enzymes; high temperatures will stop them working

Page 191 Cell division

1 a Any two from: allows organism to be larger; allows for cell differentiation; allows organism to be more complex

b i 2.0; 1.5; 1.2

ii They have a smaller surface area to volume ratio; this reduces the rate of diffusion

2 a Before cells divide, DNA replication takes place; the new cells are diploid

b The two original strands come apart; each one acts as a template attracting complementary bases

3 a The acrosome is needed to digest the egg membrane; this allows the nucleus of the sperm to enter the egg

b In the first stage there are two pairs of chromosomes; each member of the pair moves apart; in the second division there is only one chromosome from each pair and the copies move apart

Page 192 The circulatory system

1 a Disc shaped: larger surface area to take up oxygen quicker; no nucleus: more room to carry more haemoglobin/oxygen

b Haemoglobin

c It combines with oxygen in the lungs forming oxyhaemoglobin; this is reversible; oxygen is released from oxyhaemoglobin in the tissues

2 a Arteries carry blood away from the heart; veins carry blood back to the heart; capillaries are the site of exchange from the blood

b i A= artery; B= vein; C= capillary

ii Wall is only one cell thick; this makes it permeable

3 a

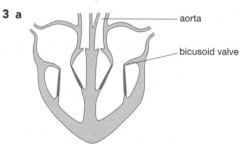

aorta

bicusoid valve

b The left ventricle has to pump the blood further; it has to generate more pressure

c Increased pressure so blood flows faster/more oxygen to the tissues

Page 193 Growth and development

1 a A bacterial cell does not have a true nucleus; or mitochondria

b It is circular rather than a strand

2 a Adolescence/puberty

b The girl is taller; because girls reach puberty/have their growth spurt younger

c Advantages: quick/not destructive; disadvantage: only measures growth in one dimension

3 a Stem cells

b Embryonic stem cells can form any type of cell; adult stem cells are more restricted

4 The parts of a plant where cells divide

Page 194 New genes for old

1 a When two individuals that are closely related mate

b Lack of genetic variation; leads to health problems

2 a Two from: many people are deficient in Vitamin A; this can be made from beta-carotene; many of these people eat large amounts of rice

b The plants might be harmful to health in the long term; they might escape into the wild and affect food chains

c The gene for beta-carotene would have to be identified; then cut out of the carrot DNA; inserted into the DNA of rice

3 a Gene therapy

b For: may cure life-threatening diseases; against: may be dangerous for the patient; some people think that it is ethically wrong to move genes between organisms

Page 195 Cloning

1 a Putting the nucleus from a body cell into an egg that has had its nucleus removed

b i The pigs have human genes; so their organs are not rejected

ii Produce large quantities of pigs with desired characteristics such as meat yield/produce large numbers of pigs which produce human proteins

2 a Given an electric shock

b Julie; because it contains her DNA

3 a Advantages: produce large numbers quickly/if the parent produces good strawberries then the offspring will also do so; disadvantage: all the strawberries will be identical and so could all be attacked by a disease

b Aseptic technique means making sure that no microbes infect the plants; the growth medium is a gel that contains all the nutrients that are needed for the plant to grow

Page 196 B3 Extended response question

5–6 marks
The answer includes an explanation of how a mutation can change a base sequence, hence the amino acid sequence and so collagen structure. It also includes an explanation of the symptoms in terms of collagen's role as a structural protein. The particular susceptibility of arteries due to the high pressure is mentioned. All information in answer is relevant, clear, organised and presented in a structured and coherent format. Specialist terms are used appropriately. Few, if any, errors in grammar, punctuation and spelling.

3–4 marks
The answer refers to mutations and changes in DNA or genes but the precise mechanism is not clear. The function of collagen as a structural protein is appreciated. For the most part the information is relevant and presented in a structured and coherent format. Specialist terms are used for the most part appropriately. There are occasional errors in grammar, punctuation and spelling.

B4 Answers

1–2 marks

An incomplete description, either showing an appreciation of mutations and their inherited nature or an understanding of the role of collagen. Answer may be simplistic. There may be limited use of specialist terms. Errors of grammar, punctuation and spelling prevent communication of the science.

0 marks

Insufficient or irrelevant science. Answer not worthy of credit.

B4 It's a green world

Page 197 Ecology in the local environment

1 a S. balanoides
  b Zonation
  c Exposure/temperature; causing drying out, some species more resistant; OR physical action of waves; some species attached more strongly

2 a Natural ecosystems have bigger diversity; native woodlands/lakes are natural ecosystems; forestry plantations/fish farms are artificial ecosystems
  b Plants depend on animals for pollination/fertilisation; animals depend on plants for food/idea of food chains; plants produce oxygen, animals use it for respiration; animals produce carbon dioxide, plants use it for photosynthesis; idea of recycling of minerals after death

3 a Capture–recapture
  b i $\frac{50 \times 60}{10}$; population size is 300
    ii Not all ladybirds are counted/population is sampled
  c No deaths, immigration, emigration; marking does not affect survival; identical technique used

(Any 4)

Page 198 Photosynthesis

1 a i $C_6H_{12}O_6$; $6O_2$
    ii Light/Sun
  b i Cell walls; starch; growth/repair; storage
    ii Energy source/respiration

2 a Plants produce oxygen
  b Isotope of oxygen $^{18}O_2$ used as part of oxygen molecule; in photosynthesis the isotope given off as oxygen gas; so water was split up by light energy not carbon dioxide

3 Heater produces warmth; and carbon dioxide; shades removed so maximum light enters; watering system provides right amount of water

4 a Carbon dioxide levels drop at midday, oxygen levels rise at midday; because carbon dioxide is used up and oxygen is released in photosynthesis; the use of oxygen and production of carbon dioxide by goldfish is constant; so it does not affect the shape of the graphs

(Any three)

  b Levels of carbon dioxide would increase; levels of oxygen would decrease; due to goldfish respiration and no photosynthesis

Page 199 Leaves and photosynthesis

1 a Shorter distance for gases to diffuse/all cells can get light for photosynthesis; variety of pigments; vascular bundles/xylem and phloem; open and close stomata
  b i Spongy mesophyll; palisade
    ii Eats cells containing chlorophyll; therefore parts will not be green

2 a chlorophyll a and b
  b Parts of light spectrum used (in photosynthesis)
  c Different pigments/more than one pigment; each absorbs different parts of the light spectrum
  d Green/about 550 nm
  e 400–500 nm approx; 640–670 nm approx
  f Little light will reach them; only blue light will reach them; little photosynthesis/poor or no growth or need very large leaves

Page 200 Diffusion and osmosis

1 a Black circles move down and mix with white circles; white circles move up and mix with black circles
  b Diffusion
  c Random movement of particles; takes place from high to low concentrations
  d A shorter distance to travel/nose nearer bottle; a higher concentration gradient/stronger perfume; a greater surface area/wider neck of bottle

2 Water; partially permeable; dilute; concentrated; random

3 a Water leaves cells; contents/cytoplasm shrinks/pulls away from cell wall
  b Cells collapse; plant wilts
  c Animal cells do not have cell walls; animal cells become creanate

Page 201 Transport in plants

1 a Vascular; xylem; phloem
  b Xylem cells are dead (phloem cells are living); have extra cell wall thickening; have a hollow lumen

2 a 2, 9
  b Leaf A (no mark), fewer stomata open; no photosynthesis/stomata close to prevent too much water loss
  c Water enters guard cells; turgidity increases; guard cells curve

3 a Water loss/transpiration
  b Leaf 4 has only lower surface exposed; it loses more weight/2.7 g instead of 1.0 g; it loses more water than Leaf 3, which has upper surface only exposed
  c Loses more weight; because loses more water; since rate of evaporation of water increased

Page 202 Plants need minerals

1 a Nitrates; phosphates; potassium; magnesium
  b Cannot make proteins; for cell growth

2 Make amino acids/proteins; make DNA/cell membranes; make chlorophyll

3 a Active transport
  b Uses energy; minerals moved against a concentration gradient; minerals are selected; uses a carrier system

C3 Answers

Page 203 Decay

1 a A temperature of 25 °C; plenty of oxygen

b Earthworms; maggots; woodlice

c For: avoids landfill/recycles minerals/saves buying fertilisers; Against: unsightly/smelly if not working properly (anaerobic respiration)/needs space or garden

d i Break up dead material; so increasing surface area for decay

ii Digest dead material; by extracellular digestion/ digesting food outside the body and absorbing it

e Increase in temperature increases growth and reproduction; of microbes/bacteria; by increasing enzyme action in respiration; extremes of temperature will slow down microbial growth; because enzymes will not work/be denatured

2 a Temperature of 5 °C slows down microbial growth/ reproduction; so slowing down decay

b Temperature of –22 °C kills most bacteria/microbes; so slowing down decay better than a refrigerator

Page 204 Farming

1 a Hydroponics

b i Better control of mineral levels; better control of diseases; can be used in areas of poor/barren soil
(Any two)

ii Lack of support for tall plants; still uses fertilisers; needs electricity supply
(Any two)

2 a Biological control

b Insecticides can enter and accumulate in food chains; insecticides may harm other useful insects; some insecticides are persistent
(Any two)

c Not native/from South America; no natural predators; lack of research on its diet/lack of trials
(Any two)

3 a To farm organically; to avoid build-up of pests

b Longer crop time/succession of crops; avoid certain times when insect pests hatch out/most abundant

Page 205 B4 Extended response question

5–6 marks

The answer includes an explanation of osmosis and the importance of the partially-permeable membrane. It correctly predicts the movement of water molecules towards the left, i.e. inside the cell. It explains the importance of the size of molecules, the large sugar molecule remaining inside the cell. The answer includes a reference to the importance of the process, i.e. uptake of water into the cell, maintaining turgor, providing water for photosynthesis as well as the cell being able to retain food molecules such as sugar. All information in answer is relevant, clear, organised and presented in a structured and coherent format. Specialist terms are used appropriately. Few, if any, errors in grammar, punctuation and spelling.

3–4 marks

The answer refers to osmosis and makes some attempt to predict movement of molecules. References to partially-permeable membrane or sugar retention may be missing. For the most part the information is relevant and presented in a structured and coherent format. Specialist terms are used for the most part appropriately. There are occasional errors in grammar, punctuation and spelling.

1–2 marks

An incomplete description of osmosis is given and only a vague reference to movement of molecules. The importance of water movement is briefly mentioned without details. Answer may be simplistic. There may be limited use of specialist terms. Errors of grammar, punctuation and spelling prevent communication of the science.

0 marks

Insufficient or irrelevant science. Answer not worthy of credit.

C3 Chemical economics

Page 206 Rate of reaction (1)

1 A rate of reaction measures how much product is formed in a fixed period of time

2 a cm^3/s

b It decreases/drops to zero

c Magnesium

3 a Slower

b The rate of reaction between 20–40 seconds gives 24 to 54 cm^3; 30 cm^3 in 20 secs; the rate is 30/20 = 1.5 cm^3/s

c 48 cm^3

4 a

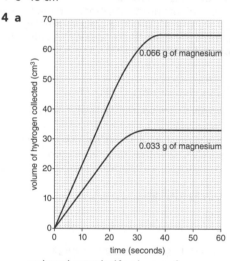

reduced rate; half volume of gas

b The rate of reaction will be slower as there are not as many particles of magnesium for the acid to make successful collisions; the volume of gas made will be half of the volume as half of the limiting reactant is used

Page 207 Rate of reaction (2)

1 a

C3 Answers

b As the temperature decreases, the particles have less kinetic energy; fewer collisions

c As the concentration is halved the particles are less crowded; (as there are fewer particles in the same area), fewer collisions take place

d The rate of reaction depends on the collision frequency; it also depends on the energy transferred during collisions; if the temperature rises collisions are more successful as the particles are more energetic and collisions are more frequent; if it is more concentrated, collision frequency rises

2 a The line levels out (no more gas is being produced)

b Construction triangle between 0 to 10 seconds; construction triangle between 20 to 30 seconds; $2.9 \ cm^3/s$; $1.2 \ cm^3/s$

Page 208 Rate of reaction (3)

1 The fine powder has a very large surface area; which provides a very large surface for contact with oxygen; which can cause an explosive reaction

2 a $CaCO_3$

b After 6 minutes

c
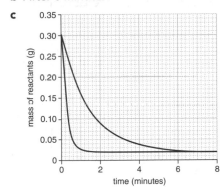

d Lump calcium carbonate has less surface area for particles of acid to collide and react with; powdered calcium carbonate has a larger surface area and can spread throughout a reaction mixture enabling greater collision frequency

e Construction triangle between 0 to 1 min;
$(0.30 - \frac{0.16}{1} = 0.14 \ g/min.$

Accept $(0.30 - \frac{0.09}{2} = 0.11 \ g/min)$

3 a A catalyst is a substance which changes the rate of a reaction; catalysts remain unchanged at the end of the reaction; only a very small amount of catalyst is needed

b i With copper at 20 seconds and with no substance at 40 seconds

ii The reaction with the copper went at a higher rate; than the reaction with no substance

iii A value between $21 \ cm^3$ and $24 \ cm^3$

Page 209 Reacting masses

1 a 5

b 74 $\quad [40 + (17 \times 2)]$

c (Formula) mass of the reactants is 103; (formula) mass of the products is 103

d (Formula) mass of the reactants is 226; (formula) mass of $Ca(NO_3)_2$ is 164; (formula) mass of the products is 226 $\quad [62 + 164]$

2 a She will need to work out how much zinc there is in the mass of zinc chloride that she needs; use excess acid; and make sure all the zinc is used up

b i $ZnCO_3 \ \rightarrow \ ZnO + CO_2$

(1 mark for all 3 symbols, 1 mark for balancing the equation)

ii The M_r of $ZnCO_3$ is 125; the M_r of ZnO is 81; 12.5 g $ZnCO_3$ decomposes to give 8.1 g of ZnO

iii $M_r \ (CO_2)$ is 44; to obtain 22 g need $\frac{125}{2}$ = 62.5 g

Page 210 Percentage yield and atom economy

1 a 28 g

b 42 g

c i Percentage yield = $\frac{\text{actual yield}}{\text{predicted yield}} \times 100$

ii 66.7 % $\quad$ Percentage yield = $\frac{28}{42} \times 100$

2 The company will want to reduce the amount of reactants wasted; reduce costs when incorrect amounts mean that not enough product is produced

3 a i The M_r of $NaNO_3$ is 85 [23 + 14 + (3 × 16)]; the M_r of H_2O is 18

ii Atom economy = $\frac{M_r \text{ of desired products}}{\text{sum of } M_r \text{ of all products}} \times 100$;

$\frac{85}{103} \times 100$; = 82.5%

b The M_r of $Ca(NO_3)_2$ is 164 [40 + 2(14 + 48)];
atom economy = $\frac{164}{200} \times 100$; = 82.0%

c The company will want to reduce the production of unwanted products; and to make the process more sustainable

Page 211 Energy

1 a Endothermic

b Exothermic

c Energy is required to break bonds; energy is released when bonds are made; in this process more energy is transferred out during bond making than is required for the bond breaking process

2 a Use a spirit burner; heat 100 g of water in a copper calorimeter; measure the mass of fuel at start and after 1 g of fuel is burnt; measure the change in temperature of the water; make sure that both tests are fair. For example, the distance between flame and the calorimeter should be the same; make sure that the tests are reliable. For example, repeat the test three times

(Any 5)

b Energy = 100 × 4.2 × 30 = 12 600 J
(1 mark for mass of 100, 1 mark temperature of 30, 1 mark for correct answer)

c Energy per gram = $\frac{\text{energy released (in J)}}{\text{mass of fuel burnt (in g)}}$

$\text{energy} = \frac{16\,800}{1.4}$

$= 12\,000 \ J/g$

d Energy per gram of Fuel A = $\frac{12\,600}{3}$; = 4200 J/g

Page 212 Batch or continuous?

1 a Pharmaceuticals need sterile environments for manufacture; the vessels can be cleaned out more times in a batch process; they are often only needed on a smaller scale as the raw materials are more difficult to source/cost more

(Any 2)

b Continuous processes can produce large quantities of products/more product made as the process can keep going day and night; the processes can be automated; there are few labour costs; raw materials can be bought in bulk so costs are reduced

(Any 2)

c Batch processes are flexible so it is easy to change from making one compound to another

d Continuous processes can be automated for 24/7 use to produce more/cheaper products

e Each batch has to be supervised, so labour costs are higher; vessels are not used efficiently because of the time spent filling and emptying reaction vessels so they are not producing chemicals continuously

2 a It costs a great deal to put in the research and development needed, it can take years to develop; countries have strict safety laws for testing a new medicine; the raw materials needed may be rare and expensive; compounds extracted from plants are difficult to find; plants contain thousands of similar chemicals, so separating the desired chemical is time consuming and expensive/raw materials are sometimes difficult to extract

(Any 3)

b Thousands of compounds often need to be tested before an effective one is found; likely compounds need to be manufactured and tested on living tissue to ensure safety; costly long-term trials are needed to identify possible side effects; lots of similar compounds need to be developed to try to reduce side effects; recommended doses need to be shown to be effective; the research needs to be independently verified; patents granted to pharmaceutical companies are sometimes not long enough for companies to recoup development costs before the patent expires and other companies can make their own version

(Any 2)

3 a The plant is crushed to break the strong cell walls; and boiled in a suitable solvent so that the chemicals can dissolve; different solvents are used to separate compounds; chromatography is used to separate and identify the compounds

b The data for chromatographic movement is the same for Z and P (2 cm in 10 mins); the melting point is slightly lower than the pure sample P; the boiling point is slightly higher than the pure sample P

Page 213 Allotropes of carbon and nanochemistry

1 a Allotropes are elements that have the same atoms that are arranged in different ways, (these are allotropes of carbon)

b They act like a cage to carry the drug and are so small they can travel around the body

2 a Diamond is hard; cut diamond reflects light from different angles

b Graphite is slippery and so when put on paper, some of the surface slides off and leaves a black mark; because it is slippery, it is also a good lubricant

c i Diamond is organised in a giant molecular structure; it does not conduct electricity as there are no free electrons

ii Graphite conducts electricity as it has unshared outer shell electrons (delocalised electrons); moving through the layers

iii It is slippery as it has strong bonds within the layers and weak bonds between the layers; so the layers move over one another easily

iv Both diamond and graphite have high melting points because the covalent bonds in the layers are strong; a lot of energy is needed to break them

3 a In both diamond and graphite, every carbon atom is joined to four others by four covalent bonds; which form in different directions

b *Features*: other compounds that have giant molecular structures are hard; have high melting points; generally do not conduct electricity; and are usually insoluble

Explanation: the substances display these properties because they generally have strong covalent bonds with no spare electrons

(1 mark for any 2 of 4 features, 1 mark for explanation)

c Catalysts can be attached to the nanotubes which have a large surface area, giving more chances that the reactants will collide with the catalyst

Page 214 C3 Extended response question

5–6 marks

Comprehensive descriptions of each experiment, including a diagram and the measurements to be made (mass of water, temperature change, mass of fuel used) and one other factor controlled. A complete explanation of formulae used;

$E = m \times c \times \Delta T$ and E per gram $= \dfrac{E}{\text{mass of fuel burnt}}$ J/g

All information is relevant, clear, organised and presented in a structured and coherent format. Specialist terms are used appropriately. Few, if any, errors in grammar, punctuation and spelling.

3–4 marks

The description of the experiment must be correct with a diagram included. Some attempt at the first calculation, but the mass chosen may be incorrect.

The information should generally be relevant and presented in a structured and coherent format. Specialist terms should mostly be used appropriately. There may be occasional errors in grammar, punctuation and spelling.

1–2 marks

An attempt must be made at a correct description of the experiment or the use of a diagram as explanation. Answer may be simplistic. There will be limited use of specialist terms. Errors of grammar, punctuation and spelling are likely to prevent effective communication of the science.

0 marks

Insufficient or irrelevant science such as repeating the question. Answer not worthy of credit.

C4 The periodic table

Page 215 Atomic structure

1 a Protons and neutrons

b

	Relative charge	Relative mass
electron	−1	0.0005 (zero)
proton	+1	1
neutron	0	1

(1 mark for each column)

c

Atomic number	Mass number	Number of protons	Number of electrons	Number of neutrons
19	39	19	19	20

d Atoms have equal numbers of positively charged protons and negatively charged electrons

e

Atomic number	Mass number	Number of protons	Number of electrons	Number of neutrons
12	24	12	10	12

2 a An element that has atoms with different numbers of neutrons/same atomic number but different mass number

b

Isotope	Electrons	Protons	Neutrons
$^{12}_{6}C$	6	6	6
$^{14}_{6}C$	6	6	8

(1 mark for each row)

c $^{35}_{17}Cl$

3 a They are arranged in atomic number order; in groups numbered with the number of electrons in the outside shell; in periods in order of how many shells the electrons are occupying

b

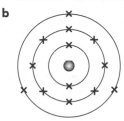

Three shells are needed as there are 13 electrons; that have a pattern 2.8.3 occupying three shells

c Chlorine

d 2.8.6

4 Not all the evidence for atomic structure was available

Page 216 Ionic bonding

1 a An atom that has extra electrons in its outer shell and needs to **lose** them to be stable. **M**; An atom that has 'spaces' in its outer shell and needs to **gain** them to be stable. **N**

b

LiF

c Atoms are stable when they have a full outer shell of electrons. Lithium has one electron in its outer shell so is more stable if this electron is lost. This leaves the atom with a positive charge

d Fluorine has an electron pattern of 2.7 so needs one more electron to become stable. This becomes a negative ion

e The positive ions of lithium are attracted to the fluoride negative ions; to make a giant ionic lattice which is solid

f $CaCl_2$

g

KCl
(1 mark for K bracket, 1 mark for Cl bracket, one mark for 1:1 ratio)

h

$$\left[K \right]^{+} \quad \left[\begin{matrix} \cdot\cdot \\ \times\times O \cdot\cdot \\ \cdot\cdot \end{matrix} \right]^{2-} \quad \left[K \right]^{+}$$

K_2O
(1 mark for K bracket, 1 mark for O bracket, one mark for 2:1 ratio)

2 a Giant ionic lattice

b When it is molten or in solution

c It contains ions; these are free to move

d A lot of energy is needed to break apart the giant ionic lattice formed by strong attractions of ions

e Magnesium transfers out two electrons not just one; magnesium ions are very small so they can get closer so the bonds are stronger

Page 217 The periodic table and covalent bonding

1 a Electron pairs are shared

b Simple molecules with weak intermolecular force

2 a

$$H \overset{\cdot\cdot}{\underset{\cdot\cdot}{\times}} F \overset{\cdot\cdot}{}$$

HF

(1 mark for H electrons, 1 mark for F electrons, one mark for 1:1 ratio)

b

$$\overset{\cdot\cdot}{\underset{\cdot\cdot}{:}} O \overset{\times}{\underset{\times}{:}} C \overset{\times}{\underset{\times}{:}} O \overset{\cdot\cdot}{\underset{\cdot\cdot}{:}}$$

CO_2

(1 mark for C electrons, 1 mark for O electrons, one mark for double bonds–four electrons.)

c It has weak intermolecular forces between its molecules which do not need much energy to separate them; there are no free electrons or ions to conduct electricity

3 a Magnesium is in group 2 as it has two electrons in its outer shell

b Both chlorine and fluorine have the same number of electrons (seven) in their outer shell

c Neon has an electron configuration of 2.8 so is in period 2 as it has two electron shells. Potassium has an electron configuration of 2.8.8.1 so is in period 4 as it has four electron shells

d Group 3

e Period 3

4 a The behaviour changed periodically/behaviour changed so that every eighth element was similar

b Electronic structure was built on electron shells that were full after every eight electrons; elements were found that filled the gaps that he predicted in his periodic table

Page 218 The group 1 elements

1 a There will be a very vigorous reaction; hydrogen will be given off; an alkaline solution will form/rubidium hydroxide will form

b Caesium; it is lower down the group/it loses its outer electron more easily

c $2K + 2H_2O \rightarrow 2KOH + H_2$
(1 mark for all four symbols, 1 mark for balancing the equation)

d Boiling point: 670–773; Atomic radius: 0.264–0.228

2 a They each have one electron in their outer shell/they each lose one electron to form an ion

b They lose one electron to form a stable electronic structure; $Na - e^- \rightarrow Na^+$ (answer accepted for any group 1 element)
(1 mark for both Na formulae, 1 mark for electron loss)

c Oxidation is the process of electron loss

C4 Answers

d The further down the group the more reactive the element; the further the outer electron shell; the easier it is to lose the outer electron

3 Dip flame-test wire into dilute HCl; dip rod into the group 1 compound and hold in flame; if flame turns yellow/red/lilac it is a compound of sodium/lithium/ potassium; other two flame colours matched with remaining two of sodium/lithium/potassium

Page 219 The group 7 elements

1

	Physical appearance	Melting point in °C	Boiling point in °C
fluorine		below −101 °C	below −36 °C
chlorine	green gas	−101	−35
bromine	orange liquid	−7	59
iodine	grey solid	114	184
astatine		above 114 °C	above 184 °C

  a *(2 marks – 3 correct, 1 mark – 2 correct)*

  b *(1 mark – 1 pair only correct or fluorine pair correct or astatine pair correct)*

2 a Sodium + bromine → sodium bromide

  b $2K + Cl_2 → 2KCl$
(1 mark for all three symbols, 1 mark for balancing the equation)

  c $Rb + I_2 → 2RbI$
(1 mark for all three symbols, 1 mark for balancing the equation)

3 a A red–brown solution

  b Displacement

  c Potassium iodide + chlorine → potassium chloride + iodine

  d Fluorine is more reactive than bromine so the fluoride ion is not displaced

  e Astatine is less reactive than chlorine so the astatide ion is displaced by chlorine as astatine

  f Group 7 elements all have seven electrons in their outer shell, so gain one to form a (singly charged) negative ion

  g $I_2 + 2e^- → 2I^-$; electrons are gained which is reduction

  h The easier it is to attract an electron into the outer shell the more reactive the element; fluorine has only two shells so the pulling power of the positive nucleus on an incoming negative electron is stronger for fluorine than for bromine which has four shells (answer accepted for any halogen element)

Page 220 Transition elements

1 a copper compounds — often pale green
iron(II) compounds — often orange/brown
iron(III) compounds — often blue

  b A substance that changes the rate of reaction but remains unchanged at the end of the reaction

  c Iron (Haber process); nickel (hardening margarine)

  d Manganese oxide and carbon dioxide

  e Copper carbonate → copper oxide + carbon dioxide

  f Blue/green solid becomes a different coloured powder (black)

  g $ZnCO_3 → ZnO + CO_2$
(1 mark for all three symbols, 1 mark for balancing the equation)

2 a Add sodium hydroxide solution to three samples; observe gelatinous precipitates; Cu^{2+} ions make a blue precipitate; Fe^{2+} ions make a pale green precipitate; Fe^{3+} ions make an orange/brown (gelatinous) precipitate (accept solid)

  b $Fe^{2+} + 2OH^- → Fe(OH)_2$
(1 mark for all three symbols, 1 mark for balancing the equation)

  c $Fe^{3+} + 3OH^- → Fe(OH)_3$
(1 mark for all three symbols, 1 mark for balancing the equation)

Page 221 Metal structure and properties

1 a Silver and copper both have higher electrical conductivity

  b Aluminium. It has a lower density than the other metals so the plane would fly more easily

  c Lower melting point

2 a The ions of the metals are held together by strong metallic bonding making them hard to separate, so a lot of energy is needed to do this

  b Metallic bonding occurs when atoms make a close-packed regular lattice; of positive metallic ions; with a 'sea' of delocalised electrons (moving through the lattice)

3 a A superconductor is a material that conducts electricity with little or no resistance

  b The potential benefits are loss-free power transmission; super-fast electronic circuits; powerful electromagnets

  c They work only at very low temperatures; this limits their use/not enough superconductors work well at 20 °C yet

Page 222 Purifying and testing water

1 a Sedimentation, filtration, chlorination

  b Chemicals are added to make solid particles and bacteria settle out; a layer of sand on gravel filters out the remaining fine particles, some types of sand filter also remove microbes; chlorine is added to kill microbes

  c Older houses sometimes have lead pipes that dissolve slowly into the water; nitrates from fertilisers get in through run-off; pesticides from spraying near rivers
(Any 2)

  d Pesticides and nitrates

  e Huge amounts of energy are needed/it is very expensive

2 a Sodium chloride; sodium iodide; sodium sulfate

  b Magnesium sulfate + barium chloride → magnesium chloride + barium sulfate

  c Precipitation

  d Sodium bromide + silver nitrate → sodium nitrate + silver bromide
(1 mark for bromide, 1 mark for the equation)

  e $MgSO_4 + BaCl_2 → MgCl_2 + BaSO_4$
(1 mark for all four symbols, 1 mark for balancing the equation)

  f $KCl + AgNO_3 → KNO_3 + AgCl$
(1 mark for all four symbols, 1 mark for balancing the equation)

P3 Answers

Page 223 C4 Extended response question

5–6 marks
A comprehensive description of the experiments using flame tests and precipitate tests with sodium hydroxide. Clear observations about the identity of each ion.

Experiments
Flame test needed.
Precipitate reaction with sodium hydroxide needed.
Use of hydrochloric acid and flame-test wire for flame tests.
Observe colour of flame/observe colour of precipitate.
Add sodium hydroxide to solution of the ion (with care).

Identification
Flame colours: lithium – red, sodium – yellow, potassium – lilac
Precipitate colours, Cu^{2+} blue, Fe^{2+} grey/green, Fe^{3+} orange/brown.
All information in the answer should be relevant, clear, organised and presented in a structured and coherent format. Specialist terms should be used appropriately. There should be few, if any, errors in grammar, punctuation and spelling.

3–4 marks
A correct description of at least one experiment. Accurate identification of the salts tested using this experiment.
The information should generally be relevant and presented in a structured and coherent format. Specialist terms should mostly be used appropriately. There may be occasional errors in grammar, punctuation and spelling.

1–2 marks
An attempt should have been made at the correct description of at least one of the experiments, or an attempt at the outcome of the results for one of the experiments. Answers may be simplistic. There will be limited use of specialist terms. Errors of grammar, punctuation and spelling are likely to prevent effective communication of the science.

0 marks
Insufficient or irrelevant science such as repeating the question. Answer not worthy of credit.

P3 Forces for transport

Page 224 Speed

1 a 80 s
  b 80 m
  c D to E
  d Speed $= \frac{40}{40}$; = 1 m/s
  e Speed $= \frac{40}{80}$; = 0.5 m/s
2 a Average speed = 12.5 m/s; time $= \frac{5000}{12.5}$; = 400 s
  b Average speed $= \frac{500}{35}$; = 14.3 m/s
  c Distance = 4500 m time = 365 s;
     average speed $= \frac{4500}{365}$; = 12.3 m/s

Page 225 Changing speed

1 a
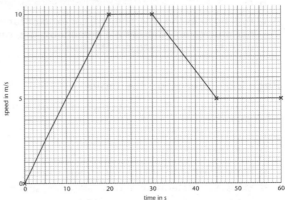

Uniform increase in speed; constant speed of 10 m/s; uniform decrease in speed; constant speed of 5 m/s
  b Area under graph

2 a
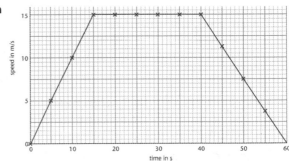

  b Change in speed = 15 – 3 = 12 m/s;
     deceleration $= \frac{12}{15}$; = 0.8 m/s^2
  c Distance travelled $= \frac{1}{2}$ (15 × 15) + (25 × 15) + $\frac{1}{2}$ (20 × 15); = 637.5m

3 Change in speed = acceleration × time; = 6 × 5 = 30 m/s is final speed

Page 226 Forces and motion

1 a Force = 500 × $\frac{40}{20}$; = 1000 N
  b Acceleration $= \frac{1250}{500}$; = 2.5 m/s^2

2 a Tiredness; age; under influence of drugs/alcohol; distracted/lacking concentration
                                                    (Any 2)
  b i Thinking distance will increase
     ii Reaction time is unchanged; so she will travel a greater distance in that time
  c Less tread/less grip; braking distance increases

3 Thinking distance – straight line
   Braking distance – curve showing 'squared' relationship; passing through (30, 13.5), (40, 24), (50, 37.5)

Page 227 Work and power

1 a Work done = 80 × 2; = 160 J
  b i Height = 80 × $\frac{2}{60}$; = 2.7 m
     ii Power = 80 × $\frac{2}{1.5}$; = 106.7 W

2 a Chris is more powerful than Abi
  b Weight = 10 × 60; 600 N
  c Power = 600 × $\frac{3}{8}$; = 225 W
  d Power = 600 × $\frac{3}{12}$; 150 W
  e 26 N/kg

3 a A

  b C

  c Fuel pollutes the environment; car exhaust fumes are harmful; carbon dioxide is a greenhouse gas; carbon dioxide contributes to climate change

(Any 3)

Page 228 Energy on the move

1 a Fewer road junctions; fewer speed changes; fewer gear changes

(Any 2)

  b Fuel used $= \frac{96}{24}$; 4

  c Renault Megane has smaller engine capacity

  d Excessive acceleration and deceleration; speed changes; braking; driving in too low a gear

(Any 3)

2 a Recharging requires electricity from power stations which do cause pollution

  b Advantage – no pollution; no batteries; does not need energy from power station

(Any 1)

  Disadvantage – not always sunny; not a constant energy source

(Any 1)

3 a Kinetic energy $= \frac{1}{2} \times 1200 \times (20)^2$; $= 240\,000$ J

  b KE proportional to v^2; so braking distance is quartered not halved

Page 229 Crumple zones

1 a Seat belt – stretches so that kinetic energy is transferred into elastic potential; crumple zones – absorb some of car's KE by changing shape on impact; air bag – absorbs some of person's KE by squashing up around them

  b Computer controls pressure on brakes; brakes are pumped; prevents locking and skidding; increases braking force just before skid

(Any 2)

2 a Force $= \frac{(25 \times 55)}{0.5}$; $= 2750$ N

  b Force $= \frac{(25 \times 55)}{0.002}$; $= 687\,500$ N

3 a Force $=$ mass $\times$ acceleration; to reduce force acceleration (deceleration) must be reduced since mass cannot change

  b Crumple zone concertinas; to reduce the car's speed more slowly

Page 230 Falling safely

1 Acceleration is independent of mass

2 a Weight acting vertically downwards; air resistance acting vertically upwards

  b Weight greater than air resistance

  c The faster she falls the greater the air resistance

  d Balanced/equal in size but opposite in direction

  e Weight is unchanged; air resistance increases suddenly

3 Greater at poles than equator; increases down a mine; decreases up a mountain; different in space or another planet/moon

(Any 3)

Page 231 The energy of games and theme rides

1 a GPE; GPE; KE; GPE

2 a C

  b GPE to KE

  c Energy is transferred as sound/heat/friction

  d Increase height of B to increase GPE; GPE is transferred to KE as the carriage falls; more KE means faster speed

3 $h = \frac{(11\,000)^2}{(2 \times 10)}$; $= 6\,050\,000$ m OR 6050 km

Page 232 P3 Extended response question

5–6 marks

A detailed description of how the forces affect the speed of descent to include; weight being greater than air resistance at the start so speed increases; air resistance increases with speed; terminal (constant) speed when air resistance equals weight; air resistance greater than weight when parachute opens so speed decreases; the parachute provides a much larger surface area and displaces more air molecules; air resistance decreases when slowing down; lower terminal speed when air resistance is again equal to his weight until he lands.

All information in answer is relevant, clear, organised and presented in a structured and coherent format. Specialist terms are used appropriately. Few, if any, errors in grammar, punctuation and spelling.

3–4 marks

A limited description of how the forces affect the speed of descent at two or three stages.

For the most part the information is relevant and presented in a structured and coherent format. Specialist terms are used for the most part appropriately. There are occasional errors in grammar, punctuation and spelling.

1–2 marks

An incomplete description of how the speed changes as he falls but with little if any reference to the forces acting.

Answer may be simplistic. There may be limited use of specialist terms. Errors of grammar, punctuation and spelling prevent communication of the science.

0 marks

Insufficient or irrelevant science. Answer not worthy of credit.

P4 Radiation for life

Page 233 Sparks

1 a i So that charge will build up on her body instead of flowing away to earth

  ii Otherwise she would get a shock

  iii Charges will begin to build up on her body

  iv Sally's hair all picks up the same charge; so it repels and stands up

  b Both balls will pick up electrons from the polythene and become negatively charged; since they have the same charge they will repel

  c The touched ball becomes charged; this induces the opposite charge on (one side) of the other ball; then the balls will attract due to these oppositely charged sides/one being charged and one uncharged; they do not touch as the force of attraction is weaker than if both balls were charged

(Any 2)

2 a i You become charged up by the friction of the rubber tyres on the road; when you touch the metal body it discharges

ii A tree could be the highest point and the charge is more likely to hit that point and jump from the tree into you

iii When you unroll cling film it becomes charged by the friction and may then attract to the uncharged parts

b The moving parts can become charged by friction if they are insulators

c The charge could travel to earth through the operators; causing a spark; rubber mats will prevent this

Page 234 Uses of electrostatics

1 a Positive

b Negative

c They have gained electrons

d Electrostatic attraction

2 a Paint particles all have same charge; so repel each other

b To attract the paint; so there is less wasted paint

c i Positive

ii The frame will start to repel the paint

3 a The shock restores a regular heart rhythm

b Hair is a poor conductor; water may conduct the charge across the chest away from where it is needed

c To avoid any other people receiving a shock

d It is safer as it is only for an extremely short time

Page 235 Safe electricals

1 a i The brightness reduces

ii Ammeter in series; voltmeter added in parallel across the bulb

iii $R = \dfrac{V}{I}$ OR $\dfrac{6}{0.25}$; = 24; Ω

b i 30×0.4; 12; V

ii There is a longer length of wire in the circuit now; the electrons have to pass through more atoms; so there are less electrons passing per second

2 Battery is DC/mains is AC; mains is high voltage/battery is low voltage

3 a i N on left; E at top; L on right

ii Earth

iii If case becomes live; current will flow down the earth wire; protecting the user

(Any 2)

b i 13 A

ii The live wire is the one at high voltage; so the fuse is placed in the live wire to break the circuit as close to this as possible

iii If a fault develops where the live wire touches the case; the case becomes 'live'; a large current flows in the earth and live wires; and the fuse blows/melts breaking the circuit

Page 236 Ultrasound

1 a i A series of compressions/high pressure areas; rarefactions/low pressure areas in the air

ii Frequency increases

b Sound above 20 000 Hz

c In a longitudinal wave the vibrations of the particles are parallel to the direction of the wave; in a transverse wave the vibrations of the particles are at right angles to the direction of the wave.

2 a Ultrasound vibrations; pass into the body to the stones; the vibrations break up the stones

b High-powered ultrasound carries more energy; stones need more energy to break them up

3 a Pulse; tissues; reflected; echoes; image; gel; probe; skin; ultrasound; reflected; skin
(2 correct = 1 mark; all correct = 6 marks)

b Density of the tissues; speed of ultrasound in the tissues

c i The gel must have a similar density to the tissue; so that the ultrasound will pass through; very little will then be reflected at the boundary
(Any 2)

ii $\dfrac{0.0002}{2}$ = 0.0001 sec; speed = $\dfrac{0.16}{0.000}$; = 1600; m/s

Page 237 What is radioactivity?

1 a

type of radiation	charge	what it is	particle or wave
alpha	+2	2 protons + 2 neutrons	particle
beta	−1	electron	particle
gamma	0	short wavelength electromagnetic radiation	wave

(1 mark each correct line or column; 3 marks total)

b i Gamma

ii Beta

2 a The time taken for the activity of a sample to drop to half of the original value; time for half of the original nuclei to decay
(Any 1)

b $5\frac{1}{2}$ years; lines drawn on graph to show this

3 a 86

b 134

c Alpha

d i $^{220}_{86}\text{Rn} \rightarrow {}^{216}_{84}\text{Po} + {}^{4}_{2}\text{He}/{}^{4}_{2}\alpha$
(All correct = 3 marks)

ii Atomic number 86 = 84 + 2; mass number 220 = 216 + 4

e A neutron changes into a proton increasing the atomic number; both are still in the nucleus so the mass number is unaffected

f $x = 131$; $y = 54$

Page 238 Uses of radioisotopes

1 a Rocks/soil; cosmic rays

b Radioactive waste from power stations; nuclear weapons testing; radioisotopes for medical uses
(Any 2)

2 a i Gamma

 ii Gamma can penetrate the pipe and soil under the surface; alpha and beta would not penetrate the pipe or soil layer

 b A Geiger counter; Geiger Muller GM tube; ratemeter/ scalar counter

 (Any 1)

 c The count rate would drop after the site of the leak

3 a Alpha radiation will be stopped by the smoke particles/beta and gamma would not

 b Without smoke, the alpha particles ionise the air; creating a tiny current that can be detected by the circuit in the smoke alarm

 With smoke, the alpha particles are partially blocked so there is less ionisation of the air; the resulting change in current is detected and the alarm sounds

4 a The activity would not have dropped enough for the difference to be measured

 b Carbon-14 is used for dating objects that were once living

 c 4000 years; lines shown on graph

Page 239 Treatment

1 a They are both ionising radiations; they can kill cancer cells

 b They cannot penetrate into the body to the site of the cancer

 c Due to their ionising abilities; they can ionise the atoms in the DNA of normal cells; and cause damage to the patient's normal cells

 d They are placed into a nuclear reactor/made to absorb neutrons

2 a A radioactive isotope which is introduced to the body; to diagnose a problem

 b Tracers will travel around the body to the site of a problem; they can be detected outside the body; this avoids having to cut the patient open

 (Any 2)

 c Gamma

 d Gamma rays are emitted by radioisotopes which can be introduced into the body; X-rays (are produced by high speed electrons hitting a metal target which cannot be done inside the body)

3 a i Gamma

 ii It can penetrate to the site of the tumour; it can kill the cancerous cells

 b This way the normal surrounding cells only receive one third of the radiation; the tumour receives the whole dose

 c The source can be rotated around the patient; delivering a constant dose of radiation to the tumour and intermittent dose to the surrounding tissue

Page 240 Fission and fusion

1 a Source of energy; water; steam; steam; turbines; generator

 (1 mark for 2 correct words)

 b Fission is the breaking down of a large nucleus; into smaller ones

 c Uranium; that has had extra neutrons added

 d i The uranium nucleus splits; releasing energy; and more neutrons

 ii The extra neutrons released are captured by more uranium nuclei causing them to split and release even more neutrons

2 a Slows down the neutrons

 b So that the neutrons are more likely to be captured by the uranium nuclei

 b The control rods can be raised or lowered to absorb more or less neutrons; controlling how many neutrons are available for capture by the uranium nuclei

3 a The joining together of two lighter nuclei to make one heavier one

 b Very high temperatures and pressures

 c There is a plentiful supply of hydrogen/fuel (in sea water); waste products of fusion are less harmful than fission

 d The experimental results cannot be reproduced by other scientists

Page 241 P4 Extended response question

5–6 marks

A detailed description of how ultrasound could produce an image including its reflection from different tissue boundaries, how the reflection depends on tissues having different densities and the different speed of the ultrasound in the different media, ultrasound echoes being used to build up the picture of her lungs and detect the tumour without the need for surgery.

The alternative methods, i.e. X-rays and gamma radiation cause ionisation in cells and are potentially harmful especially to the foetus whose cells are rapidly dividing. Ultrasound is safe as a scanning method even with unborn babies.

All the information in the answer is relevant, clear, organised and presented in a structured and coherent format. Specialist terms are used appropriately. There are few, if any, errors in grammar, punctuation and spelling

3–4 marks

A limited description of how the ultrasound method produces the image, lacking in specific details.

Insufficient detail in the comparison of ultrasound with alternative methods.

For the most part the information is relevant and presented in a structured and coherent format. Specialist terms are used for the most part appropriately. There are occasional errors in grammar, punctuation and spelling.

1–2 marks

An incomplete description of how ultrasound produces an image, with little if any reference to the advantages it may have.

Answer may be simplistic. There may be limited use of specialist terms. Errors of grammar, punctuation and spelling prevent communication of the science.

0 marks

Insufficient or irrelevant science. Answer not worthy of credit.